Foams

Symposium on Foams, 1975

ORGANIZED BY THE

SOCIETY OF CHEMICAL INDUSTRY
President: E. L. Streatfield

COLLOID AND SURFACE CHEMISTRY GROUP
Chairman: Professor K. S. W. Sing

SYMPOSIUM COMMITTEE

Dr M. M. Breuer (*Chairman*)
Dr D. Seaman (*Secretary*)
Dr R. J. Akers (*Editor*)

Mr S. A. Mitchell
Dr A. L. Smith
Dr T. Walker

Foams

Proceedings of a Symposium
organized by the
Society of Chemical Industry,
Colloid and Surface Chemistry
Group, and held at Brunel University,
September 8–10, 1975

edited by

R. J. AKERS
University of Technology
Loughborough, Leicestershire, England

1976

ACADEMIC PRESS
LONDON NEW YORK SAN FRANCISCO
A Subsidiary of Harcourt Brace Jovanovich, Publishers

Academic Press Inc. (London) Ltd
24–28 Oval Road
London NW1

US edition published by
Academic Press Inc.
111 Fifth Avenue,
New York, New York 10003

Library of Congress Catalog Card Number: 76-47157
ISBN: 0-12-047350-X

Printed in Great Britain by
Page Bros (Norwich) Ltd,
Mile Cross Lane, Norwich

Contributors

W. G. M. Agterof
Van't Hoff Laboratory for Physical and Colloid Chemistry, University of Utrecht, Padualaan 8, Utrecht, The Netherlands

C. Axberg
The Swedish Institute for Surface Chemistry, Drottning Kristinas väg 45, S-114 28 Stockholm, Sweden

J. S. Burton
Berk Pharmaceuticals Ltd, Godalming, Surrey GU4 8HE, England

R. Buscall
Department of Pharmacy, University of Aston in Birmingham, Gosta Green, Birmingham B4 7ET, England

J. G. Corrie
Fire Research Station, Borehamwood W06 2BL, Hertfordshire, England

S. C. Cribbs
Research Chemistry Branch, Whiteshell Nuclear Research Establishment, Atomic Energy of Canada Limited, Pinawa, Manitoba ROE 1LO, Canada

E. J. Derderian
Ames Laboratory, USERDA, Iowa State University, Ames, Iowa 50010, USA

R. B. Donaldson
Unilever Research, Port Sunlight Laboratory, Wirral, Merseyside L62 4XN, England

D. Exerowa
Bulgarian Academy of Sciences, Institute of Physical Chemistry, Sofia 13, Bulgaria

J. A. De Feijter
Van't Hoff Laboratory for Physical and Colloid Chemistry, University of Utrecht, Padualaan 8, Utrecht, The Netherlands

H. M. Fijnaut
Van't Hoff Laboratory for Physical and Colloid Chemistry, University of Utrecht, Padualaan 8, Utrecht, The Netherlands

S. Friberg
The Swedish Institute for Surface Chemistry, Drottning Kristinas väg 45, S-114 28 Stockholm, Sweden

D. E. Graham
Biosciences Division, Unilever Research Laboratory, Colworth/Welwyn, The Frythe, Welwyn AL6 9AG, Hertfordshire, England

R. S. Hansen
Ames Laboratory, USERDA, Iowa State University, Ames, Iowa 50010, USA

R. W. Huddleston
Unilever Research, Port Sunlight Laboratory, Wirral, Merseyside L62 4XN, England

B. T. Ingram
Procter & Gamble Technical Centre, Newcastle upon Tyne NW12 9TS, England

Khr. Khristov
Bulgarian Academy of Sciences, Institute of Physical Chemistry, Sofia 13, Bulgaria

J. W. Mansvelt
Lenderink & Co. BV, Schiedam, P.O. Box 126, The Netherlands

E. Matijević
Institute of Colloid and Surface Science and Department of Chemistry, Clarkson
College of Technology, Potsdam, New York 13676, USA

J. B. Melville
Unilever Research, Port Sunlight Laboratory, Wirral, Merseyside L62 4XN, England

G. Nishioka
Chemistry Department, Rensselaer Polytechnic Institute, Troy, New York 12181, USA

R. Österlund
The Swedish Insitute for Surface Chemistry, Drottning Kristinas väg 45, S-114 28,
Stockholm, Sweden

R. H. Ottewill
School of Chemistry, University of Bristol, Bristol BS8 1TS, England

I. Penev
Bulgarian Academy of Sciences, Institute of Physical Chemistry, Sofia 13, Bulgaria

M. C. Phillips
Biosciences Division, Unilever Research Laboratory, Colworth/Welwyn, The Frythe,
Welwyn AL6 9AG, Hertfordshire, England

A. Prins
Unilever Research, Vlaardingen, P.O. Box 114, The Netherlands

V. W. Punton
Wiggins Teape Research and Development, Butler's Court, Beaconsfield, Buckinghamshire HP9 1RT, England

M. J. Quinn
Research Chemistry Branch, Whiteshell Nuclear Research Establishment, Atomic
Energy of Canada Limited, Pinawa, Manitoba ROE 1LO, Canada

K. Roberts
The Swedish Institute for Surface Chemistry, Drottning Kristinas väg 45, S-114 28
Stockholm, Sweden

R. T. Roberts
The Brewing Industry Research Foundation, Nutfield, Redhill, Surrey RH1 4HY,
England

E. L. J. Rosinger
Research Chemistry Branch, Whiteshell Nuclear Research Establishment, Atomic
Energy of Canada Limited, Pinawa, Manitoba ROE 1LO, Canada

S. Ross
Chemistry Department, Rensselaer Polytechnic Institute, Troy, New York 12181, USA

J. V. Russo
2 Hexham Gardens, Isleworth, Middlesex TW7 5JR, England

N. H. Sagert
Research Chemistry Branch, Whiteshell Nuclear Research Establishment, Atomic Energy of Canada Limited, Pinawa, Manitoba ROE 1LO, Canada

H. Saito
The Swedish Institute for Surface Chemistry, Drottning Kristinas väg 45, S-114 28 Stockholm, Sweden

D. Segal
School of Chemistry, University of Bristol, Bristol BS8 1TS, England

A. L. Smith
Unilever Research, Port Sunlight Laboratory, Wirral, Merseyside L62 4XN, England

A. Vrij
Van't Hoff Laboratory for Physical and Colloid Chemistry, University of Utrecht, Padualaan 8, Utrecht, The Netherlands

Contents

Plenary Lecture

Problems in foam origin, drainage and rupture

ROBERT S. HANSEN and EDMOND J. DERDERIAN

*Ames Laboratory, USERDA, Iowa State University,
Ames, Iowa 50010, USA*

Summary

The development of a polyhedral foam is illustrated by an analysis of bubble birth, growth, and motion in a carbonated beverage.

The gross features of polyhedral foam drainage can be modelled in terms of flow of a viscous fluid between plates, with plates corresponding to surfactant monolayers of varying rigidity and the flow driven by pressure gradients due to gravity and forces due to Plateau border suction. The film elastic modulus is an important flow parameter. Problems of the model stem from complicated boundaries, fluctuations in thickness, and possibly from anomalous viscosities at small thicknesses.

Soap films can thin to equilibrium thicknesses at which van der Waals stresses favouring thinning are balanced by double-layer repulsion forces. While this interpretation is clear conceptually, there has been some controversy about the magnitude of the Hamaker constant inferred from equilibrium film thickness measurements, and the manner in which foams stabilized by nonionic surfactants provide the analogue of double-layer repulsion is unclear.

Spontaneous rupture of soap films can be modelled in terms of fluctuations in thickness, and these can be treated as resulting from thermally generated capillary waves, i.e. surface phonons. Translation of this model into rupture rates is still at a rather crude level, and methods for comparing fluctuations resulting from surface phonons with those resulting from environmental fluctuations are needed.

Marangoni flows appear to be critically important both to the spontaneous healing of thin spots in stable films and in foam destruction by antifoaming agents. While a number of cases of Marangoni flow can be quantitatively interpreted with no adjustable parameters, the flows encountered in foam problems are more complex and adequate quantitative modelling, particularly in the antifoaming problem, is still lacking.

1 Introduction

If we consider foams broadly as dispersions of gas in liquid, we can quickly divide foams into two classes depending on whether the volume fraction of dispersed gas is small or large. The first class is well illustrated by carbonated beverages (excluding the foamy heads or "collars" formed, for example, on beer and ale). The second class is well illustrated by just such foamy heads, by soap foams, by whipped cream, etc.

In dilute foams we are concerned with isolated bubbles. Typical physical chemistry of such systems can be illustrated by carbonated beverages. In such beverages, carbon dioxide is dissolved in an aqueous solution at a pressure of about three atmospheres. The Bunsen coefficient k_B (volume of gas dissolved by unit volume of liquid) is about one for carbon dioxide in water at 15°C, so when container pressure is released the beverage should release an amount of (gaseous) carbon dioxide equal to twice the volume of the liquid. Only the nucleation step (normally heterogeneous) in this release process is at all challenging. Once a bubble is nucleated it will grow by diffusion; its rate of capture of carbon dioxide is

$$Q = 4\pi D C_0 r \tag{1}$$

where C_0 is the excess CO_2 concentration above saturation, D the diffusion coefficient of carbon dioxide in water, and r the instantaneous radius of the bubble. The approximate volume growth rate of the bubble is $(RT/P)Q$, where P is the pressure (approximately one atmosphere) within the bubble; from this, the stated solubility, and Henry's law we deduce that its size satisfies the law

$$r^2 = k_B P^{-1}(P_i - P)\, Dt \tag{2}$$

where P_i is the pressure prior to pressure release. Hence for $P_i = 3$ atmospheres, $P = 1$ atmosphere, $D = 10^{-5}\,\text{cm}^2\,\text{s}^{-1}$, $r = 10^{-2}$ cm after 5 seconds. The bubble will rise towards the surface at a velocity v given by Stokes's law

$$6\pi\eta r v = (4/3)\pi r^3 \rho g \tag{3a}$$

or
$$v = (2/9)\rho g r^2/\eta \tag{3b}$$

where ρ and η are the solution density and viscosity respectively, and g the acceleration of gravity. The growing bubble will travel in time t a distance s given by

$$s = \int_0^t v\, dt \tag{4a}$$

$$= k_B \rho g (9\eta P)^{-1}(P_i - P)Dt^2 \tag{4b}$$

using v from equation (3b) with r^2 given by equation (2). With the values of P_i, P, k_B and D previously given, $\rho = 10^3\,\text{kg m}^{-3}$, $g = 10\,\text{m s}^{-2}$, $\eta = 10^{-3}$ N s m^{-2} $s \approx (2/9)t^2$ (s in cm, t in s). The maximum lifetime of a bubble in a

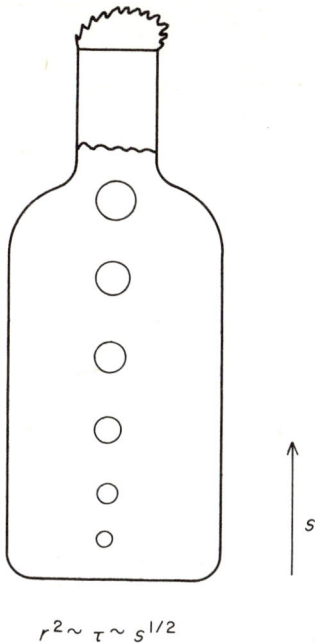

$$r^2 \sim \tau \sim s^{1/2}$$

FIG. 1. Nucleation, growth and rise of bubbles in a dilute foam.

bottle 25 cm high will hence be about 7·5 s and the mean bubble lifetime about 5 seconds. Figure 1 depicts the life of a bubble in such a system.

The foregoing calculations provide a time scale for estimating the likelihood of colloidal processes occurring within dilute foams. Let n be the number of bubbles cm^{-3} in a dilute foam; if their coalescence is diffusion controlled, their rate of disappearance will be

$$-dn/dt = 4kT\eta^{-1}n^2 \tag{5}$$

and their number will be halved in a time $\eta/4kTn$. This half-life proves to be about an hour for $n = 5 \times 10^7$ bubbles cm^{-3}, and this number of bubbles of radius 10^{-2} cm would occupy about $250\ cm^3$. One can immediately conclude that bubbles in a dilute carbonated beverage will almost inevitably rise to the top of the liquid without coalescence, since the mean time for the former process is several orders of magnitude less than the half-life for the latter process. A quick glance at equation (4b) indicates that if the overpressure ratio $(P_i - P)/P_i$ is reduced by a factor 10^4, the residence time will be increased by a factor 10^2, but that coalescence prior to escape from the dilute foam is still improbable.

Except for the nucleation phenomenon, the foregoing illustration indicates a straightforward physical chemistry for dilute foams produced by saturating a pure solvent with a gas and then partially releasing the pressure. Dilute two-component foams generated by other means (e.g. bubbling, beating) do not appear to offer any new problems. A surfactant may serve to reduce bubble growth (in gas release generation) because its adsorbed film on the bubble serves as a partial barrier to transport either directly or through a large elastic modulus; this would alter the quantitative estimates reached earlier, but not the qualitative conclusions.

Let us now consider the properties of the concentrated foam represented, for example, by the collar formed by bubbles rising to the top of a bottle of beer. An important principle of foam stability is illustrated by a simple mechanical model in Fig. 2. Imagine two bubbles blown from a common

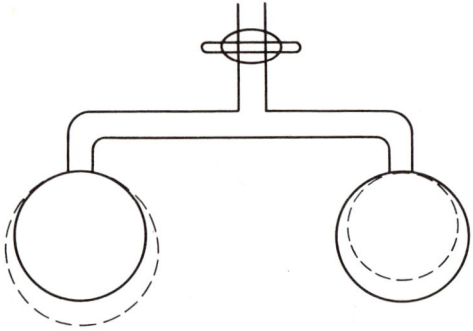

Mechanical equilibrium is unstable unless $2\epsilon > \gamma$

(Gibbs, 1878)

Fig. 2. Stability conditions for connected bubbles.

source as shown, and the stopcock closed. According to the Laplace equation, the pressure inside a spherical bubble is greater than that outside by an amount

$$P - P_0 = 2\gamma/r. \tag{6}$$

Since the outside pressure is the same for both bubbles, if the tensions are equal the bubble radii must be equal for the two inside pressures to be equal, as is required for mechanical equilibrium. But if the tensions are equal and constant the equilibrium attained with equal radii is not a stable one; if a small amount of gas is transferred from right bubble to left bubble the radius of the former increases and that of the latter decreases, as illustrated. Hence, according to equation (6) the pressure of the former decreases, that of the latter increases, and still more gas will flow from right to left; the left bubble will hence finally disappear. Whenever an equilibrium is unstable, as in the present case, a fluctuation will grow (at least initially) exponentially with time. The time required for the fluctuation to double will of course depend on the transport mechanism, but it is important to recognize that processes such as diffusion and convection can perfectly well serve the purpose of the connecting tube in Fig. 2. There is therefore no way for bubbles to coexist in stable equilibrium if their tensions are constant and there is any mechanism for transport of the gaseous component between bubbles. For this reason pure liquids do not foam.

For the two bubbles in Fig. 2 to be in stable equilibrium, it is plain that transfer of a small amount of gas from right to left must be associated with a pressure increase on the right and a pressure decrease on the left, so that gas will flow back to restore the original equilibrium. This will happen if

$$\mathrm{d}p/\mathrm{d}r > 0, \tag{7a}$$

i.e. if

$$2r^{-2}(-\gamma + r\mathrm{d}\gamma/\mathrm{d}r) > 0 \tag{7b}$$

or if

$$\mathrm{d}\gamma/\mathrm{d}\ln r > \gamma. \tag{7c}$$

Since the area of the bubble $A = 4\pi r^2$, this can also be written

$$2\mathrm{d}\gamma/\mathrm{d}\ln A > \gamma. \tag{7d}$$

We define the quantity $\varepsilon = \mathrm{d}\gamma/\mathrm{d}\ln A$ as the surface elastic modulus, and therefore have

$$2\varepsilon > \gamma \tag{7e}$$

as the condition for stability equilibrium of a family of coexisting bubbles. It is also a requirement for stable equilibrium of a single bubble in its surrounding liquid. Equations equivalent to (7d) and (7e) were first given by Gibbs[1] almost a century ago.

Presuming the stability condition equation (7e) is satisfied, a concentrated foam will initially consist of a dense collection of bubbles, and we can estimate the initial drainage rate from arguments presented by Mysels, Shinoda and Frankel.[2] The flow rate per cm Q of liquid of viscosity η, density ρ flowing downward under the influence of gravity between two parallel plates separated by a distance δ is readily obtained by classical hydrodynamics as

$$Q = \rho g \delta^3/12\eta \tag{8}$$

and its mean velocity is

$$Q/\delta = \rho g \delta^2/12\eta. \tag{9}$$

For water at 25°C the mean velocity is about 5 mm min^{-1} if $\delta = 10$ μm and 0·05 mm min^{-1} if $\delta = 1$ μm. The mean velocity of water in a capillary of diameter δ is 0·375 that given by equation (9). Hence, even the most slowly draining water foams (those characterized by rigid adsorbed films around the bubbles) will drain so that the thickness of the water film separating the bubbles (and diameters of columns of water between collections of bubbles) will be reduced to about 10 μm in a minute or so, but by the time these thicknesses are reached drainage has become quite slow. For variable foam heights and liquid viscosities the time required to reach the 10 μm average thickness is roughly proportional to the height and to the viscosity.

Mysels, Shinoda, and Frankel[2] show that, for a supported film with no influx of liquid from the top, the conservation equation

$$\partial\delta/\partial t = -\partial Q/\partial z \tag{10}$$

together with equation (8) leads to

$$\delta^2 = 4\eta z/\rho g t \tag{11}$$

for variation in thickness of the film with time and distance from the top. Equation (11) is a stable solution to equations (8) and (10), but it is not adapted to arbitrary initial conditions and the model does not indicate intermediate configurations, for example, in the transition from rectangular to parabolic cross-section illustrated in Fig. 2. Experimentally, they find the film after some time has a shape approximately given by this equation; the parabolic film predicted is actually supported by a thin black film as shown in Fig. 3, and the distance z must be measured from the top of the parabolic portion. The black film is assumed[2] to result from marginal regeneration (to be discussed later) but there is not yet a model giving its length as a function of time and film parameters.

Since the interstitial water drains rather rapidly to thicknesses of the order 10 μm, and since foam bubbles most commonly have diameters of 100 μm or larger, the foam must achieve by drainage a structure in which bubbles are separated from each other by thin walls of water; if the pressures within two adjoining bubbles are equal (as would generally be the case if their radii before contact were equal) this wall will be flat; were all bubbles initially equal in size, individual bubbles would take polyhedral shapes with flat walls. The junctions of these walls furnish additional problems in equilibria and stability analysed by Plateau[3] and Gibbs;[1] at equilibrium the line of intersection can have no resultant force perpendicular to it. The cross-section illustrated on the left side of Fig. 4 satisfies this condition (presuming

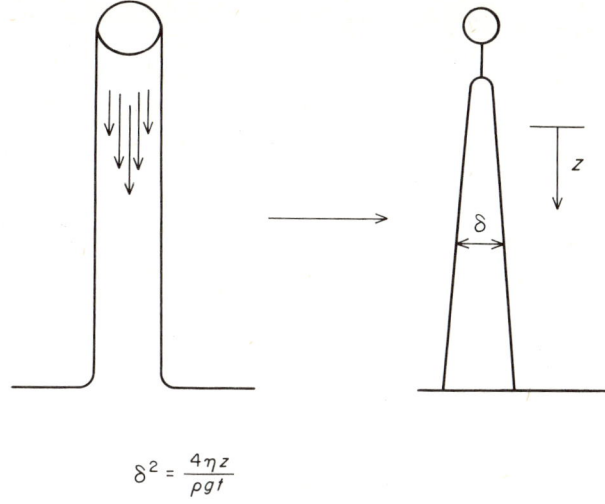

$$\delta^2 = \frac{4\eta z}{\rho g t}$$

FIG. 3. Drainage of liquid between rigid soap film boundaries. (From Mysels *et al.*, 1959.)

the tensions equal), but the equilibrium is unstable with respect to the transition from four films intersecting along a line to two lines each corresponding to three intersecting films at 120° angles (as shown on the right-hand side of Fig. 4). In general, polyhedral foams are characterized by nearly flat walls, with three walls meeting at 120° angles along lines, and four lines meeting at tetrahedral angles at points for reasons of mechanical equilibrium and stability similar to those illustrated.

The foam walls are of finite thickness, and hence so are the "lines" along

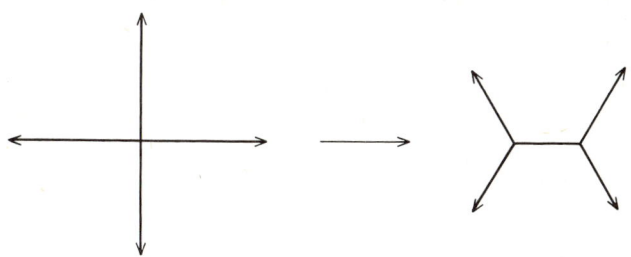

Mechanical instability of 4 lamellae intersecting in a line

FIG. 4. Instability of intersecting films.

which they intersect. Figure 5 illustrates an important detail of this intersection. The films must curve in the neighbourhood of the intersection to satisfy the meniscus equation. The pressure P'' inside the intersection or Plateau border region must be less than that outside, P', by a difference

$$P' - P'' = \gamma/r, \tag{12}$$

where r is the radius of curvature of the Plateau border. Presuming the walls at some distance from the border to be flat as illustrated, the pressure inside the walls there must be P', and so the Plateau border exercises a suction on

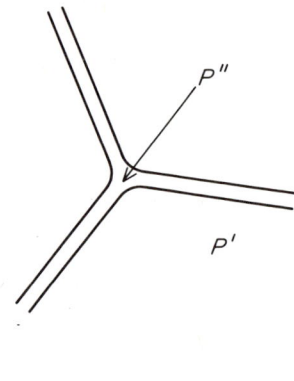

$$P'' < P'$$

Plateau border suction

Fig. 5. Mechanism of border suction.

liquid in the walls of the film. The order of magnitude of this suction can be estimated; presumably P' is approximately atmospheric pressure and P'' approximately $P' - \rho gh$, where h is the height above bulk liquid surface level. Note that this provides an estimate of the radius of curvature of the Plateau border as $r = \gamma(\rho gh)^{-1}$, about 500 μm for $h = 1$ cm, that the maximum internal dimension of the border cross-section is also in this range, and therefore is about a hundredfold greater than the wall thickness after the films have drained a minute or two.

The Plateau border suction establishes a pressure difference between border and foam walls, but if the wall centres are indeed flat the same argument indicates that there is no pressure gradient to cause a flow within these flat portions. Suppose the walls to be curved in such a way as to provide a constant pressure gradient between borders and wall centres; where x_o is the distance between borders this pressure gradient would be approximately $2\rho gz/x_0$. The discussion following equation (9) indicated that drainage between flat parallel plates under a pressure gradient ρg could be expected to reduce film thickness to 10 μm in a minute or so. As the ratio z/x_0 in a foam is of the order 100 the Plateau border suction drainage mechanism will lead to a film thickness of about 1 μm in a minute or so even if the foam-stabilizing surfactant forms a rigid adsorbed film.

Equation (9) shows the mean fluid velocity between parallel rigid plates to be proportional to the square of their separation, and so if the rigid adsorbed film model indicates a time scale of minutes for reduction of wall thicknesses to 1 μm it implies a time scale of hours for reduction to 0·1 μm. In this time scale thinning of isolated films by evaporation can be very noticeable; for example, diffusion of water vapour across a partial pressure gradient of 1 torr cm^{-1} at 25°C would provide a thinning rate of about 0·15 μm min^{-1}. Prins and van den Tempel[4] point out that equalization of temperature and chemical potentials of volatile components should occur within a closed system within seconds, so thinning by evaporation should not occur in the foam interior. An important further mechanism can occur when the adsorbed film is mobile. This is the marginal regeneration mechanism studied by Mysels, Shinoda and Frankel[2] and illustrated in Fig. 6. The top lamella is shown with two curved borders (for example, Plateau borders) and the same pressure drop Δp at each. If a thickness variation develops as shown, there is a net force equal to the product of Δp and the difference in cross-sections between thick and thin films, so that the thick film disappears

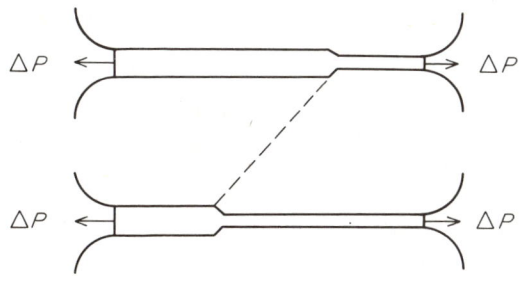

Marginal regeneration in lamellae with mobile films

Left thickness > right thickness

Force left > force right

Junction moves left

FIG. 6. The mechanism of marginal regeneration. (From Mysels *et al.*, 1959.)

into its border and the thin film is withdrawn from its border. The films are regarded to be moving as slabs except in the region of border curvature. In this manner little patches of thin film are generated at Plateau borders, and (in the case of large suspended soap films) their upward motion by convection can readily be observed through interference patterns.

Now the order of magnitude of the force causing the marginal regeneration flow, supposing the original film to have a thickness δ, the thick and thin products $\frac{3}{2}\delta$ and $\frac{1}{2}\delta$ respectively, is $\rho g z \delta$ or (for $z = 1$ cm, $\delta = 1$ μm) about 10^{-4} N m^{-1}. In the absence of compensating forces this would, of course, lead to accelerated flow. Compensating forces limiting the velocity exist in the viscous drag generated in pulling the lamella out of the Plateau border (approximately $\eta v / r_p$ or $0.2v$ for $r_p = 500$ μm as the approximate mean radius of the border). In the example given, this drag would limit the velocity to 0.5 cm s^{-1}. A more severe limitation is likely to be furnished by surface tension variations. The flow illustrated in Fig. 6 evidently requires a source of surfactant at the right boundary and a sink for it at the left. Suppose for illustration that both are provided by the liquid in the two borders, with surfactant diffusing from solution to new surface at the right boundary and dissolving from film into solution at the left boundary. For this process to proceed at an appreciable rate the surface concentration at the right boundary must be less than, and at the left boundary greater than, its equilibrium concentration and the surface tension at the right boundary must be less than that at the left boundary, opposing the illustrated flow. If the surface tension difference equals the force difference resulting from border suction and differing thicknesses, the flow will be steady. Even with a modest elastic modulus (5×10^{-2} N m^{-1}) and reasonably soluble surfactant (10^{-2} molar solution), diffusion from border interior could keep the surface tension difference to 10^{-4} N m^{-1} in the previous example only at film velocities of 0.01 cm s^{-1}. It is also quite likely that marginal regeneration will tend to occur in such a way that "sinks" for surfactant in the generation of one thin area will become "sources" for the generation of another thin area. If the border volume is small and the surfactant of low solubility, the total amount of surfactant in the border may sharply limit the area of surfactant-stabilized lamella that can be drawn from it.

As lamellae stabilized by surfactant films become thinner, repulsion of electrical double layers and thinning pressure arising from van der Waals attraction become increasingly important. Overbeek[5] gives a concise presentation of the pressure resulting from double-layer repulsion; with the approximate result (for 1–1 electrolyte)

$$P_{dl} = 64 \gamma^2 \, nkT \exp(-d/\lambda_D). \tag{13a}$$

where $\gamma = \tanh(e\Psi/kT)$, n is the counterion concentration (ions cm^{-3}), d is the separation between plates, and λ_D is the Debye length. Mysels and Jones[6] have verified the form and approximate magnitude of this pressure for $0.0025 \, \text{mol dm}^{-3}$ sodium tetradecyl sulphate solutions, using a surface potential Ψ of $100 \, \text{mV}$. With this value $\gamma = 0.76$, and if C is the molar concentration, $\lambda_D \approx 0.3 \, \text{nm} \, C^{-\frac{1}{2}}$. Then

$$P_{\text{dl}} \approx 9 \times 10^7 \, C \exp(-d/\lambda_D) \tag{13b}$$

in N m^{-2}. For $C = 0.0025 \, \text{mol dm}^{-3}$, $\lambda_D = 6 \, \text{nm}$, the double-layer repulsive pressure is of the order of $100 \, \text{N m}^{-2}$ (i.e. of the order of border suction) when $d \approx 47 \, \text{nm}$, and increases by a factor 10 for each 14-nm decrease in lamella thickness. The van der Waals attraction between walls of a lamella leads to a thinning pressure

$$P_{\text{vdW}} = A/6 \, \pi d^3 \tag{14}$$

as discussed by Overbeek,[7] who suggests a value $5 \times 10^{-20} \, \text{J}$ for the Hamaker constant A. In this case the thinning pressure is about $100 \, \text{N m}^{-2}$ when $d \approx 30 \, \text{nm}$. The thinning and repulsive pressures, with the parameters given, would be equal when $d = 113 \, \text{nm}$ and this would be the equilibrium film thickness in the absence of other pressures (e.g. border suction). The net disjoining pressure is given by

$$\pi(h) = P_{\text{dl}} - P_{\text{vdW}}. \tag{15}$$

A typical film equilibrium condition requires the disjoining pressure to be equal to the border suction, and this normally corresponds to a thickness in the black film range.

Vrij,[8] and Vrij and Overbeek[9] have shown that thin films can develop instabilities leading to rupture for suitable film thicknesses and instability wavelengths. The subsequent detailed analysis of waves in thin liquid films is given by Lucassen et al.[10] and by Vrij et al.,[11] and we shall pursue their analysis briefly. Liquid velocities within a flat film with air on both sides are derived from the potential functions

$$\Phi = (A \cosh kz + A' \sinh kz) \exp(ikx + iwt) \tag{16}$$

$$\Psi = (B \cosh mz + B' \sinh mz) \exp(ikx + iwt) \tag{17}$$

according to

$$v_x = -\partial\Phi/\partial x - \partial\Psi/\partial z \tag{18}$$

$$v_z = -\partial\Phi/\partial z + \partial\Psi/\partial x \tag{19}$$

where $w = 2\pi v$, $k = 2\pi/\lambda$, v is the frequency of the disturbance, λ is its wavelength, and m is defined by

$$m^2 = k^2 + i\rho w/\eta. \tag{20}$$

The dispersion equation relating w to k is obtained by requiring the motions to satisfy normal and tangential stress boundary conditions at the film surfaces. w is treated as complex

$$w = w_0 + i\beta \tag{21}$$

so that β is a time-damping coefficient (if positive) or growth coefficient (if negative). In general the boundary conditions result in a number of roots w for each value of k, and to each root there corresponds a definite ratio of the four coefficients A, A', B, and B' which defines the motion of liquid in the film. If gravity is neglected, however, the roots separate into two sets. One set involves only the terms whose coefficients are A and B', the second only terms whose coefficients are A' and B. The first set involves only symmetric motions, in the sense $v_z(z) = -v_z(-z)$, $v_x(z) = v_x(-z)$ where z is the direction of the normal to the top film surface, $z = 0$ is the midplane of the unperturbed film, and $-x$ is the direction of wave propagation (in the plane of the film). The second set involves only antisymmetric motions, $v_z(z) = v_z(-z)$, $v_x(z) = -v_x(-z)$. The two motions are illustrated in Fig. 7. Only the symmetric

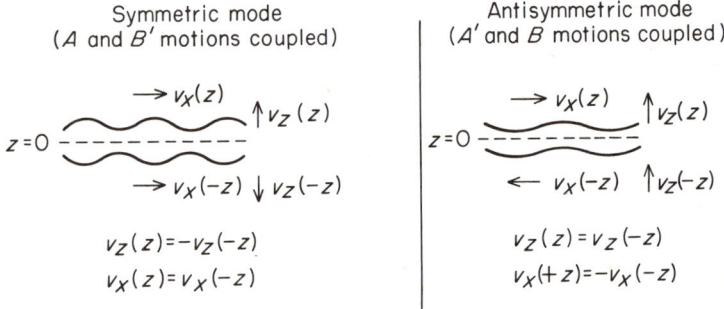

FIG. 7. Symmetric and antisymmetric motions in soap films.

motions lead to film thinning, and we shall therefore limit our attention to them.

The general solutions for the dispersion equation are tedious and best obtained numerically using a computer. If $kh \ll 1$ (and this is the condition of greatest interest for foam rupture phenomena), approximate solutions can be fairly easily obtained. Where

$$\gamma_{\text{eff}} = \gamma - 2k^{-2}\, d\pi/dh \tag{22}$$

and $\varepsilon \gg 0.25\, \gamma_{\text{eff}}\, k^2 h^2$, the smallest root in w is pure imaginary

$$w = ik^4 h^3\, \gamma_{\text{eff}}/24\, \eta \tag{23}$$

and corresponds to a periodic damping if γ_{eff} is positive and to growth if γ_{eff} is negative. If we suppose the disjoining pressure arises solely from the van der Waals thinning pressure, we can use equations (14) and (15) to show that

$$\gamma_{eff} = \gamma - A/\pi k^2 h^4 \tag{24}$$

and therefore is zero if

$$k_{crit} = (A/\pi\gamma h^4)^{\frac{1}{2}}. \tag{25}$$

A disturbance in this mode will hence damp if $k > k_{crit}$ and grow if $k < k_{crit}$. The maximum growth rate occurs when $k = k_{crit}/\sqrt{2}$, and for representative values of parameters $\beta \approx -1$ (so the disturbance increases its amplitude by a factor e each second) when $h \approx 30$ nm. Where γ_{eff} is given by equation (24), which neglects double-layer repulsion, this mode plainly provides a mechanism for film rupture, and Vrij and Overbeek[9] have shown how to estimate the time scale for this process. If equation (24) is modified to include double-layer repulsion, consideration of equations (13a), (15), and (22) indicates that for representative values of parameters, γ_{eff} will go through a minimum and a maximum before decreasing to $-\infty$ (an unrealistic figure reflecting neglect of contact repulsion in equation (14)) as film thickness is steadily decreased. This plainly complicates the analysis of instabilities by (functionally) introducing an activation energy to the thinning process. An analysis incorporating double-layer repulsion appears possible, but so far as I am aware it has not yet been executed.

Two other roots providing dispersion equations are approximately

$$w = (2\,i\eta k^2/\rho) \pm [(\gamma_{eff}\,k^4 h/2\rho) + (2\varepsilon k^2/\rho h) - (4\eta^2 k^2/\rho^2)]^{\frac{1}{2}}. \tag{26}$$

For positive values of γ_{eff} and ε both roots correspond to damped periodic motion, but for negative γ_{eff} one of these roots will correspond to a periodic growth if

$$(\gamma_{eff}\,k^4 h/2\rho) + (2\varepsilon k^2/\rho h) < 0, \tag{27}$$

i.e. if

$$-(\varepsilon/\gamma_{eff}) < \tfrac{1}{4}k^2 h^2. \tag{28}$$

This will require very small values of ε, and in this mode the liquid velocity is nearly constant through the film cross-section, the surface film also moves with this velocity, and the motion is principally in the plane of the film (longitudinal waves). Conceivably this mode may play a role in the marginal regeneration phenomenon previously discussed. The characters of motions associated with equations (23) and (26) for various parameter values have been extensively investigated by Vrij et al.[11]

Scheludko[12] has found it possible to treat the thinning of small (few tenths of a mm radius) circular films according to a theory of Reynolds[13] according to which

$$dh^{-2}/dt = 4P/3\eta\pi^2 \tag{29}$$

where P is the pressure difference between the film and its border (e.g. the border suction). Vrij and Overbeek[9] have combined this with the maximum growth rate of fluctuations to be expected from equation (23) to estimate the critical thickness of films such as Scheludko[12] considered, and obtained results in the right order of magnitude. It is rather remarkable that an essentially linearized instability theory, neglecting double-layer repulsion (which should have been small in the experiments to which the theory was applied) should follow so well the growth of an initial fluctuation by a factor of several hundred.

References

1. Gibbs, J. W. (1948). *In* "Collected Works", Vol. I, pp. 237ff, esp. Eq. 531. Yale University Press, New Haven.
2. Mysels, K. J., Shinoda, K. and Frankel, S. (1959). *In* "Soap Films, Studies of Their Thinning and a Bibliography". Pergamon Press, New York.
3. Plateau, J. A. F. (1873). *In* "Statique Experimentale et Theorique des Liquides Soumisaux Seules Forces Moleculaires". Gauthier-Villars, Paris.
4. Prins, A. and van den Tempel, M. (1970). *In* "Thin Liquid Films and Boundary Layers" (No. 1, p. 20, Special Discussions of the Faraday Society).
5. Overbeek, J. Th. G. (1952) *In* "Colloid Science" (H. R. Kruyt, Ed.), Vol. I, pp. 252–255. Elsevier, Amsterdam.
6. Mysels, K. J. and Jones, M. N. (1967). *Discuss. Faraday Soc.* **42**, 42.
7. Overbeek, J. Th. G. (1960). *J. Phys. Chem.* **64**, 1178.
8. Vrij, A. (1966). *Discuss. Faraday Soc.* **42**, 23.
9. Vrij, A. and Overbeek, J. Th. G. (1968). *J. Amer. Chem. Soc.* **90**, 3074.
10. Lucassen, J., van den Tempel, M., Vrij, A. and Hesselink, F. Th. (1970). *Proc. Kon. Ned. Akad. Wetensch.* Ser. B, **73**, 109.
11. Vrij, A., Hesselink, F. Th., Lucassen, J. and van den Tempel, M. (1970). *Proc. Kon. Ned. Akad. Wetensch.* Ser. B, **73**, 124.
12. Scheludko, A. (1957). *Kolloid-Z.* **155**, 39.
13. Reynolds, O. (1886). *Phil. Trans. Roy. Soc. London*, **177**, 157.

Discussion

Russo I would like to point out that in cake batters which are usually Bingham plastics (i.e. non-Newtonian) Handlemann *et al.*[1] have shown that rise of bubbles obeys Stokes's law.

Hansen Stokes's law as used in our equation (3a) asserts that the viscous drag on a sphere of radius r moving with velocity v through a medium of viscosity η is $6\pi\eta rv$. It is derived by continuum hydrodynamics for a Newtonian liquid, and there is no theoretical basis for expecting it to represent the motion of a sphere through a non-Newtonian liquid. The work of Handlemann *et al.* cited does indeed discuss the use of Stokes's law to describe the motion of bubbles in Newtonian liquids, and their equation (6) is equivalent to our equation (3a). They point out on the following page of their paper that cake batters are neither homogeneous nor Newtonian, and that therefore "batter properties governing bubble movement act in a more complicated way". They then make the comment that "equation (6) will also apply to cake batters" if the velocity is made a proper function of η. This statement is misleading, and apparently has misled Russo. It does *not* mean that Stokes's law governs the bubble motion, and indeed their equations (7) and (8) show that they are taking the drag force as $Yf(r)$, where Y is the yield value and $f(r)$ is an unknown function of the radius, rather than $6\pi\eta rv$ as given by Stokes's law.

Padday (*Kodak Ltd, Harrow, England*) The force of attraction between two bars of liquid separated by a parallel gap is only equal to the thinning force in a film of equal thickness to the gap, when the forces acting arise from dispersion interactions only. Water films are reinforced by hydrogen bonding forces which make the thinning force greater than that between two bars of water separated by a space. Thus one has to apply Hamaker and Lifshitz to calculations with water films with greater precaution.

Hansen Padday is correct in recommending caution in calculation of thinning forces by the Hamaker or Lifshitz models. The refinements to which he alludes would affect the quantitative aspects of the calculations (e.g. magnitudes of A in equations (24) and (25) but not the qualitative results).

Israelachvili *et al.*[2] have reviewed the van der Waals force problem, and their equations permit focusing of issues. For two semi-infinite slabs of material 1 separated

[1] Handleman, A. R., Conn, J. F. and Lyons, J. W. (1961). *Cereal. Chem.* **38**, 294.

[2] Israelachvili, J. N. and Tabor, D. (1973). *In* "Progress in Surface and Membrane Science" (Eds J. F. Danielli, M. D. Rosenberg and D. A. Cadenhead), Vol. 7, pp. 2–55. Academic Press, New York and London.

by material 3 of thickness d their equations (53) and (55) show that the force per unit area for *nonretarded* van der Waals force is the same if media 1 and 3 are interchanged, but this is not true for equation (56) governing the retarded force. In both cases the equations are derived for isotropic media. As they are based on macroscopic theory it would appear that all microscopic information (including bonding and hydrogen bonding) is included in the frequency dependent dielectric constants, and the limitations due to molecular character of the media would only become important when the thickness d is not large compared to molecular dimensions.

1

Foaming behaviour of partially miscible liquids as related to their phase diagrams

SYDNEY ROSS and GARY NISHIOKA

Chemistry Department, Rensselaer Polytechnic Institute, Troy, New York 12181, USA

Summary

The phase diagrams of partially miscible liquids in systems of two, three, or more components can be used to predict the foaming behaviour of compositions close to the critical point (two components) or plait point (three or more components). Surface activity is the precursor of incipient phase separation and this distinctly appears as an increasing foaminess of the unsaturated solutions as they approach in composition to the solubility curve. When the system separates into two conjugate solutions, one of them has a positive and the other a negative spreading coefficient, each with respect to its conjugate. The former, when dispersed, acts as a defoamer to its conjugate by virtue of this property; the latter, dispersed but lacking the ability to spread, does not affect the foaminess of its conjugate. Within a narrow range of compositions on the tie-line between the two immiscible conjugates, emulsion inversion occurs: what was previously the dispersed phase becomes the continuous phase and vice versa. The conjugate of lower surface tension, which when dispersed acts as a foam inhibitor, when continuous acts as a foamy matrix; the conjugate of higher surface tension, which when it is the continuous phase is defoamed by its conjugate, when dispersed cannot reciprocate the defoaming action. The foam behaviour of the heterogeneous system therefore switches from unfoamable to foamable at or near the total composition where the emulsion inversion occurs. These effects are illustrated by descriptions of the foaming behaviour of various two- and three-component systems as related to their phase diagrams.

1 Introduction

Conventional amphipathic solutes such as soaps or synthetic detergents are not unique in displaying surface activity. Certainly such a solute, whose molecular structure is planned to be amphipathic by a balance of hydrophilic and lipophilic moieties, is thereby assured of any required degree of interaction with the solvent; but other solutes may well exist that have the requisite degree of interaction for surface activity although their molecular structure does not proclaim their amphipathy even to experienced chemists. The surface activity of a solute is primarily the result of a weak interaction with the solvent, which interaction must of course still be sufficient to dissolve the solute but need not be greater than the least degree required to do so. A propensity towards phase separation is therefore a general guide to surface-active behaviour. Indeed a prescient but passing and incidental remark made by Irving Langmuir could have been seized at the time by an alert reader to stimulate something like the present series of experimental researches and so anticipate our finding by fifty years. Langmuir[1] wrote: "Undoubtedly, in mutually saturated liquids, especially near the critical temperature, the conditions are favourable for orientation and segregation of the molecules in the liquid." This hint was not taken up by anyone, so far as we know, nor did we ourselves notice it until after our work was well under way. Langmuir was unaware of the implications of his incidental comments; but it probably would not have surprised him to learn that surface activity displayed by solutions or mixtures can be related to their phase diagram, especially at compositions close to the critical temperature (for two-component systems) or at compositions close to the plait-point concentration (for three or higher component systems). The thesis that surface activity is the precursor of incipient phase separation is distinctly demonstrated by our reported observations[2] of increasing foaminess of the unsaturated solutions as they approach in composition to the solubility curve, with maximum foaminess near the critical solution point.

The foaminess of the homogeneous solutions is not the only sign of the surface activity of the solute. When the solubility limit is exceeded a few drops of a conjugate solution separate out as a dispersed second phase. The newly separated phase will act to destroy the foam of the parent phase if it meets the following conditions:

a. It is present in so small a quantity as to make it the dispersed phase of the unstable "emulsion" that is continuously regenerated during the agitation of the system by dynamic foaming.

b. It has a surface tension lower than that of the matrix.

Those conditions can be satisfied by one or the other conjugate solutions when present at a low relative amount, which results in a rather dramatic, abrupt transition of the system, on the separation of a new phase, from a highly foamable solution to an unfoamable mixture. So sensitive to the first separation or, alternatively, absorption of a second phase is the loss or, alternatively, appearance of foaminess that it offers a better indication of the solubility limit than does the visual observation of a cloud point.

2 Theory of foam stabilization and antifoaming action

In describing the foam behaviour of liquid systems as related to their phase diagrams we have to assume the reader's familiarity with the mechanisms of foam stabilization and foam inhibition.[3, 4, 5]

A film of liquid tends to contract because of its surface tension unless that tendency is counteracted by the presence of a surface-active component, which cannot be transferred from the surface to the bulk solution without the expenditure of work. If the work required is more than the work done by the surface on contracting, the contraction is inhibited and the liquid film is thermodynamically stable. But it is not necessarily mechanically stable against external stresses such as those imposed by gravity or capillary forces, which pull the liquid downward or into regions of convex curvature, respectively. The film could not exist for more than a moment unless those forces were opposed by another force that checks the flow of the liquid. The opposing force is the higher surface tension that is created when the film begins to flow. A film composed of pure liquid produces a higher surface tension at a newly formed surface only very briefly, of the order of microseconds, and so such a film can have but a very brief life. However, if the film has an adsorbed layer of solute on its surface then a higher surface tension may develop at a newly formed surface and persist for several microseconds until the adsorbed layer of solute is restored by diffusion from the bulk solution. But events do not wait for that relatively slow diffusion to occur. Where a higher tension appears, contraction of the surface at that point immediately ensues; surrounding areas of lower-tension surface are pulled in towards the high-tension spot until the difference has been effaced. The movement of the surface from areas of lower to areas of higher surface tension is accompanied by the flow of thick layers of underlying fluid, amounting to several microns in depth, so that drainage or capillary flow of the film is balanced by a counterflow of liquid at the surface. The restoration of the film by this mechanism is called the "Marangoni effect";[6] it was recognized also by Gibbs.[7]

The mechanism of foam inhibition by an insoluble agent added for the purpose also depends on a Marangoni effect, if that term is defined to refer to flow of liquid due to local differences in surface tension. The action of such an agent arises from its ability to spread spontaneously over the surface of the foamy liquid. This purely mechanical action ruptures the foam film. The lateral flow of the spreading liquid communicates a shearing stress to the liquid underneath so that the substrate, to a depth of several microns, is carried away by viscous drag as the liquid on top advances. A single drop of the agent, once it has arrived at the surface of the liquid film, performs in effect like a Venturi pump, ejecting on every side all the liquid lying beneath it and causing the film to rupture by the agitation produced by its action. If the matrix is a foamable liquid and has formed a stable liquid film the spreading action of the agent will destroy it. Obviously the process just described can occur only when relations obtain between the various surface and interfacial tensions. The agent will spread spontaneously if the value of the spreading coefficient S be positive, where S is given by

$$S = \sigma_m - \sigma_a - \sigma_{int}$$

where σ_m is the surface tension of the foamy matrix, σ_a is the surface tension of the foam-inhibiting agent, and σ_{int} is the interfacial tension between them. If S is negative the insoluble drop of liquid may enter the surface without spreading on it, or may not even be able to enter the surface. This happens if the interfacial tension between the two immiscible liquids is too great. Foam inhibition is then greatly curtailed or does not occur at all.

3 Materials and their sources of supply

Benzene	C_6H_6	Fisher Scientific Co.	Certified reagent grade
2,6-Dimethyl heptan-4-ol (di-isobutyl carbinol)	$[(CH_3)_2CHCH_2]_2\text{-}CHOH$	Union Carbide	Industrial grade
Ethanol (absolute)	CH_3CH_2OH	United States Industrial Chemical Co.	Reagent grade
12-Ethane diol (ethylene glycol)	$HOCH_2CH_2OH$	Fisher Scientific Co.	Certified reagent grade
4-Butoxy-ethan-2-ol (ethylene glycol monobutyl ether)	$CH_3(CH_2)_3\text{-}O(CH_2)_2OH$	Fisher Scientific Co.	Certified reagent grade
Methyl acetate	CH_3COOCH_3	Fisher Scientific Co.	Certified reagent grade

4 Apparatus and procedure

The foam stabilities were measured with the dynamic foam meter shown in Fig. 1. Nitrogen gas was bubbled through 25 or 50 cm³ of a suitable liquid at measured rates V/t cm³ s⁻¹ (V is gas volume, t is time) and the steady-state volume v_0 of foam was read. The ratio $v_0 t/V = \Sigma$ is almost independent of V/t

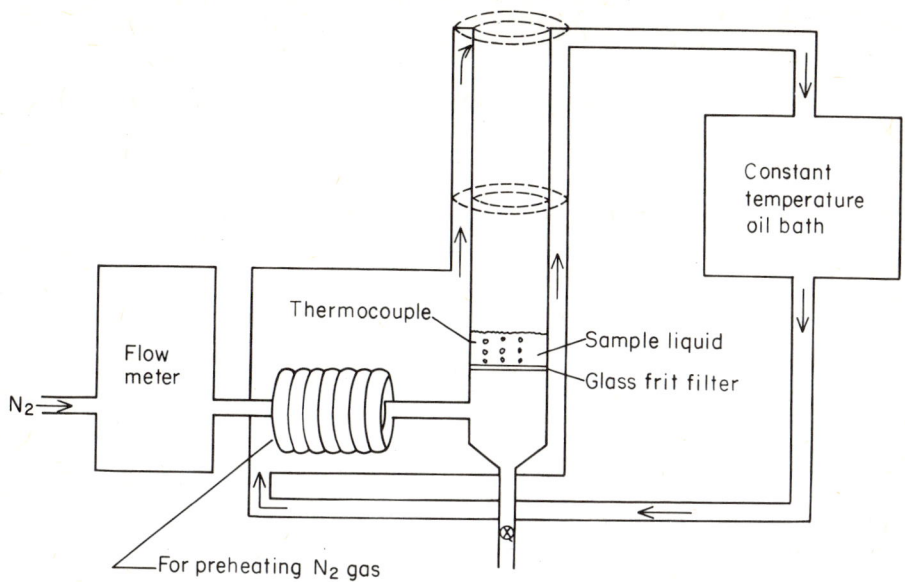

FIG. 1. Dynamic foam meter, equipped for temperature control of entering gas and of the entire foam column. Not shown is the heating tape used to counteract thermal losses of the oil in the upper portions of the column.

and may be used as a measure of foaminess.[8] Σ is, for instance, about 6 s for 1 per cent aqueous l-butanol; about 140 s for 0·04 per cent solution of egg albumin at pH 4·8.

Solutions were foamed in a column 70 cm high with a diameter of 3 cm. Temperature was controlled by means of an oil circulating from a thermostat. The cooling of the oil as it passed through the jacketed column was counteracted by heating tapes wound in a wide spiral around the outside of the apparatus. Thermocouples placed inside the oil jacket allowed proper adjustment of the current in the heating tapes to ensure uniformity of temperature throughout the column. The temperature of the foamed solution was read directly by a thermocouple placed inside the sample space of the foam meter. The sample temperatures were stable to within two-tenths of a Celsius degree.

5 Foaminess of homogeneous solutions near their critical or plait point[2]

Results for the two component-system diisobutyl carbinol + ethylene glycol are reported in Fig. 2 as interpolated lines of equal foam stability (*isaphroic* lines, from *aphros* the Greek word for foam) superimposed on the

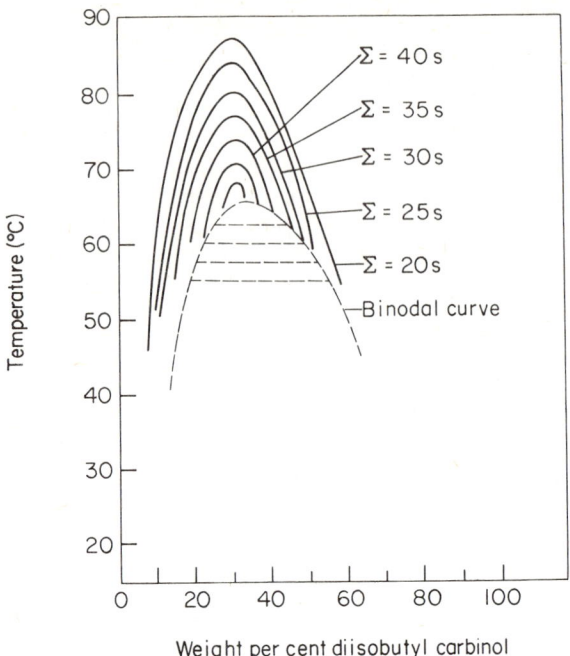

Fig. 2. Phase diagram and interpolated isaphroic contours of the two-component system, diisobutyl carbinol and water, showing maximum foaminess near the critical-solution temperature.

phase diagram. The isaphroic lines centre about the critical-solution point as a maximum and decrease in value the farther they are from it. Mixtures of composition falling within the two-phase region were separated into their conjugate solutions and their foaminess determined individually. The foaminess of the unseparated mixture was not determined with this system.

Foam stabilities ($\Sigma > 5$ s) at 20°C for three-component systems are shown in Figs 3 and 4 for the benzene–ethanol–water system and the ethylene glycol-*n* butanol-water system respectively. The data are reported as interpolated isaphroic curves. Superimposed on these graphs are the binodal curves[9, 10] and tie lines for these systems. Solutions of composition falling within the two-phase region of each diagram do not foam. No plait-point data were available for these systems at 20°C but its approximate position on the binodal curve can be observed by eye.

From these diagrams one notices a maximum of foaming at the plait point, with a gradual decrease in foaming of solutions in the one-phase region with increasing distance from the plait point, and an absence of foam within the two-phase region. For these systems, compositions on one side of the binodal

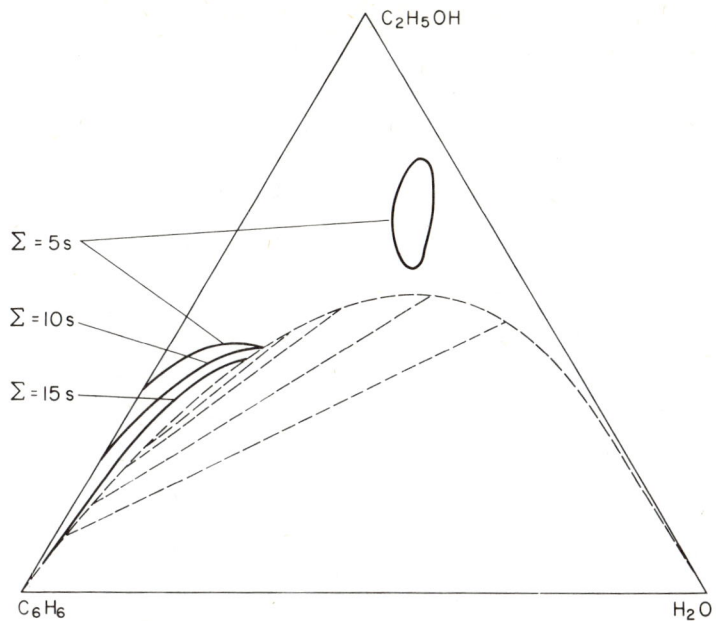

FIG. 3. Phase diagram and interpolated isaphroic contours of the three-component system, ethanol–benzene–water, at 20°C.

curve foamed while those on the other side did not. In the ethylene glycol-*n* butanol–water system, the composition at the aqueous extremity of the tie line foamed while its conjugate composition at the organic extremity did not. Addition of small portions of the organic phase to the foaming aqueous phase produced a non-foaming system, showing that the organic phase acts to defoam its aqueous conjugate. This result explains the absence of foam within the two-phase region: mixtures within this region contain an intrinsically foamable solution along with its inhibitor.

The explanation of the defoaming action of one conjugate solution on the other is to be found in their relative surface and interfacial tensions near the critical point, where the interfacial tension between the conjugates approaches zero and the two surface tensions approach equality. The former effect is found to precede the latter.[11] The spreading coefficient of A on B is defined as

$$S = \sigma_B - (\sigma_A + \sigma_{A/B}) \qquad (1)$$

where σ_A, σ_B and $\sigma_{A/B}$ designate the surface tensions of A, B and the interfacial tension of A/B respectively. As $\sigma_{A/B}$ approaches zero more rapidly than σ_A

B

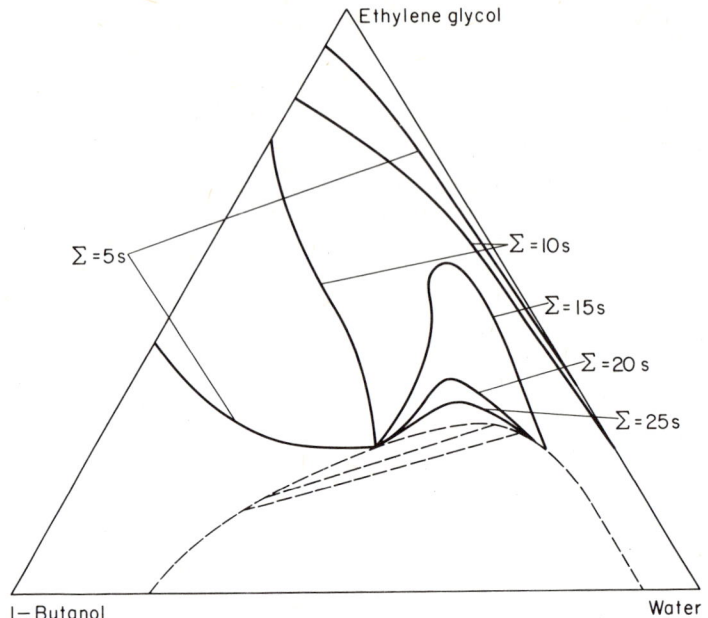

FIG. 4. Phase diagram and interpolated isaphroic contours of the three-component system, ethylene glycol–butanol–water, at 20°C.

becomes equal to σ_B, the value of S in the vicinity of the critical point is approximated by

$$S = \sigma_B - \sigma_A \tag{2}$$

If A and B represent the two conjugate solutions, then S is positive when $\sigma_B > \sigma_A$ and S is negative when $\sigma_A > \sigma_B$, i.e. the solution with the lower surface tension will spread spontaneously on a surface of its conjugate. The behaviour is exactly what is required of a foam inhibitor. Conjugate solutions that lie farther from the critical point do not behave in this way and such two-phase mixtures would not necessarily be defoamed.

The current literature on foaming problems that occur during fractionation or distillation of liquids emphasizes the surface tension relation of solutions containing volatile constituents, and even promulgates a dogmatic rule of thumb, as follows: If the loss of volatile constituents causes the surface tension of the remaining solution to increase, then anticipate a foam problem; if the loss of volatile constituents causes the surface tension to decrease, then no foam problem need be feared.[12] So runs the dictum, which is based on the mechanism of bubble stabilization by the Marangoni effect, resulting from local changes in surface tension.

The normal behaviour of liquids is such that, in general, involatility and high surface tension go together; so that, although exceptions exist (the polydimethyl siloxanes being the best known of these exceptions), the usual behaviour of solutions leads to the surface tension of the remaining solutions being higher after the loss of its more volatile constituents. According to the rule cited above, that should lead to foaming within a fractionation tower. But it does not always do so in practice. The incidence of the surface tension properties that one is warned against is therefore more to be taken as an alert to a situation that is potentially rather than necessarily troublesome. The converse conclusion, namely that a reduction of the surface tension of the remaining solution after the loss of some of the more volatile constituents can be interpreted as an all-clear signal for the nonoccurrence of foaming, is indeed corroborated by the small body of existing data on the subject. The evidence pertaining to the foregoing rule is that it is inconclusive when it proffers a positive warning about an impending foam problem, and conclusive only when it indicates a safety signal. The relative frequency of receiving the inconclusive positive warning is, however, much higher than the frequency with which one will receive an "all-clear".

The usefulness of the rule is much reduced by the lack of surface tension data under conditions of temperature and varying concentrations within the column. These data are not readily obtained by laboratory measurements, requiring special equipment suitable for conditions prevailing inside the column.

A more useful and more far-reaching rule is to be obtained from the present work. Foaming of a solution reaches its maximum under conditions of temperature and concentration where a transition into two separate liquid phases is imminent. Two-component systems show maximum foam stability at the temperature and composition of the critical point, but only as long as the system is maintained as an homogeneous one-phase solution. Should the slightest degree of phase separation occur, one of the separated liquid phases acts as a defoamer for its foamy conjugate, *even though the more volatile constituent has a lower surface tension than the less volatile constituent.* Here is a case, therefore, where the former rule about surface tensions would sound an unnecessary alert: an additional factor that the rule does not provide for has supervened and overturned its prediction.

Three-component systems show maximum foam stability at a given temperature at compositions near that of the plait point, but again only when the system is an homogeneous single phase. As before, the slightest degree of separation of liquid phases produces a conjugate solution that can defoam its foamy conjugate. It is highly probable that polycomponent systems would behave in a similar way, and show a maximum foam stability where separation of phase is imminent, and also show defoaming action once the slightest degree of phase separation occurs.

In the light of these conclusions some prior observations of foams are seen to be more significant. Bolles[13] describes a foaming problem encountered in the industrial-scale separation of propylene from dimethyl formamide. Although no phase diagram is given, various statements in this report suggest that the abrupt decrease of the foam stability with increasing temperature between 51 and 54°F might be due to the separation of a liquid phase acting as a defoamer to its foamy conjugate solution. In general in extractive distillation, in which a solvent is introduced, troublesome foaming is so common an occurrence that in the design of extractive-distillation columns excess capacity is always built in: experience has indicated that a foaming problem is likely to arise. We point out as significant that in those solutions phases appear and disappear at various stages inside the column; therefore, a knowledge of the phase diagram of the systems to be treated in an extractive-distillation process is the best guide to locate potential sources of troublesome foam.

Other prior sets of observations are also of interest. When piperidine is added to a foamy cyclohexane-nitrobenzene solution, the foaminess is depressed.[14] At the temperature of these experiments (0°C) the ternary system is homogeneous, but the suggested explanation is that the miscibility of C_6H_{12} and $C_6H_5NO_2$ is enhanced by $C_5H_{11}N$, i.e. the composition of the mixture lies farther away from the phase-separation boundary. Apparently an opposite effect occurs in the homogeneous ternary solution of nitrobenzene, p-cymene, and cyclohexane at 0°C where maximum foaminess is observed at 60:25:15 parts. Decreased miscibility is the probable explanation here. Regions of higher foaminess in homogeneous solutions lie near phase separations, so presumably at lower temperatures this ternary system would show partial miscibility. Again, in agreement with expectation, the foaminess of solutions of anhydrous methanol and p-cymene was raised by the addition of water as long as the ternary system remained homogeneous.[15]

6 Foaminess of heterogeneous mixtures near their critical point

The heterogeneous portions of these phase diagrams denote two conjugate solutions in equilibrium at any temperature. Depending on their interfacial and surface tensions one such conjugate A can spread on the other B if it meets the condition that $S_1 > 0$, where

$$S_1 = \sigma_B - \sigma_A - \sigma_{A/B}$$

The reciprocal condition, i.e. that B can spread on A, depends on $S_2 > 0$, where

$$S_2 = \sigma_A - \sigma_B - \sigma_{A/B}$$

The two spreading coefficients S_1 and S_2 cannot both be positive; one can be positive and the other negative, or they can both be negative. Near the critical point, where $\sigma_{A/B}$ tends to zero, the spreading coefficients are expressed by simpler equations:

$$S_1 = \sigma_B - \sigma_A$$

$$S_2 = \sigma_A - \sigma_B$$

From these equations it is immediately obvious that either S_1 or S_2 is positive,

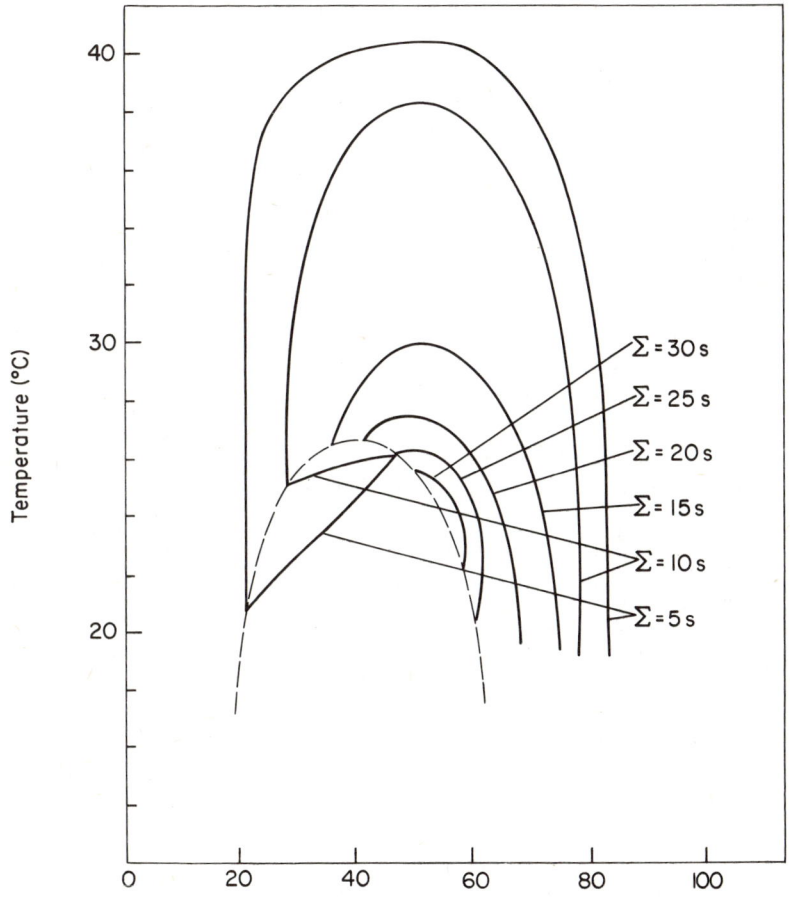

FIG. 5. Phase diagram and interpolated isaphroic contours of the two-component system, methyl acetate and ethylene glycol.

but not both. Therefore, for conjugate solutions near the critical point, the one of lower surface tension spreads on its conjugate, but the one of higher surface tension does not spread on its conjugate.

Foam-inhibiting action depends on the dispersed antifoam spreading spontaneously on the foamable solution.[3, 4] Thus one conjugate solution will act as a foam-inhibiting agent for its conjugate partner, but the action cannot be mutually reciprocal. If foam measurements are made of a series of solutions and mixtures whose compositions span the phase-separation boundary of the phase diagram, the foam stability would be expected to increase continuously up to the point of phase separation, after which it may take one of the two following courses: if the newly separated, dispersed phase has a lower surface tension than that of the solution, the foam will be inhibited, but if the newly separated, dispersed phase has the higher surface tension it cannot spread, it cannot even "enter"[16] the surface as that requires the medium to withdraw on the close approach of a dispersed drop to the surface, and so it has no inhibitory action on the foaminess of the solution. If a composition near one end of a tie line shows one of these effects, the composition near the other end will show the other. An understanding of these principles helps us to interpret foam behaviour that would otherwise appear puzzling.

The behaviour patterns described above are shown by the partially miscible liquids methyl acetate (MA) and ethylene glycol (EG). Figure 5 reports isaphroic contours for this system as a function of temperature, superimposed upon the binodal curve of the phase separation, which has a critical point at 27°C and 38 per cent EG by weight. The maximum of foaming does not correspond to the critical point but occurs at 24°C and 50 per cent EG by weight. The foaminess of the heterogeneous mixtures is never great and occurs only at total compositions of less than 50 per cent EG by weight. The foaminess of a range of compositions at 20°C is shown in Fig. 6. The surface tensions of the pure components and the conjugate solutions at 20°C are reported in Table 1.

TABLE 1

Composition	$\sigma \times 10^3/\mathrm{N\,m}^{-1}$
Pure MA	24·1
20 per cent EG (conjugate A)	25·4
61 per cent EG (conjugate B)	26·2
Pure EG	46·9

The data in Table 1 show that, throughout the whole range of bulk composition, the surface is richer in MA than is the bulk solution.

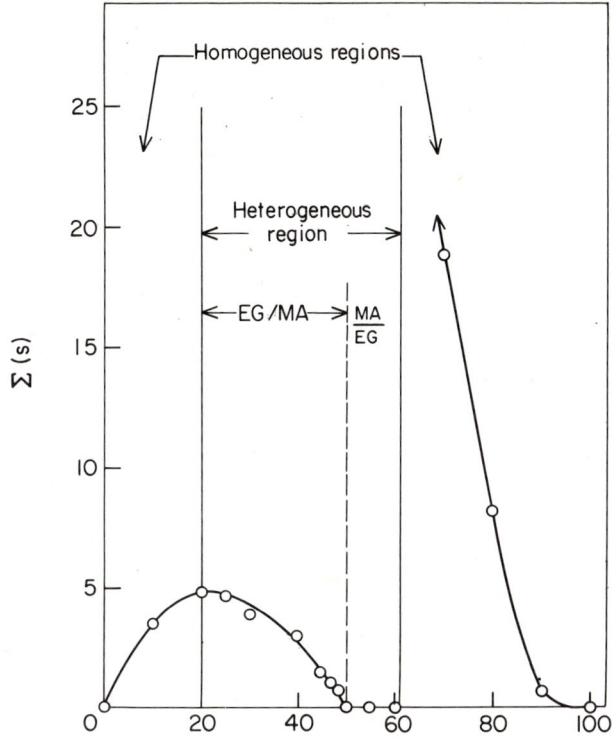

FIG. 6. Variation with composition of foam stabilities of two-component solutions and mixtures at 20°C (methyl acetate and ethylene glycol).

The interfacial tension between the conjugates is unmeasurably small ($\sigma_{A/B} < 10^{-4}\,N\,m^{-1}$) at 20°C. The spreading coefficients are therefore:

$$S_1 = +0.8 \times 10^{-3}\,N\,m^{-1}$$
$$S_2 = -0.8 \times 10^{-3}\,N\,m^{-1}$$

The coefficient S_1 refers to the spreading of conjugate A on B, and S_2 refers to the spreading of conjugate B on A. The expected behaviour, therefore, is that conjugate B will be defoamed by A, and that conjugate A will not be defoamed by B. Figure 6 shows that this is indeed the case. The defoaming of conjugate B is very notable, as the foaminess of B reaches a high value in the homogeneous region, which is immediately reduced to zero by the first separation of its immiscible conjugate. Conjugate A also foams in the homogeneous region but the foaminess suffers only a slow decline as conjugate B separates out. The slow decline of foaminess is partly the result of replacing the amount of the matrix (A) with dispersed B, and partly due to the instability of the dynamic emulsions EG/MA and MA/EG, both of which may exist transiently because of the agitation of the mixture during the foam measurement. Above 50 per cent, however, the emulsion type MA/EG predominates; and with it the foam-inhibiting action of conjugate A.

Another example of essentially the same foam behaviour as reported in Fig 6 is provided by results published by Prigorodov[17] for the partially-miscible liquid polymers polydiethylsiloxane (PES) and polymethyl-γ-trifluoropropylsiloxane (FS), which have surface tensions of 26.8×10^{-3} and 18.8×10^{-3} N m^{-1} respectively at 20° C. The phase transitions were observed by cloud points at 1·5 per cent FS and 93 per cent FS, but the change in foaminess, a more sensitive indicator of phase separation, alters the former value to 0·8 per cent FS, which we have adopted in Fig. 7, our report of Prigorodov's

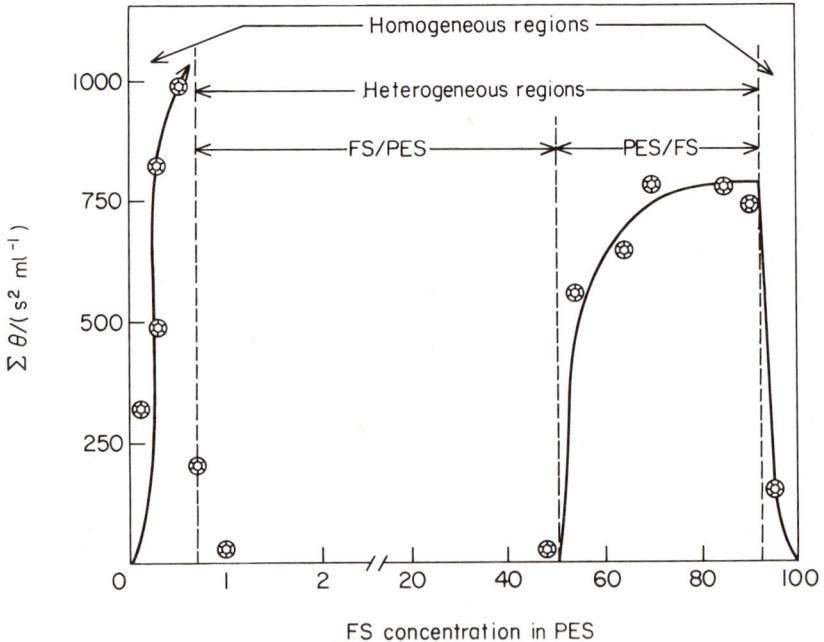

FIG. 7. Variation of the foam stability with composition of polymer mixtures at 20°C.

data. Reading the diagram from left to right, PES solutions of low FS concentration show increasing foaminess with addition of solute until the separated conjugate phase causes defoaming. The heterogeneous mixtures remain defoamed until the emulsion inverts at 50 per cent FS. The new matrix, consisting of 93 per cent FS, is not defoamed by the PES solution of higher surface tension, and its foaminess continues to increase throughout the heterogeneous region as the amount of foamy matrix increases. Past the point of immiscibility, the solutions of PES in FS become progressively less concentrated and finally reach the zero foaminess of the pure solvent.

We cannot but agree with the reviewer who found Prigorodov's own explanation of those phenomena "not entirely clear".[18]

Acknowledgement

The authors gratefully acknowledge permission from Fractionation Research Incorporated, South Pasadena, California, to publish some of the foregoing data, and from the American Chemical Society to reprint Figs 1, 2, 3 and 4.

References

1. Langmuir, I. (1925). *Colloid Symp. Monogr.* **3**, 62.
2. Ross, S. and Nishioka, G. (1975). *J. Phys. Chem.* **79**, 1561.
3. Ewers, W. E. and Sutherland, K. L. (1952). *Aust. J. Sci. Res.* **5A**, 697.
4. Ross, S. (1967). *Chem. Eng. Progr.* **63** (9), 41.
5. Bikerman, J. J (1973). "Foams", Ch. 1. Springer-Verlag, New York.
6. Marangoni, C. (1871). *Nuovo Cimento*, (2) **5–6**, 239; (1878). (3) **3**, 97, 193.
7. Gibbs, J. W. (1906). "Scientific Papers", Vol. 1, pp. 300–314. Longmans Green, London.
8. Bikerman, J. J. (1938). *Trans. Faraday Soc.* **34**, 634.
9. Bancroft, W. D. and Hubard, S. C. (1942). *J. Amer. Chem. Soc.* **64**, 349.
10. Matsumoto, K. and Sone, S. (1956). *J. Pharm. Soc. Jap.* **76**, 475.
11. Donahue, D. G. and Bartell, F. E. (1952). *J. Phys. Chem.* **56**, 480.
12. Zuiderweg, F. J. and Harmens, A. (1958). *Chem. Eng. Sci.* **9**, 89.
13. Bolles, W. (1967). *Chem. Eng. Progr.* **63**, (9), 48.
14. Teitelbaum, B. Ya. and Ganelina, S. G. (1952). *Kolloid. Zh.* **14**, 267.
15. Teitelbaum, B. Ya., Ganelina, S. G. and Gortalova, T. A. (1950). *Bull. Kazan Branch Acad. Sci. USSR, Ser. Chem. Sci.* **1**, 105.
16. Robinson, J. V. and Woods, W. W. (1948). *J. Soc. Chem. Ind.* **67**, 361.
17. Prigorodov, V. N. (1970). *Kolloid. Zh.* **32**, 793; *Colloid J. USSR*, **32**, 662 (English translation).
18. Vincent, B. (1973). *In* "Specialist Periodical Reports, Colloid Science", Vol. 1, p. 251. The Chemical Society, London.

Discussion

Friberg In your Fig. 5 you would expect maximum foam stability at the maximum temperature for phase separation if the factor you have referred to is the decisive one.

Ross and **Nishioka** Yes, but the relative stabilities of foams are seldom determined by a single decisive factor.

Breuer (*Brunel University, England*) Could you explain the difference between the behaviour of systems with a lower and an upper consolute point.

Ross and **Nishioka** By your use of the term "consolute point" I take it that you refer to three (or more) component systems. I know of no experimental data on foam stabilities that refer to such systems with lower consolute points.

Laughlin (*Procter & Gamble Co., Cincinnati, Ohio, USA*) Have foaming experiments been done on systems which display a lower critical temperature rather than an upper.

Ross and **Nishioka** We are aware of at least one system (water and ethylene glycol monobutyl ether) with a lower critical temperature whose foam behaviour differs from systems with an upper critical temperature in that the approach to immiscibility is not accompanied by an increase in foam stability. Such behaviour may be caused by an endothermic reaction between the two components, leading to a phase separation that is not presaged by increasing surface activity. But we cannot say whether all systems with lower critical temperatures behave in this way. As for surface-active nonionic solutes containing a polyethylene-oxide moiety, these are well known to be profoamers in water at temperatures below their cloud-points; but we do not know how relative foam stabilities of such solutions change as the temperature is raised towards that point.

2

Foam stability and association of surfactants

S. FRIBERG and H. SAITO

The Swedish Institute for Surface Chemistry,
Drottning Kristinas väg 45,
S-114 28 Stockholm, Sweden.

Summary

Foam stability was determined using different phases of surfactant association structures and mixtures of them.

It was shown that liquid crystals may have a pronounced influence on the stability of foams if spreading conditions are satisfactory. In such cases the spreading of one phase on another in a thin film give rise to a spontaneous thickening of the thin film.

1 Introduction

The stability of foams[1] has in general been related to the colloid stability of thin films[2,3,4] by which the distance relations of the mutually independent Van der Waals' attraction potential and the electric double-layer repulsion potential are dominant factors.[5,6]

In this paper we will examine the influence of the introduction of a third phase with a liquid crystalline structure on foam stability. It was suggested several years ago[7-12] that water, surfactants and amphiphiles, in combination, could profoundly affect the stability of emulsions and foams, but the first reports in which well-defined and separated phases were mixed to establish critical conditions for the stability are comparatively recent.[13,14,15]

This paper deals specifically with investigations on the role of liquid crystals in foam stability and the application of these results to the action of foam-breaking compounds.

2 Materials and methods

Cetyl trimethyl ammonium bromide, CTAB, hexanol and twice distilled water were used.

The foams were prepared by shaking a volume of 5 ml of the mixture in a 15-ml stoppered test tube until no liquid was visible in the lower part of the tube. For very unstable foams no observation of bubble size was possible; for those foams with a time for half-height stability of a few minutes the shaking time was adjusted to give bubble sizes of approximately 1 mm diameter.

3 Results

The phases in the system have been described earlier by Ekwall and his co-workers.[13] They are illustrated in Fig. 1, showing one isotropic solution with normal micelles (L_1), one isotropic solution with inverse micelles (L_2) and two liquid crystalline phases of which one (N) has a lamellar structure.

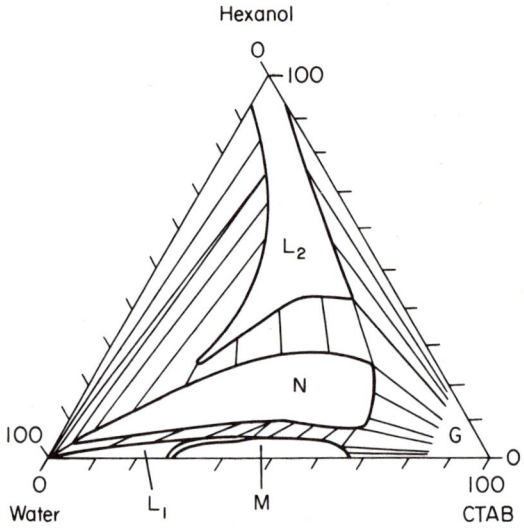

FIG. 1. Phase diagram of ternary system water, cetyltrimethylammonium bromide and *n*-hexanol. L_1 and L_2, aqueous and alcohol solution; M and N, liquid crystalline phases; G, solid CTAB.

Different phases gave foams of different stability and the differences were distinct. The times for the height of the foams to be reduced to half the initial value are given in Table 1.

4 Discussion

The results illustrate the importance of the effect of different association structures on the stability of foams. In the present case it appears obvious

TABLE 1

Half-height times for foams

Phases	Time
L_2	< 30 s
$L_{1\ (c\ <\ cmc)}$	1–2 h
$L_{1\ (c\ <\ cmc)}$	5–25 h
N	No foams
$L_{1\ (c\ >\ cmc)} + L_2$	< 30 s
$L_{1\ (c\ >\ cmc)} + L_2 + N$	10–60 min
$L_2 + N$	No foams
$L_1 + N$	25–30 h

that foam stability is principally due to the properties of the isotropic solutions. Isotropic solutions containing inverse micelles give extremely unstable foams and also have a detrimental effect when combined with the aqueous phase, reducing the times for stability from hours to seconds.

The liquid crystalline phases examined in these experiments do not form foams. They have recently been shown[14] to form reasonably stable thin films but in the experiments cited the films were carefully prepared to avoid vibration. It is reasonable to postulate that sudden thinning from very thick films (≈ 0.1 mm) to black films observed for liquid crystalline phases[14] is unlikely to be reproduced under the violent conditions characteristic of the present experiments.

When combined with the aqueous phase the liquid crystalline phase gives enhanced foam stability; this was also the case when both aqueous and alcoholic micellar solutions were present. The contribution of the liquid crystalline phase was, however, limited to improvement of foam stability; when combined with the alcoholic solution containing inverse micelles no foams at all could be produced.

The present results confirmed an earlier opinion[15] that the liquid crystal mainly operates as stabilizer localized at the junctions of the thin films in the foam. Its function appears to be two-fold: its high viscosity reduces liquid drainage through the junctions, and it serves as a reservoir of surfactants of optimal composition to stabilize the foam.

If the ratio between interfacial tensions in the air, isotropic liquid solution and liquid crystalline phase obey a relationship $\gamma_{1.\,cr./air} + \gamma_{1.\,cr./i.s.} > \gamma_{i.s./air}$, the liquid crystal may spread over the thin film. Such spreading has recently been observed experimentally by Roberts, Axberg and Saito.[16]

It appears probable that further information about spreading conditions in these systems will give a rational explanation of the present empirical information about liquid crystals and foam stability. The notion that the interfacial energies are a decisive factor is to some extent supported by the fact that the systems investigated so far have shown a stabilizing effect of the liquid crystal only in combination with one of the liquid phases.

The solubility of the surfactant in the liquid phase has been demonstrated not to be decisive; in one case[17] the surfactant and the hydrocarbon were completely miscible with each other in all proportions and the liquid crystalline phase stabilized foams when combined with water which did not dissolve the emulsifier. Examples of an opposite effect have been published.[18]

The results point to an explanation of the fact that identical molecules can act either as foam stabilizers or foam breakers. Our investigations show that when hexanol is mixed with a solution such that the resultant system consists of liquid crystalline and aqueous phases foam stabilization occurs. If the alcohol to surfactant ratio is high enough to give a combination of the liquid crystal with the solution containing inverse micelles, the foam is destabilized. This is especially the case when the alcohol is added drop-wise to the foam.

Acknowledgement

The authors wish to express their gratitude for financial support from the Swedish Board for Technical Development.

References

1. Prince, L. M. (1967). *J. Colloid Interface Sci.* **23**, 165.
2. Rosano, H. L., Peiser, R. C. and Eydt, A. (1969). *Rev. Tran. Corps. Goas.* **16**, 249.
3. Shah, D. O. and Hamlin, R. M. (1971). *Science*, **171**, 483.
4. Winsor, P. A. (1948). *Trans. Faraday Soc.* **44**, 376.
5. Palit, S. R., Moghe, V. A. and Biswas, B. (1959). *Trans. Faraday Soc.* **55**, 463.
6. Shinoda, K. (1967). "Solvent Properties of Surfactant Solutions", p. 4. Marcel Dekker, New York.
7. Gillberg, G., Lehtinen, H. and Friberg, S. (1970). *J. Colloid Interface Sci.* **33**, 40.
8. Ekwall, P., Mandell, L. and Fontell, K. (1970). *J. Colloid Interface Sci.* **33**, 215.
9. Shinoda, K. and Kunieda, H. (1973). *J. Colloid Interface Sci.* **42**, 381.
10. Rosano, H. L. and Gerbacia, W. E. (1972). Proc. VI Intern. Conf. Surf-Active Subst. Zürich. In press.
11. Corkill, J. M., Goodman, J. F. and Haisman, D. R. (1961). *Trans. Faraday Soc.* **57**, 821.
12. Balmbra, R. R., Clunie, J. S., Goodman, J. F. and Ingram, B. T. (1973). *J. Colloid Interface Sci.* **42**, 225.
13. Mandell, L. (1963). *Finska Kemistsamfundets. Medd.* **72**, 49.
14. Friberg, S., Saito, H. and Lindén, S. (1974). *Nature*, **251**, 494.
15. Saito, H. and Friberg, S. (1975). "Liquid Crystals" (S. Chandresekhar, Ed.). Raman Research Institute (1975).
16. Roberts, K., Asberg, C. and Saito, H. *Nature.* In press.
17. To be published.
18. Ahmad, S. I. and Friberg, S. (1971). *Acta Polytech. Scand.* Ch. 102.

Discussion

Exerowa In the work of Professor Friberg it is shown that a liquid crystalline phase does not form a foam but does form stable black films. However a mixture of liquid crystalline phase and isotropic solution gives a very stable foam—stable for 30 minutes. In this situation does one find black films? In many experiments we have proved that the formation of stable foams is connected with the formation of black films. At low foaming agent concentrations black films are not present and foam life is short whilst at concentrations sufficient for foam formation, black films are formed and the foam is stable. All the evidence indicates that these phenomena would seem to be caused by particular properties of the adsorbed layer. Is it that in the presence of a liquid crystalline phase a "surface" mechanism exists or are the high stabilities observed in these cases due to "bulk" effects?

I believe that it is the former mechanism.

Friberg The liquid crystal itself will not form a stable foam because its thin films thin differently from films from a solution. The thinning takes place momentaneously from a thick film to a black film. Such a black film is hence obtained only under special conditions (no vibrations and temperature control) which are not found in the formation of a foam.

The liquid crystal is collected to the borders in the foam and improves foam stability by two mechanisms: (a) it reduces the drainage rate from the foam; and (b) it serves as a reservoir of an optimal mix of emulsifiers to form a layer structure at an interface. Its action should be viewed as a "surface" mechanism.

Barber (*Shell Research, B.V., Amsterdam, The Netherlands*) You presented a sequence of two slides as evidence that the liquid crystalline phase, introduced into a Plateau border spread into the adjacent films, thus stabilized the foam. How can you be sure that the liquid crystalline phase spread into the films and not merely along the Plateau border network?

Friberg The spreading of material from a liquid crystal onto a thin film has also been observed directly on a macroscopic thin film adhering to a glass frame.

Padday The passage of the liquid crystal into the flat part of the thin foam film from the Plateau border increases the total thickness of the film and thereby reduces the total effective disjointing pressure. Is this process at equilibrium and if so was the equilibrium approach from both sides?

Friberg The passage of the liquid crystal onto the flat part of the foam film is a spreading of a new phase onto the liquid phase and the discussion about total effective disjoining pressure has to consider a three-layer structure at nonequilibrium. The free

37

interfacial energy between the liquid and the liquid crystal is certainly small compared to the differences in the tension towards air. Equilibrium cannot be obtained in a thin-film system of this kind; even a pure liquid crystal will thin in air to a black film.

3

Emulsion foam killers in foams containing fatty and rosin acids

KELVIN ROBERTS, CLAES AXBERG and ROLF ÖSTERLUND
The Swedish Institute for Surface Chemistry,
Drottning Kristinas väg 45, S-114 28 Stockholm, Sweden

1 Introduction

Though the stability of both foams and thin films have been extensively studied, and the forces which stabilize thin films are well understood both qualitatively and quantitatively,[1,2] the "breaking and prevention" of foams is less well understood. It is known, however, that addition of small quantities of specific agents to foaming systems can cause reduction in the stability of foams formed. These agents can be divided into two types. The first, foam breakers, are added to existing foams, and they are generally considered to act in the form of small droplets, which spread on the foam lamellae, so that the spreading liquid film carries with it a layer of the underlying liquid which comprises the foam lamella. This means that the lamella is thinned and the foam breaks.

Foam preventatives, on the other hand, are generally believed to adsorb at the air–water interface in preference to the film stabilizing surfactants. These molecules which have a strong affinity for the air–water interface do not have the capacity for stabilizing foam, and foam prevention is achieved.

More recently, it has been shown that the presence of a liquid crystalline phase in equilibrium with an aqueous micellar solution of surfactant improves the stability of the foams formed from the surfactant solution. Addition of a reversed micellar solution (an organic solution) capable of solubilizing the liquid crystalline phase to such a foam causes foam breaking.[3,4]

This paper is concerned with a new and previously unreported mechanism of action for foam preventers. In certain practical systems (e.g. sulphate mill back-waters) foam prevention is carried out in systems containing natural surfactants in alkaline solution by employing foam preventers containing an oil, an alcohol and a fatty acid. The mechanism of action of alcohol in foam prevention and foam killing has been discussed previously,[3, 5, 6] but the role of fatty acids in such systems is far from clear. In particular, it would be expected that the acids would saponify rapidly, and the soaps thereby produced aid foam stabilization. It would also be expected that such saponification would produce a spontaneous emulsification.[7, 8] This paper is concerned with studies determined to elucidate the role of such acids in foam prevention. As a model foaming system we have used a solution of 80 g l^{-1} abietic acid and 50 g l^{-1} myristic acid at pH 10, and 60°C. As foam preventer we have used decane containing nonionic surfactant.

2 Experimental

Foaming experiments were carried out in a thermostated column described elsewhere.[9] Air was introduced via a sinter at a rate of 2 ml min^{-1}, controlled by a flowmeter. The abietic acid used was Fluka (practical grade) and myristic acid was used as fatty acid and obtained as analytical grade from Merck. Berol 259, supplied by Berol Kemi AB, Sweden, is a nonyl phenyl polyoxyethylene monoglycol ether. Solutions containing 80 g l^{-1} abietic acid and 50 g l^{-1} myristic acid were made up and thermostated at 60°C. Air-flow was started, and foam height measured as a function of time up to 10 minutes. The air-flow was then switched off, and the foam collapse followed as a function of time. A series of solutions of foam preventer were made up in decane, containing increased amounts of Berol 259 from 20 g l^{-1} to 200 g l^{-1}. 0·2 per cent of each of the foam preventer solutions were added to the foaming solution prior to onset of foaming, and a foaming "run" carried out as described previously. In addition, the interaction of the foam preventer solutions with the alkaline foaming solution was studied by placing the two solutions in contact under a microscope and the thickness of the emulsion layer formed spontaneously was determined.

Following the microscope studies, the maximum solubilization capacity of the foaming solution for the various foam preventers was determined using standard techniques.[9, 10] An estimation was made of the relative rates of solubilization of the emulsions initially formed under standard mixing conditions corresponding approximately to the mixing conditions in the foaming cell.

Cine-microphotography of the foam formed in the presence of preventer was carried out.

3 Results

In Fig. 1 is shown the data obtained for foaming of solutions containing

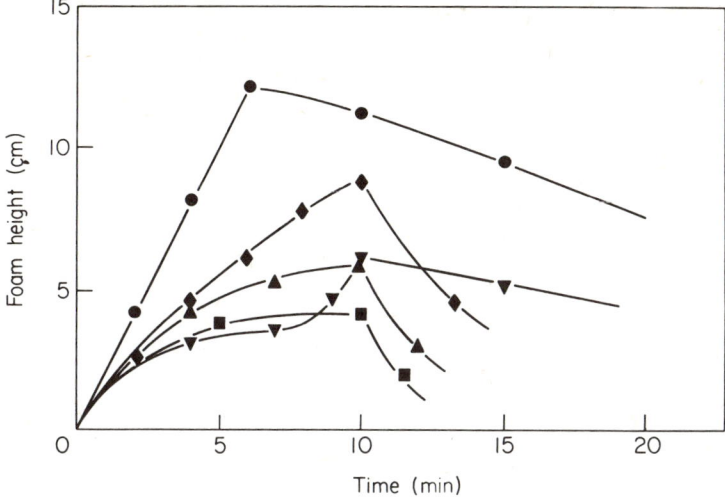

FIG. 1. Foam height as a function of time (from $t = 0$) for solutions of 50 g l^{-1} myristic acid, 80 g l^{-1} abietic acid, at pH 10 (NaOH for pH adjustment). Various foam preventers (0·2 per cent of total volume), added 15 s prior to foaming. ●, No foam preventer; ◆, decane added; ▲, decane containing 20 g l^{-1} Berol 259 added; ■, decane containing 100 g l^{-1} Berol 259 added; ▼, decane containing 200 g l^{-1} Berol 259 added.

80 g l^{-1} abietic acid and 50 g l^{-1} myristic acid at pH 10, with the same amounts of different foam preventers added. Air-flow begins at time $= 0$ and the air supply is switched off after 10 min (except in Run 1).

In Run 1, the foaming solution without added preventer is employed. The foam height obtained in the absence of foam preventer is 12 cm after 6 minutes. As the foam then neared the top of the column, the air supply was switched off and the height decrease followed (18 min to 6 cm). Run 2 shows the foaming behaviour of the solution to which decane alone has been added. Foam height is still increasing after 10 min (9 cm), and the half-time for decay is 3·25 minutes. On adding 20 g l^{-1} Berol 259 to the foam preventer, the foam height is 6 cm after 10 min and the half-life of the foam 2 minutes. An increase in concentration of Berol 259 in the foam preventer to 100 g l^{-1} results in a decrease of foam height to 4 cm after 10 min, with half life of 1·5 minutes. On adding 200 g l^{-1} Berol 259 to the foam preventer, foam height is 3 cm after 6 min, but begins to increase after 8 min to 6 cm after 10 minutes. The half-life for decay is 20 min, similar to the original foam before adding preventer. The relative slopes of the foam height against time curves indicate that the foam stability is highest with no foam preventer added, and the lowest for the foam preventer containing 100 g l^{-1} Berol 259.

Microphotographs of the foam preventers in contact with the foaming solution containing $20 \, gl^{-1}$ and $100 \, gl^{-1}$ Berol 259 in contact with the foaming solution are shown in Figs 2 and 3 respectively. Figure 3 shows the spontaneous formation of an emulsion at the oil–water interface. The foam preventer solution containing $200 \, gl^{-1}$ Berol 259 also showed spontaneous

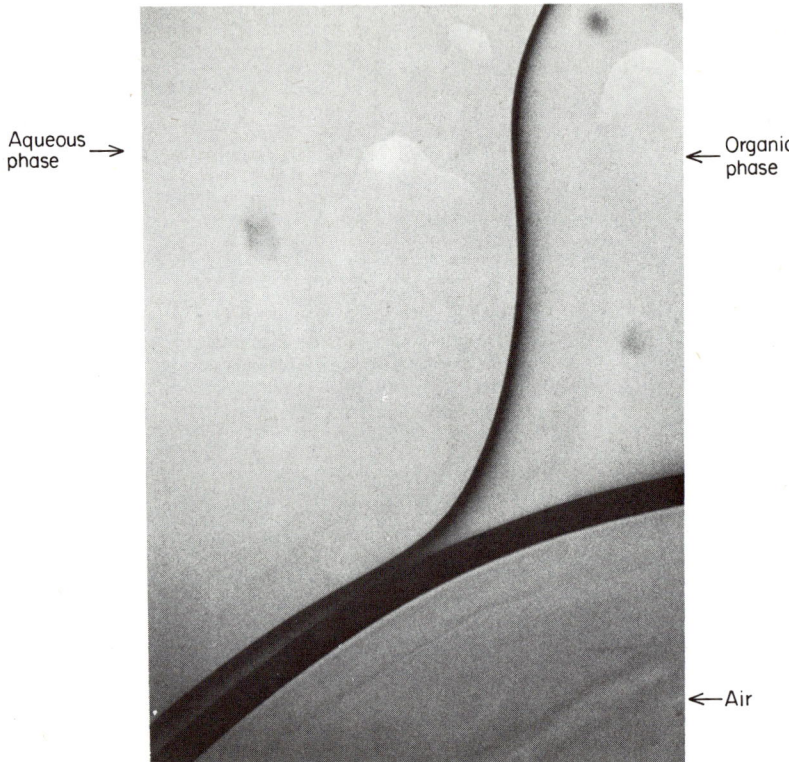

Aqueous phase →

← Organic phase

←Air

FIG. 2. Microphotography (70 ×) of the organic–aqueous interface for phases of the following compositions. Aqueous phase: $50 \, gl^{-1}$ myristic acid; $80 \, gl^{-1}$ abietic acid; pH 10 (NaOH). Organic phase: decane containing $20 \, gl^{-1}$ Berol 259.

emulsification. The preventers containing 2 per cent Berol 259 shows no spontaneous emulsification but emulsifies, of course, under the mechanical agitation in the foaming cell.

Solubilization studies indicated that a $2 \, mll^{-1}$ addition of the various foam preventers to the foaming solution under stirring, complete solubilization occurred within 20 min, though the preventer containing 200 g l^{-1} Berol 259 dissolved within about 10 minutes.

The cine-microphotography of the foam containing foam preventer containing 100 g l^{-1} surfactant is shown in Figs 4, 5 and 6. The film speed is

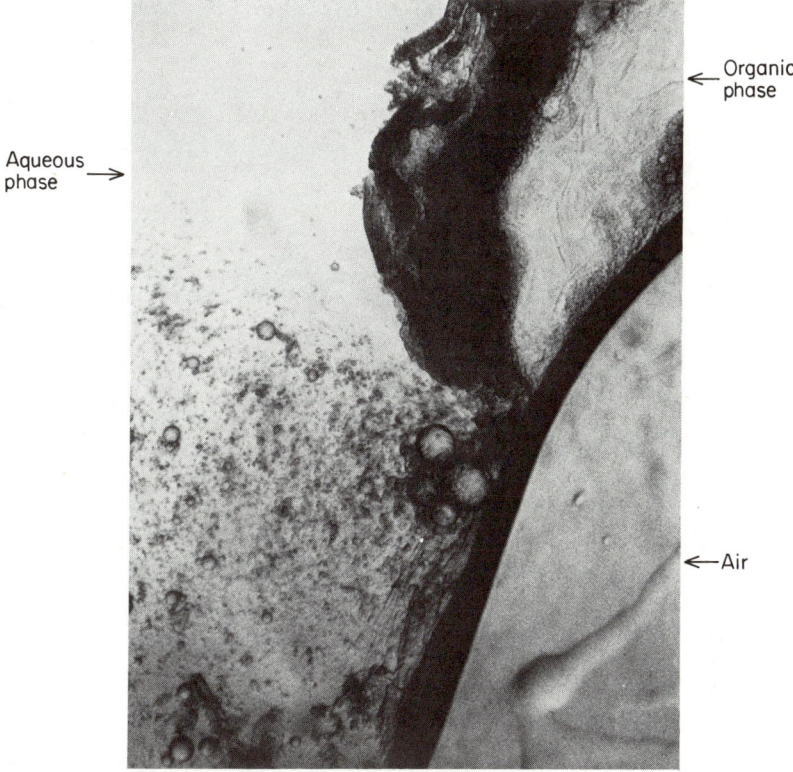

FIG. 3. As Fig. 2, but with the following phase compositions. Aqueous phase: 50 g l^{-1} myristic acid; 80 g l^{-1} abietic acid; pH 10 (NaOH). Organic phase: decane containing 100 g l^{-1} Berol 259.

64 frames per second. During the sequence, the emulsion drop which finally rests in the centre of the picture migrates through the film, and when the drop reaches the middle of the lamella shown the lamella bursts leaving the drop. In view of the small thickness of the glass cell used, c, 1 mm, the drop falls to the lower cover glass and remains in focus.

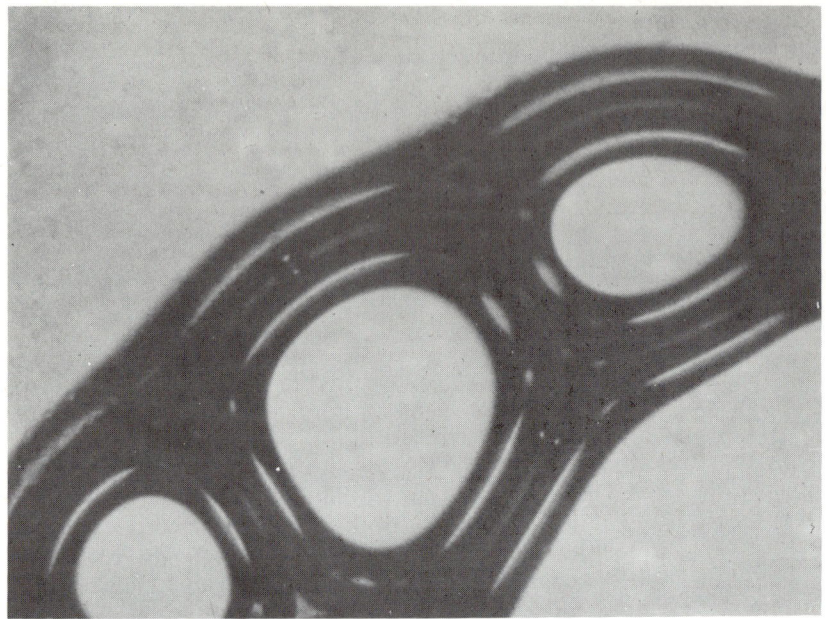

FIG. 4

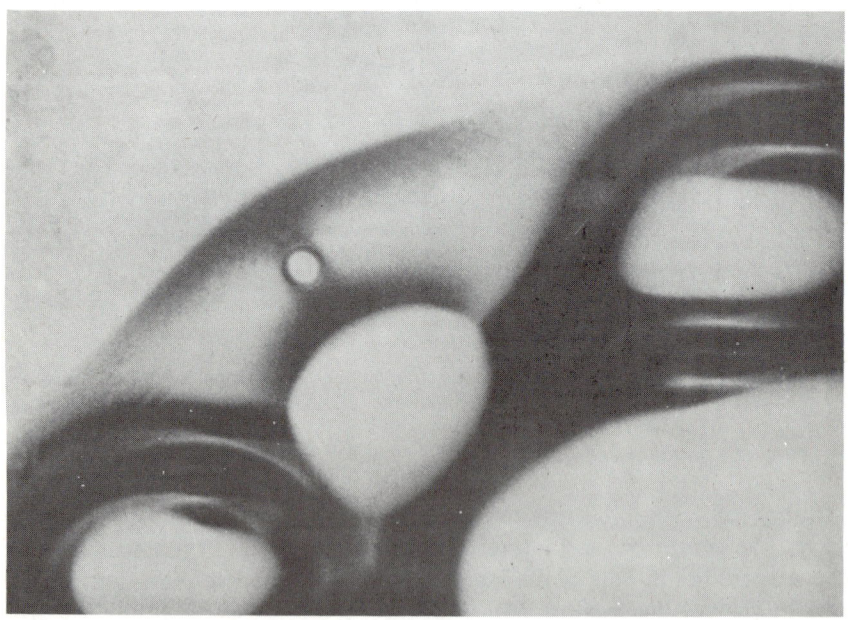

FIG. 5

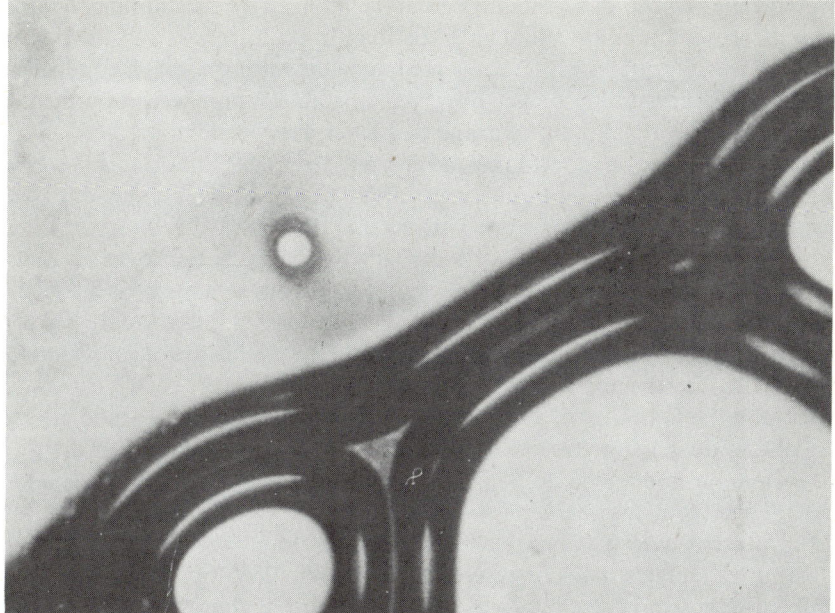

FIG. 6

FIGS 4–6. Sequence of microphotographs (60 ×) from cinefilm 64 frames per second of emulsion drops in foam lamella. Foaming solution: 50 g l^{-1} myristic acid; 80 g l^{-1} abietic acid; pH 10 (NaOH). Foam preventer: (0·2 per cent added) decane containing 100 g l^{-1} Berol 259.4. Drop in middle of lamella; 5. Lamella breaks, foam lamellae rearrange; 6. Lamella broken, drop remains.

4 Discussion

The experiments here reported indicate that addition of organic solutions containing nonionic surfactants can be used to prevent foam formation in alkaline solution containing sodium myristate and sodium abietate. The fact that foam preventers either spontaneously or mechanically emulsify in the foaming solution and then dissolve in the solution is significant. The life time of the emulsion drops (i.e. the time for solubilization) appears to be of the same order as the persistence of the foam prevention effect. The rapid increase in foam height observed with the foam preventer containing 200 g l^{-1} surfactant after 8 min indicates a reduction in foam preventing capacity. On discontinuation of the air-flow, the foam formed is shown to be as stable as the original foam before preventative is added.

It has recently been established that in systems containing an anionic surfactant, an amphiphilic substance, and water, the stability of foams formed from micellar solutions systems are greatly enhanced when a liquid crystalline phase is present in equilibrium with the foaming solution.[3,10] In the regions of the phase diagram in which a reversed micellar solution was in equilibrium with a liquid crystalline phase, however, unstable foams were obtained. In connection with these investigations by Friberg, it was demonstrated that in a three-phase region containing a normal micellar solution, a reversed micellar solution and a liquid crystalline phase, stable emulsions were formed,[11,12] and that foaming within the three-phase region produced stable foams with moderate life times. This indicates that an organic phase, if coated with a liquid crystalline phase, does not act as a foam preventative.

With this as background, the present experiments point strongly towards the conclusion that the transient existence of emulsion drops is of decisive importance for the foam-preventing effect. Emulsion drops which do not dissolve in the foaming solution, and which are protected by a liquid crystalline phase, do not function as foam inhibitors, whereas emulsion drops in the system here reported do.

Evidence from microcinephotography must always be treated carefully. In the present case it is necessary for depth of focus reasons to use a thin cell, and wall effects must be carefully considered. Two possible conclusions can be drawn from the cinephotographs. The first is that the emulsion drop in the photograph actually destroys the foam lamellae by an interaction at the lamella–air interface or the organic–aqueous interface. The second is that the same drop destroys the foam lamella by interaction at the lamella–glass interface. In the latter case, the drop would be expected to spread on the glass and thereby displace the lamella liquid, causing thinning and breaking. Two facts are against this interpretation. The first is that the drop does not continue to spread on the glass after the lamella has broken, as indicated by the fact that the drop does not decrease in size. (This sequence was followed for several minutes and the drop size remained constant.) The second is that it is unlikely that the drop, which according to the second hypothesis spreads on glass, should do so in the middle of one particular lamellae, when it has been observed in a previous photograph to migrate under foam drainage through several lamellae and plateau borders.

This argument, though indirect, strongly suggests that the first hypothesis, that the drop destroys the foam lamella by interaction at the lamella–air interface or organic–aqueous interface is correct. This hypothesis is readily understandable in surface chemical terms. The experimental investigations demonstrate that the spontaneously formed emulsion drops have a short life-time, since they dissolve into the micellar abietate and myristate solution. This means that a concentration gradient must exist between the emulsion drops and the foaming solution, and therefore a free energy difference exists between the drop and the material of the foam lamellae. It is reasonable then to expect that when such a drop reaches the air–lamella interface, the difference in free energy will be expressed as a difference in surface tension, and, since the drop is dissolving slowly in the material of the lamella, this will create a surface tension gradient in the lamella, operating radially out from the drop. This surface tension gradient would be expected to act as a force radially from the drop, and to lead to rupture of the lamella. This idea is supported by the fact that when the lamella breaks, the drop does not move, which indicates the foam-breaking force acts evenly round the drop. A similar surface tension gradient could also be brought about by a changed surfactant concentration in the neighbourhood of the lamella–air interface caused by the dissolving of the drop.

These investigations have demonstrated that it is possible to obtain a foam-preventing action by adding to a micellar foaming solution an organic solution containing a surfactant, in such a concentration that the organic phase spontaneously emulsifies, and then dissolves over a period of minutes to hours. The emulsion drops can then be carried by the foam-forming air stream into the foam, and, on draining, disrupt the foam lamellae at the air–lamella interface. Formation of stable emulsions containing liquid crystalline phases as stabilizing agents result in a lack of foam prevention,[3] as does the solution of the emulsion droplets including micellar foaming solution.

Acknowledgement

The authors wish to thank Professor Stig Friberg for advice and discussions, and the Swedish Board for Technical Development for financial support.

References

1. Kitchener, J. A. (1964). *In* "Recent Progress in Surface Science", Vol. 1. Academic Press, New York and London.
2. Clunie, J. S. *et al.* (1971). *In* "Surface and Colloid Science", (E. Matijević, Ed.), Vol. 3. Wiley Interscience, New York.
3. Saito, H. and Friberg, S. (1974). *In* Proceedings of the 25th Anniversary Conference. Raman Research Institute, Bangalore.
4. Friberg, S., Saito, H. and Lindén, S. (1974). *Nature* (*London*), **251**, 494.
5. Ross, S. and Butler, J. N. (1956). *J. Phys. Chem.* **60**, 1255.
6. Ross, S. and Bramfitt, T. H. (1957). *J. Phys. Chem.* **61**, 1261.
7. Ruschak, K. J. and Miller, C. A. (1972). *Ind. Eng. Chem. Fundam.* 534.
8. Miller, C. A. and Serwen, L. E. (1970). *J. Colloid Interface Sci.* **33**, 360.
9. Roberts, K., Österlund R. and Axberg, C. To be published.
10. Friberg, S. and Ahmad, S. I. (1971). *J. Colloid Interface Sci.* **35**, 175.
11. Friberg, S., Mandell, L. and Larsson, M. (1969). *J. Colloid Interface Sci.*, **29**, 155.
12. Friberg, S. and Mandell, L. (1970). *J. Amer. Oil. Chem. Soc.* **47**, 149.

Discussion

Hansen By what mechanism does your emulsion droplet lead to film rupture?

Roberts, K. We believe that the emulsion droplet causes film rupture via a Marangoni effect induced by the concentration gradient between the emulsion drop and the material comprising the foam lamellae. This would be expected to give rise to a surface tension gradient and thereby cause rupture.

Prins According to your explanation, the spreading of emulsion droplets over the surface of a film causes the collapse of this film. Friberg and Saito, however, consider that the spreading of liquid crystalline phase over the film surface is responsible for the film stability. Can you comment on this contradiction?

Roberts, K. The Marangoni effect responsible for breaking of the foam with an emulsion droplet occurs very rapidly. The time scale is far less than 1/100 second. The spreading of a liquid crystalline phase over the film surface is a completely different phenomenon, and the film thickening thereby produced occurs over seconds or minutes. Further, since the micellar solution from which the foams formed as described by Friberg and Saito are in thermodynamic equilibrium with the liquid crystalline phases which spread on the foam lamellae, no major thermodynamic driving forces exist to promote film instability.

Friberg If the Marangoni effect is responsible for the bursting of a film (foam breaking) then a spray of droplets should be as effective in foam breaking.
   Spontaneous emulsification produces finer droplets than are normally generated by a spray. In addition, the practical systems have an uneven supply of air, as well as mixture of foam and cellulose fibres. It is probable that the selectivity of spontaneous emulsified "foam killer" droplets is an asset in this last respect.

Roberts, K. Spontaneous emulsification produces finer droplets than are normally generated by a spray. In addition, the practical systems have an uneven supply of air, as well as mixture of foam and cellulose fibres. It is probable that the selectivity of spontaneous emulsified "foam killer" droplets is an asset in this last respect.

Padday and **Blake** (*Kodak Ltd. Harrow, England*) We would like to suggest that the bubble of antifoam material only acts when it enters the film–air interface. We have shown experimentally that a monolayer spreads very quickly by the Marangoni effect. In spreading, the monolayer then carries a substantial amount of the underlying liquid before it, so that an area is successively thinned until rupture occurs.

Roberts, K.   The interactions which you describe would explain the phenomena we observed and are in line with the previous comments and questions in this discussion.

Russo Is the 10 per cent of anionic detergent an optimum both from the working and economic point of view?

Roberts, K.   The 10 per cent anionic detergent concentration is not optimal from an economic viewpoint. The economic optimum must be determined for the particular system in question.

4

Dynamic surface properties and foaming behaviour of aqueous surfactant solutions

A. PRINS
Unilever Research, Vlaardingen, P.O. Box 114,
The Netherlands

Summary

It is known from experience that the amount of foam which can be produced from an aqueous surfactant solution depends both on the manner of foam formation and on the amount and nature of the surfactant. Up till now, attempts to explain this foaming behaviour in terms of equilibrium properties such as surface tension and adsorption have, however, failed.

As the production of foam is a dynamic process, the assumption is made that dynamic surface properties are involved. It is further assumed that the amount of foam is determined by the stability of the foam films during foam production.

A mechanism is introduced which explains how the stability of foam films is related to the way the foam is made and the surface dilational viscosity. Differences in foaming behaviour which are observed when various surfactant solutions are subjected to different foaming techniques are explained by this theory.

1 Introduction

The production of foam from an aqueous surfactant solution requires some form of agitation, such as shaking, whipping, beating or bubbling. It is known from experience that the amount of foam produced depends on the nature and amount of the surfactant and on the way the foam is generated, vigorous shaking of a nonionic surfactant solution in a partly filled bottle produces only a small amount of foam whereas gentle bubbling results in abundant foaming.

We have investigated the relation between agitation and foaming behaviour by analysing the foam-making process in detail. During foaming, agitation is responsible for the introduction of gas bubbles into the liquid.

In general the amount of gas brought into the liquid increases with the amount of agitation. On the other hand, mechanical agitation has a destabilizing effect on foam already formed. The ultimate amount of foam produced depends upon the equilibrium between the rates of foam production and breakdown.

Differences in foaming behaviour between various surfactant solutions subjected to different amounts of agitation will be explained on the basis of this principle.

Briefly, the theory considers that agitation causes local expansions (and contractions) of the thin liquid films between the bubbles. The connected increase in the film surface tension is estimated and compared with the measured increase in surface tension resulting in instantaneous film rupture. It is found that at a critical degree of agitation, where a further increase in agitation causes a considerable reduction of the foam volume, the increased film surface tension is in satisfactory agreement with the film rupture surface tension. This is held to prove that beyond critical agitation, foam collapse takes place as a result of the increased surface tension due to expansion of films.

2 Experimental

2.1 MATERIALS

Technical-grade nonionic surfactant Nonylphenol 14 EO (EO = ethylene oxide groups) was used without further purification. Water was distilled after being de-ionized.

All experiments were carried out at room temperature and atmospheric pressure.

2.2 MEASUREMENT OF THE FOAMING BEHAVIOUR

Foam was generated from 0·275 l of aqueous surfactant solution by means of a rotating wire cage. The foam volume in the stationary state was measured as a function of the velocity of the wire cage.

2.3 MEASUREMENT OF THE DYNAMIC SURFACE PROPERTIES

The surface of the surfactant solution was expanded in a shallow vessel (length 50 cm, width 6 cm, depth 1 cm) by moving three barriers attached to an endless belt over the surface of the solution in the direction of the arrow (Fig. 1). The surface tension was measured continuously by means of a roughened glass plate suspended from a transducer. The surface tension reached in the stationary state was measured as a function of the surface expansion rate.

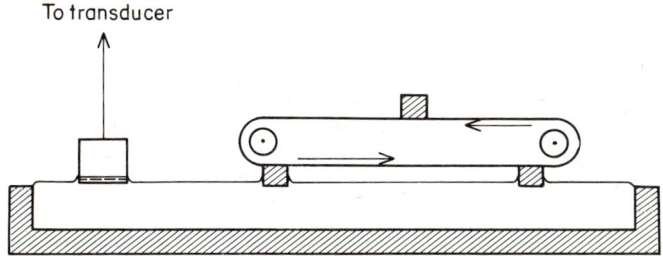

FIG. 1. Schematic representation of the experimental set-up for measuring the surface tension as a function of the rate of expansion.

2.4 MEASUREMENT OF THE FILM RUPTURE SURFACE TENSION

Films of the surfactant solution were prepared using a horizontal glass frame (Fig. 2).[1]

A small amount of the solution was brought into contact with two parallel glass rods A and B, situated at a distance of about half a millimeter. Due to

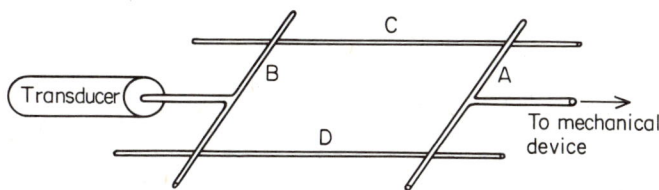

FIG. 2. Horizontal glass frame for measuring the film rupture surface tension.

capillary suction the space between rods A, B, C and D was completely filled with solution.

Rod A, in contact with rods C and D, was moved mechanically at a velocity of $0.2\ mm\ s^{-1}$ in the direction of the arrow. In this way an expanding film was formed between rods A, B, C and D.

Rod B, connected to a force transducer responding to horizontal forces only, was separated from rods C and D by a distance of 0.001 cm. This enabled us to continuously measure the surface tension of the expanding film. The surface tension measured just before the film ruptured was taken as the surface tension of film rupture.

To ensure a saturated water vapour atmosphere, the whole apparatus was enclosed in a Perspex box containing a layer of surfactant solution at the bottom.

3 Results

3.1 FOAMING BEHAVIOUR

The amount of foam measured as a function of the velocity of the wire cage
is given in Fig. 3. The maximum amount of foam appears to increase with
the surfactant concentration. At low velocities of the wire cage, the foam

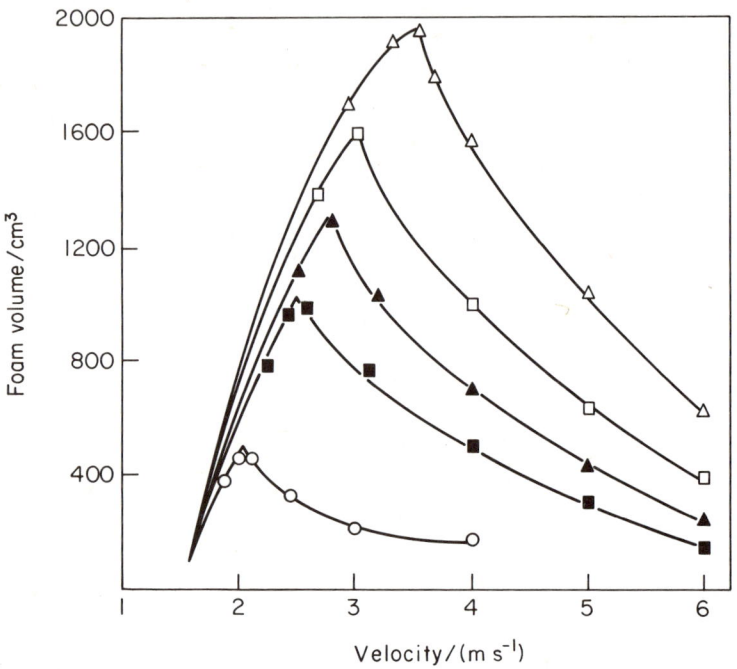

FIG. 3. Foam volume as a function of the velocity of the wire cage for various surfactant concentrations. $\triangle$, $0.2\,g\,l^{-1}$; $\square$, $0.1\,g\,l^{-1}$; $\blacktriangle$, $0.05\,g\,l^{-1}$; $\blacksquare$, $0.03\,g\,l^{-1}$; $\circ$, $0.01\,g\,l^{-1}$.

volumes of all the systems follow the same trend. At a well-defined higher
velocity, however, the foam volume suddenly decreases. The velocity at
which the sudden decrease takes place increases with the surfactant concentration. This velocity is taken as the critical value beyond which mechanical
agitation causes rupture of foam films resulting in lower foam volumes.

3.2 DYNAMIC SURFACE PROPERTIES

The surface tension as a function of the rate of surface expansion for the
various surfactant concentrations is given in Fig. 4. The surface tension

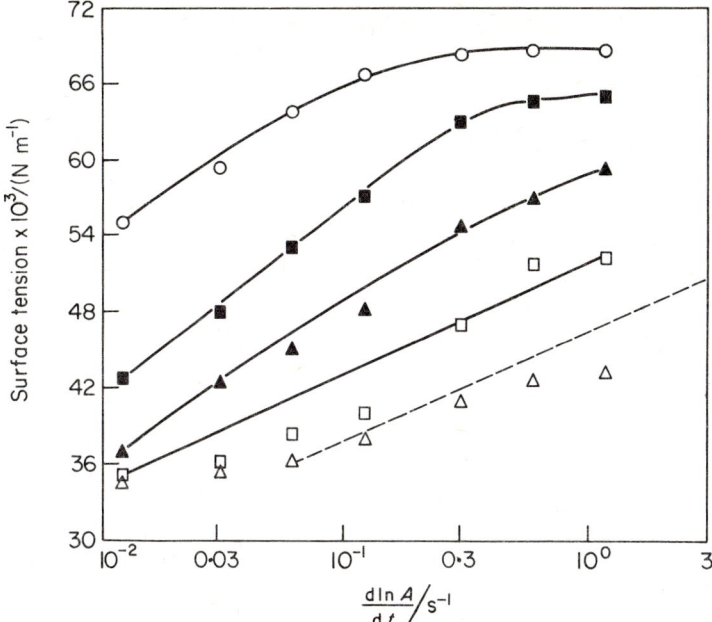

FIG. 4. Surface tension as a function of the rate of surface expansion for various surfactant concentrations. △, 0·2 gl⁻¹; □, 0·1 gl⁻¹; ▲, 0·05 gl⁻¹; ■, 0·03 gl⁻¹; ○, 0·01 gl⁻¹.

increases with the rate of expansion, as was shown previously by Van Voorst Vader *et al.*[2] Higher surface tensions are found with lower surfactant concentrations.

For low surfactant concentrations, the curves bend at higher expansion rates, obviously due to the surface tension of aqueous surfactant solutions being restricted to values below $72 \times 10^{-3}\,\mathrm{N\,m^{-1}}$.

3.3 SURFACE TENSION OF FILM RUPTURE

The film rupture surface tension curve A, Fig. 5, decreases as the surfactant concentration is increased. However, the difference between the film rupture surface tension A and the equilibrium surface tension B is almost independent of the surfactant concentration.

4 Discussion

During the production of foam, agitation causes accelerations of the liquid volume elements. The associated pressure fluctuations Δp, as given by

C

Bernoulli's law, take place in a time Δt given by the characteristic dimension of the wire cage L used for the agitation. Consequently

$$\frac{\Delta p}{\Delta t} = -\frac{\rho v^2}{2\Delta t} = -\frac{\rho v^3}{2L} \tag{1}$$

where ρ is the density of the liquid and v is the peripheral velocity of the wire cage.

The gas bubbles are also subjected to these pressure fluctuations and they adjust their volumes in order to remain in mechanical equilibrium with their surroundings.[3] For the sake of simplicity, it will be assumed that the bubbles contain only air, at a pressure which equals the pressure of the surrounding liquid and which behaves according to Boyle's law at constant temperature:

$$p\frac{dV}{dt} + V\frac{dp}{dt} = 0, \tag{2}$$

where V is the volume of the bubble. Consequently

$$p\frac{dR}{dt} + \frac{R}{3}\frac{dp}{dt} = 0, \tag{3}$$

where R is the radius of the bubble.

The change in surface area of the bubble, dA, follows from equation (1) and equation (3):

$$\frac{d \ln A}{dt} = \frac{1}{3}\frac{\rho v^3}{Lp}. \tag{4}$$

The associated change in surface tension is in principle given by the value of the surface dilational viscosity[2,4] which depends on the amount and

TABLE 1

Film rupture surface tension and the surface tension corresponding to the rate of surface expansion at the outset of foam collapse, for various surfactant concentrations

Surfactant concentration $g\,l^{-1}$	Critical rate of expansion at the outset of foam collapse s^{-1}	Surface tension at the critical rate of expansion $(N\,m^{-1} \times 10^3)$	Surface tension at film rupture $(N\,m^{-1} \times 10^3)$
0·01	0·68	69	68
0·03	1·2	65	64
0·05	1·35	62	62
0·1	2	56	56
0·2	3·2	52	51·5

nature of the surfactant. For the purpose of the present investigation, however, the increased surface tension is measured directly as a function of the rate of expansion (Fig. 4).

Application of equation (4) requires the value of a characteristic dimension L. The rather arbitrary assumption will be made that L equals the diameter of the wire cage: 4 cm. The critical velocity of the wire cage beyond which the foam volume suddenly decreases, is now equated by means of equation (4) to the critical rate of expansion of the bubble surface (Table 1). The values of the increased surface tensions which correspond to these expansion rates are obtained from Fig. 4 and are given in the third column of Table 1. For the highest concentration used (0·2 g l^{-1} extrapolation of the curve in Fig. 4 was necessary. This was done by assuming that this curve runs parallel to the other curves.

During agitation, the thin liquid films between the bubbles are also subjected to increased surface tension. The surface tension at which film rupture takes place may be obtained from Fig. 5 and is given for the various

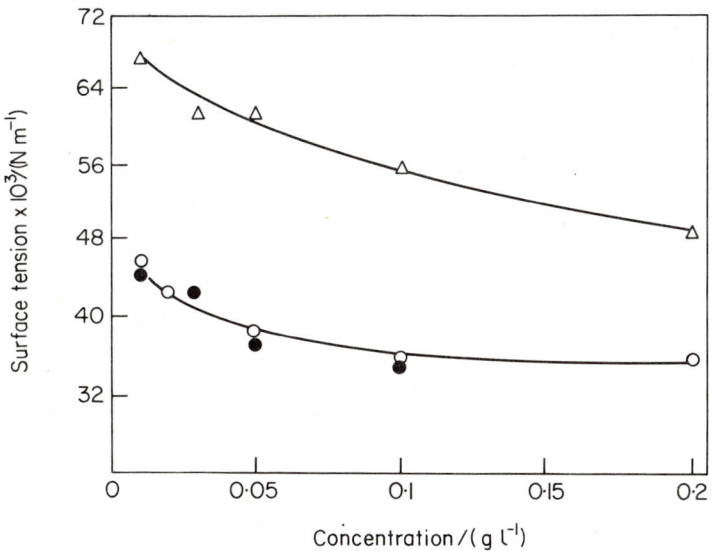

FIG 5. Equilibrium surface tension of film (●) and bulk solution (○) and film rupture surface tension (△) as a function of the surfactant concentration.

surfactant concentrations in the last column of Table 1. The agreement between the film rupture tension and the surface tension reached during foam making at the outset of foam collapse is taken as an argument in favour of the present mechanism, which explains the relation between foaming behaviour and the way of foam making.

It is realized that in the present study properties of surfaces which have different histories are compared; for instance the bubble surface during foaming, the surface of the bulk solution in the shallow vessel and the surface of the single film. The composition of a liquid surface may depend on the way it has been made, and therefore one might expect that these surfaces have different properties. However, several observations suggest that the surface properties of the systems in the present investigation are not strongly dependent on their history. These observations are:

i. the surface tension of the nondisturbed single film equals the equilibrium surface tension of the bulk solution in the shallow vessel;

ii. the surface tension required for film rupture is not changed when the rate of expansion of the film is increased by one order of magnitude.

Acknowledgement

Thanks are due to Mr J. C. van de Pas and to Mr B. J. Bel for their skill in carrying out the experiments.

References

1. Krotov, V. V. and Rusanov, A. I. (1962). *Kolloid Zh.* **34**, 297.
2. Van Voorst Vader, F., Erkens, Th. F. and Van den Tempel, M. (1964). *Trans. Faraday Soc.* **60**, 1.
3. Minnaert, M. (1933). *Phil. Mag.* **16**, 235.
4. Lucassen-Reynders, E. H. and Lucassen, J. (1970). *Advan. Colloid Interface Sci.* **2**, 347.

Discussion

Exerowa I believe that it would be interesting to extend the physical treatment of the results presented by Prins on the stability of foams and the nature of the adsorbed layer. We found that the same surface tensions corresponded to equal values of surface concentration. In this context it would be justifiable to interpret the results in terms of the mean surface concentration. On this basis the results of Prins agree very well with earlier work of Sheludko and Polikarova, as shown by the simple experiments in which a fixed volume of known concentration of sodium laurate solution was formed into a bubble until it burst.

It was shown that rupture occurred at the same surface concentration ($3 \cdot 6 \times 10^{18}$ molecules m^{-2}) independent of initial solution concentration and for different critical film thickness. In the same way the life of microscopic foam films may be interpreted in terms of surfactant concentration as demonstrated in our paper.

Beyond a particular surfactant concentration the life of a thin film is observed to increase markedly. For the formation of a stable film it is necessary that the surface concentration of surfactant be sufficient to form a compact adsorbed layer; e.g. NaLS: $\Gamma = 2 \cdot 6 \times 10^{18}$ molecules m^{-2}, F = 0·39 nm^2; OP2O: $\Gamma = 1 \cdot 7 \times 10^{18}$ molecules m^{-2}, F = 0·59 nm^2.

These results prove categorically that the nature of the adsorbed layer is responsible for the existence of very stable foams.

Prins The interesting agreement between our results and the results of Scheludko and Polikarova may lead to the conclusion that the adsorption plays an important role in the mechanism of foam collapse as suggested by Exerowa. More quantitative confirmation of this statement cannot be obtained from our results because the nonionic surfactant was of technical grade and in this case adsorption data cannot be obtained in a simple way from surface tension measurements.

Breuer Is there any dependence of the bubble size on the rate of beating?

Prins The smallest bubbles are found at that rate of beating which corresponds with the maximum in foam height. It is well known that bubbles are split up into smaller ones when velocity gradients of sufficient magnitude are generated in the surrounding liquid. This explains why by increasing the velocity of the whipping rod the bubble size decreases. Beyond the maximum of the foam height, however, higher velocity of the whipping rod causes the collapse of the film between the bubbles which results in an increase bubble size.

Zichy *(ICI Plastics Division, Welwyn Garden City, England)* Could you explain the details of the stirrer you use in your experiments. Are the bubbles formed at the tip of the blade at the point of maximum verticity, or are bubbles formed at a range of radii, so that at higher speeds the stirrer tip destroys some of the bubbles formed at smaller radii.

Prins The effect of the detailed geometry of the stirrer on the foaming behaviour has not been investigated in a systematic way. The present theory accounts for the mechanical agitation exerted on the foam by the rotating rod. By increasing the velocity of the whipping rod the mechanical agitation for foam collapse is reached first at the tip of the stirrer. Under these conditions it is indeed to be expected that bubbles formed at smaller radii of the stirrer are destroyed at the tip of the stirrer.

Epstein *(White Sea and Baltic Co. Ltd., South Woodford, London, England)* The results shown in Fig. 3 imply that for a given concentration of surfactant there is an optimum speed of stirrer for a maximum yield of foam. Does this optimum exist independently of the kind and detailed mechanical design of the apparatus used to produce the foam, for example, stirring, shaking, blowing a gas, etc.? If so would the graph be the same for various ways of mixing?

Assuming that the aim of stirring is to produce maximum amount of foam, is it possible to establish some general recommendations other than trial and error?

Prins In the theory the amount of foam is related to the amount of mechanical agitation and not to the way in which the agitation is exerted on the system. Foaming experiments, in which nonionic solutions have been shaken in partly filled bottles at various velocities, have confirmed that the foaming behaviour is qualitatively the same as that obtained with a Kenwood whipping machine.

A very small amount of mechanical agitation is exerted on the system when foam is made by gentle bubbling of gas through the liquid. In agreement with expectations a large amount of foam is produced out of the nonionic solutions under these conditions. The "general recommendation" for the production of the maximum amount of foam by stirring follows the theory: measure the collapse surface tension of a single film and measure the rate of expansion of the surface of the solution $d \ln A/dt$ which is required to reach the collapse surface tension. By introducing this value of $d \ln A/dt$ in equation (4) the maximum value of the velocity of the stirrer v can be calculated which corresponds to the maximum amount of foam.

5
Equilibrium soap films

B. T. INGRAM

Procter & Gamble Technical Centre,
Newcastle upon Tyne, NE12 9TS, England

Summary

A review is given of the ways in which equilibrium soap films have been used to study the intermolecular forces contributing to foam stability and to provide systems for investigating theories of colloid stability in general. The two types of equilibrium soap film—the "common black" and "Newton black"—are discussed and the problem of interpreting measurements made on the latter is highlighted.

Two experiments with Newton black films are described. In the first, film thickness was measured as a function of water vapour pressure giving a measure of the repulsion force operating in the film. In the second, the effect of polymeric additive (polyethylene glycol) was investigated and the large increases in the absolute value of the film's excess tension produced by the higher molecular weight polymers tentatively attributed to increased van der Waals attraction caused by polymer adsorption in the film.

1 Introduction

Equilibrium soap films have received a considerable amount of scientific attention in recent years[1,2] because, apart from their interest for those specifically studying aqueous foams, they provide fairly simple experimental systems for investigating the various long-range intermolecular forces involved in colloid stability in general. Equilibrium soap films—the term "soap" is usually taken to include films containing any kind of surfactant, not exclusively alkyl carboxylates—are usually so thin (< 20 nm thick) that they reflect very little light and are therefore known as "black" films.

The structure of the equilibrium soap film has been determined by a variety of techniques.[2] Radiotracer experiments using labelled surfactants have shown[3,4] that the surfactant concentration in black films corresponds to that of two close-packed monolayers. It is therefore assumed that the film has the sandwich-like structure of a thin aqueous layer between two surfactant monolayers, although there is evidence[5] that some films may have multilayer structures during drainage. The aqueous layer thickness has been measured using infrared absorption[6] and the total film thickness obtained by a combination of reflectance and Brewster angle measurements[7] and by ellipsometry[8] assuming a symmetrical, isotropic, three-layer optical model. The use of such a model for interpreting reflectance measurements is supported by the agreement between film thicknesses obtained in this way with those determined independently using low-angle X-ray scattering.[9]

It should be mentioned that the structural similarity between the soap film and the repeat unit of lamellar mesomorphic (neat) phase—the liquid crystalline phase often found at high solute concentration in aqueous surfactant solutions—has prompted some investigators[10, 11] to compare film thickness with lamellar unit repeat distances for the same surfactant system. Differences were found between these, however, but were partly ascribed to the larger compressive forces expected to act on a single film in air compared to one in a multi-lamellar structure.

Two main types of equilibrium soap film have been recognized. The "common" or "first" black film has a thickness which decreases mono-tonically with increasing electrolyte concentration and is usually much thicker than the "Newton" or "second" black film which has a thickness almost independent of electrolyte concentration and only slightly greater than the thickness of its two surfactant monolayers. Both types of film can occur with the same surfactant by changing the electrolyte concentration, the transition from one to the other usually being indicated by a large, abrupt change in equilibrium thickness. Films containing ionic surfactant have been found[12, 13] to exhibit a transition from common to Newton black film as the electrolyte concentration is increased, whereas with a nonionic surfactant[14, 15] the reverse has been found.

An explanation for the two types of black film has been sought in con-siderations of the various intermolecular forces likely to be operating in the film,[1, 2] for example, van der Waals attraction due to dispersion forces, repulsion arising from the overlap of diffuse ionic layers which extend from each surface into the aqueous core as a result of the surface charge caused either by the ionized state of the surfactant itself or by inorganic ions preferentially adsorbed,[15] and a much shorter range repulsion caused by interaction between the assemblies of solvation molecules associated with the surfactant. The resultant force acts perpendicularly to the film surface to resist thinning and, as a force per unit area, is often called the "disjoining" pressure. In an equilibrium film, where it is usually much smaller than the separate contributing forces, it is balanced by the hydrostatic or capillary pressure. The common black film is thought to be stabilized by a balance between the electrostatic double layer repulsion and van der Waals attraction and is thus an example of a system to which the DLVO theory of colloid stability[16] applies. The effect of electrolyte concentration on its thickness is therefore attributed to the reduction in double-layer repulsion with increasing ionic strength. The forces stabilizing the Newton black film, however, are less well understood. It seems that the Newton black film has a more organized structure than the common black film. Conductivity experiments[17,18,19] have indicated that the ions in the aqueous layer of a Newton black film are less mobile than in common black films. Newton black film formation is very dependent upon the exact type of ion in the aqueous phase[12]—some ions promote Newton black film formation while others inhibit it. These observations suggest that the aqueous core of a Newton black film is very unlike the bulk aqueous solution and, in films containing salt additives, may actually have a quasi-crystalline structure.

The potential energy of a soap film decreases as the film thins until it reaches a minimum at equilibrium. It seems probable that in some cases two minima, corresponding to the energy states of the Newton and common black films, can coexist in the free energy versus thickness curve for a film, and the transition from one state to the other is thus abrupt and rather like a phase transition. The depth of the energy minimum is related to the "excess tension", $\Delta\sigma$,[2,14] which is the difference between the total film tension and twice the surface tension of the bulk solution, and is always negative. $\Delta\sigma$ has been measured directly using a modified Wilhelmy plate technique[20,21] and also determined indirectly from measurements of the contact angle between the film and its bulk solution.[22,23,24] It has been related[2,25,26] to the disjoining pressure and composition of the film via a Gibbs–Duhem equation:

$$d(\Delta\sigma) = h\,dP_D - \Delta s\,dT - \sum_i \Delta\Gamma_i\,d\mu_i, \tag{1}$$

h is film thickness, P_D is disjoining pressure, T is temperature, μ_i is the chemical potential of component i in the bulk solution, Δs is the difference between the excess entropy of a unit area of film and twice the excess entropy of a unit area of solution surface (the excesses being relative to bulk solution of the same volume), and $\Delta\Gamma_i$ is the difference between the excess concentration of i per unit area of film and twice its excess at the solution surface. (This equation has been used[26] to obtain the excess salt concentrations in films from determinations of the variation of $\Delta\sigma$ with electrolyte concentration.) Since P_D is the sum of the different intermolecular forces in the film and each force is a function of film thickness we then have two relationships:

a force balance

$$P_{D,eq} = \sum_j P_j(h_{eq}), \tag{2}$$

where $P_{D,eq}$ is the equilibrium disjoining pressure (equal to capillary or hydrostatic pressure) and $P_j(h_{eq})$ represents the jth type of contributing force at the equilibrium thickness, h_{eq}; and an energy balance

$$\Delta\sigma = h_{eq}P_{D,eq} - \int_{\infty}^{h_{eq}} \sum_j P_j(h)\,\mathrm{d}h \tag{3}$$

which is obtained by integrating equation (1) at constant temperature and solution composition. With common black films, where the absolute value of $\Delta\sigma$ increases as ionic strength increases and thickness decreases, these equations have been applied using the assumption that only dispersion and double-layer forces contribute to the disjoining pressure and there has, on the whole, been quite good agreement between experiment and theory.[5,26,27] (The same assumption has also been used quite successfully to interpret light-scattering measurements on common black films.[28]) However, with Newton black films the situation is very different. $\Delta\sigma$ is found to be very sensitive to both the concentration and type of additive.[1,5,26,27] Very large changes in $\Delta\sigma$ occur over quite narrow ranges of electrolyte concentration without much change in film thickness, and order of magnitude differences have been found[27] between observed values of $\Delta\sigma$ and values calculated using DVLO assumptions, the latter values usually being gross under-estimates.

It is apparent that there are large gaps in our understanding of the factors which affect the stability of very thin equilibrium soap films, particularly the structure of their aqueous core, and much more study of Newton black films is warranted. To conclude this short review of equilibrium soap films then, I describe two pieces of previously unreported work concerning Newton black films. In the first experiment, the thickness of a large planar film formed from an aqueous solution of the anionic–cationic complex surfactant n-decyl trimethylammonium n-decyl sulphate, $C_{10}^+C_{10}^-$, was measured at different water vapour pressures. In the second, the effect of polyethylene glycols on the excess tensions of films stabilised by n-decyl methyl sulphoxide was investigated.

2 Experimental

2.1 VAPOUR PRESSURE EXPERIMENTS

The preparation of the $C_{10}^+C_{10}^-$ and the technique used for measuring film thickness have been described previously.[6,14] The films were formed by completely withdrawing a large, rectangular, glass frame from an aqueous solution containing $C_{10}^{+}C_{10}^{-}$ at a concentration of 0.45 mol m^{-3} inside a glass vessel submerged in a water bath held at 298.15 ± 0.01 K. The surfactant solution was then replaced by various salt solutions chosen to give relative water vapour pressures down to 0.97. The thickness, monitored by reflectance, decreased with decreasing water vapour pressure. Constant thickness was achieved within about five hours and was maintained thereafter for many hours. (Even the thinnest film lasted for over ten hours.) The thickness was calculated from the reflectance by assuming that the thick, draining film had a three-layer optical structure consisting of two outer layers with refractive index of 1.41 and thickness of 1.45 nm enclosing an aqueous layer with refractive index of 1.33, but that the equilibrium film was best represented by a single isotropic layer with refractive index 1.41. (This assumption

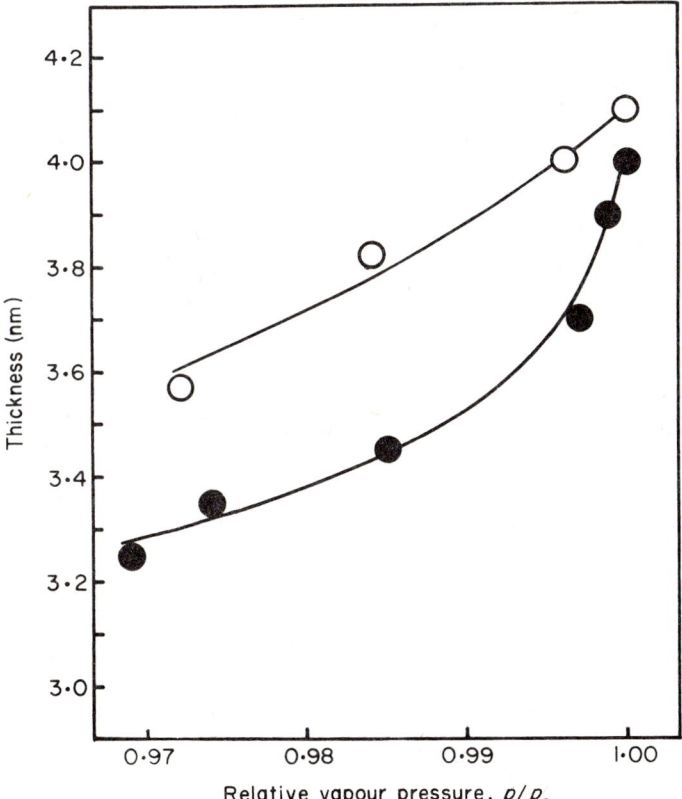

FIG. 1. Film thickness and lamellar liquid crystal spacing in the $C_{10}^{+}C_{10}^{-}/H_2O$ system, 298 K. Film (○); neat phase (●).

resulted in a correction of about -1.1 nm being applied to the equivalent water thickness.) The uncertainty inherent in such a calculation of film thickness and doubts about whether the film could really be in equilibrium with the vapour prompted us to use the structural similarity between the film and the lamellar liquid crystal[10] and compare the film thickness with the unit spacing of $C_{10}^+ C_{10}^-$ neat phase.

The neat phase spacing for a range of compositions (about 60 per cent to 80 per cent $C_{10}^+ C_{10}^-$) was measured by low-angle X-ray diffraction[29] and the water activity in the neat phase was determined isopiestically.[30] Figure 1 shows the equilibrium film thickness and neat phase spacing as a function of relative water vapour pressure. (The film thickness had a reproducibility of about ± 0.2 nm and the neat phase spacing about ± 0.05 nm.)

2.2 POLYMER ADDITIVE EXPERIMENTS

The preparation of the n-decyl methyl sulphoxide and the techniques for measuring the thickness and excess tension of the film have been described in detail previously.[15] The excess tension was measured with a micro-balance[20] and also calculated from the contact angle which was obtained by the diffraction technique.[24, 31] The intensity of the diffraction pattern was monitored with a photomultiplier rather than recorded photographically as in earlier work.[15]

The polyethylene glycols (PEG) were BDH materials which were used without purification. Because these had Gaussian distributions of molecular weight some measurements were made using a polyethylene glycol which had been made by the Williamson ether synthesis and contained only a single species of molecular weight 810.

The films were all formed from solutions containing 3.0 mol m^{-3} decyl methyl sulphoxide.

3 Results and discussion

3.1 VAPOUR PRESSURE EXPERIMENTS

The van der Waals compression in a soap film is probably an order of magnitude greater than in the neat phase bilayer[32] so that we would expect the film to be slightly thinner than the neat phase unit, although, because the force of repulsion would undoubtedly increase greatly with decreasing thickness, the difference in thickness would not be expected to be large. We can see from Fig. 1 that in fact the film seems to be slightly thicker. However, the uncertainty regarding the appropriate optical model for the film makes this unlikely to be significant.

If the assumption is made that the surfactant layers are unaltered by the dehydration process, the disjoining pressure can be calculated from the relationship[25]

$$P_D = -(RT/\overline{V}) \ln (p/p_0). \tag{4}$$

R is the gas constant, $\overline{V}$ is the molar volume of water in the aqueous layer and is taken to have the same value as in bulk water, and p/p_0 is the relative water vapour pressure. Figure 2 shows P_D as a function of the thickness of

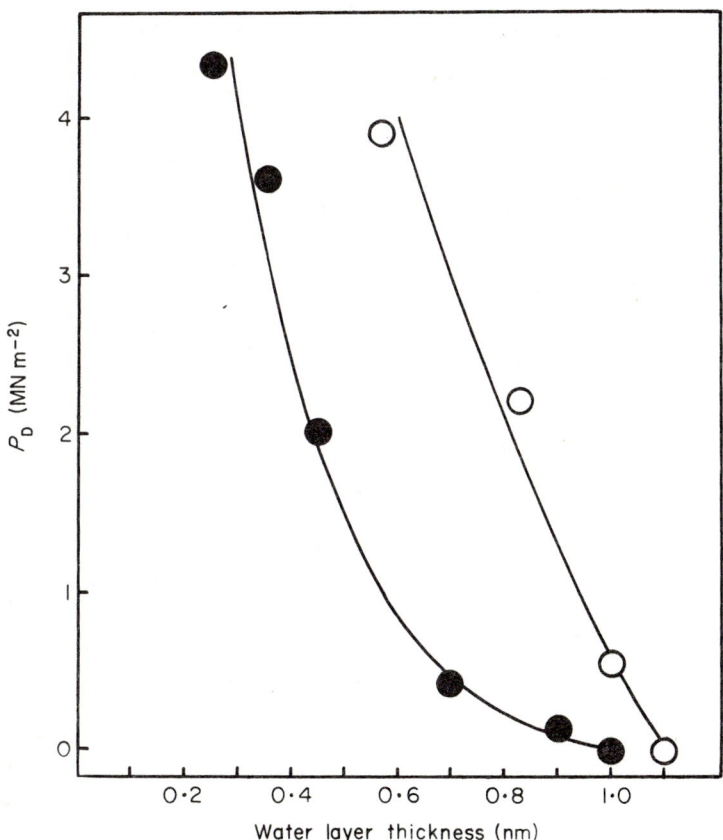

FIG. 2. Disjoining pressure isotherm. Film (○), neat phase (●).

the water layer which was obtained by subtracting 3·0 nm for the surfactant bilayer thickness (i.e. crystal unit spacing) from the total thickness. The steep rise in P_D as the water layer thinned is clearly seen.

P_D is, of course, the nett force acting on the film and includes forces of attraction, such as dispersion forces and perhaps electrostatic interaction between the two mosaic-like planes of surfactant ions, as well as any forces of repulsion. No attempt is made here to obtain a theoretical P_D but any such attempt would have to take account of the highly structured nature of the water—the orientation of the water dipoles must be considerably affected by the close proximity of the surfactant ions—and might perhaps best be carried out using a Monte Carlo computer simulation.[33]

3.2 POLYMER ADDITIVE EXPERIMENTS

Figure 3 shows the excess tensions of films containing PEGs of various

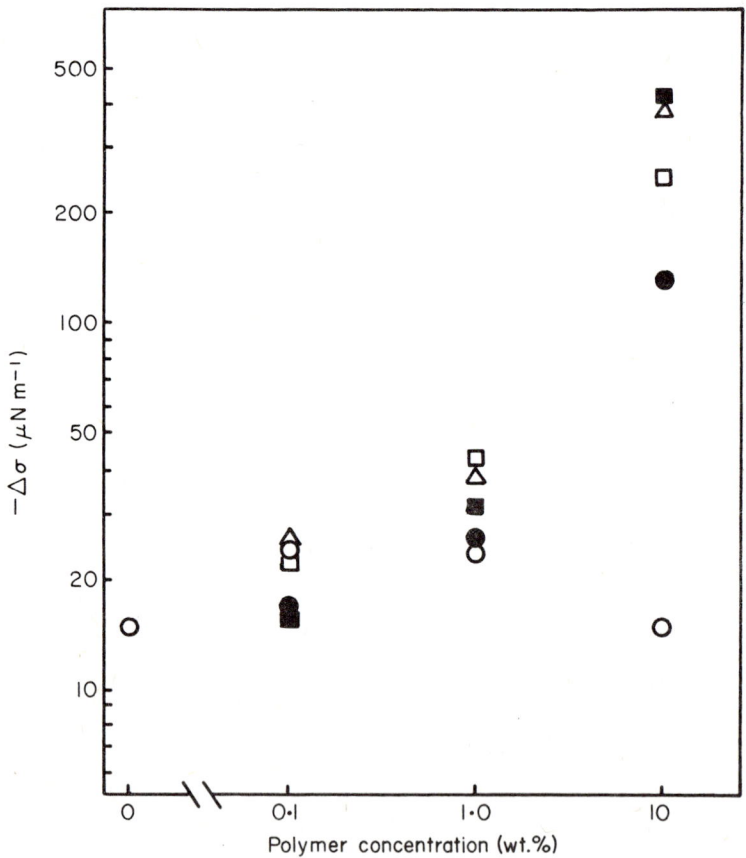

FIG. 3. Excess tensions of aqueous decyl methyl sulphoxide films containing polyethylene glycol, 298 K. PEG 106 (○), PEG 400 (●), PEG 1500 (□), PEG 4000 (■), PEG 6000 (△).

average molecular weights at 298 K. The relationship $\Delta\sigma = 2\gamma (\cos\theta - 1)$, in which γ is the surface tension of the film-forming solution, was used to obtain the excess tension from the contact angle, θ. There was no significant difference between values obtained in this way and those measured directly with the modified Wilhelmy plate method.[20]

The film thickness was $4\cdot6 \pm 0\cdot3$ nm in the absence of polymer but seemed to be slightly increased when polymer was present. However, since the maximum increase was only about 15 per cent (for PEG 6000 at 10 per cent concentration) this may not have much significance, particularly in view of the uncertain effect of polymer on the film's optical properties. (In calculating the thickness it was assumed that the film's aqueous layer had the same refractive index as the bulk solution. This would underestimate the refractive index of a layer containing adsorbed polymer and hence overestimate the thickness.)

The absolute value of $\Delta\sigma$ increased with polymer concentration and this increase became greater as the polymer molecular weight was increased. The addition of sodium chloride at concentrations up to 1 kmol m^{-3} had little effect on either $\Delta\sigma$ or thickness, and the single species PEG 810 gave $\Delta\sigma$ values close to those interpolated for that molecular weight from the values given by the commercial polymers.

The excess concentration of polymer in the film compared with that adsorbed at the surface of the bulk solution can be calculated from equation (1) if we assume that only the chemical potential of the polymer varied. (P_D was very small and virtually constant throughout these experiments.) There are insufficient results for an accurate calculation but we can see that $\Delta\Gamma_{polymer}$ is positive and increases with increasing concentration and molecular weight. Thus for PEG 6000 at 10 per cent concentration it is at least 60 nmol m^{-2}. It is perhaps not surprising that there is more high molecular weight polymer in the film than at the surface of the solution since it would seem to be fairly easy for polymer to be trapped by the very rapid thinning which precedes the formation of the Newton black film. It is not entirely clear, on the other hand. how polymer can produce such large $|\Delta\sigma|$ values.

The lack of sensitivity to ionic strength suggests that electrostatic forces do not play an important part. However, adsorbed polymer would increase the effective thickness of the film's surface layers and this could cause a large increase in the van der Waals energy of a very thin film.[34] Rough calculations indicate that the observed $\Delta\sigma$'s would probably need fairly densely packed polymer layers up to about 1 nm thick at each film surface, but these dispersion force calculations take no account of probable steric interactions between the polymer layers and an adequate explanation for the effect of polymer awaits a more detailed theoretical analysis.

Acknowledgement

I am indebted to Mr R. R. Balmbra for carrying out the measurements on the neat phase.

References

1. Sheludko, A. (1967). *Advan. Colloid Interface Sci.* **1**, 391.
2. Clunie, J. S., Goodman, J. F. and Ingram, B. T. (1971). *In* "Surface and Colloid Science" (E. Matijević, Ed.), Vol. 3, p. 167. John Wiley, New York.
3. Corkill, J. M., Goodman, J. F., Haisman, D. R. and Harrold, S. P. (1961). *Trans. Faraday Soc.* **57**, 821.
4. Jones, M. N. and Ibbotson, G. (1970). *Trans. Faraday Soc.* **66**, 2396.
5. Lyklema, J. and Bruil, H. G. (1971). *Nature phys. Sci.* **233**, 19.
6. Corkill, J. M., Goodman, J. F., Ogden, C. P. and Tate, J. R. (1963). *Proc. Roy. Soc.* **A273**, 84.
7. Corkill, J. M., Goodman, J. F. and Ogden, C. P. (1965). *Trans. Faraday Soc.* **61**, 583.
8. Den Engelsen, D. and Frens, G. (1974). *J. Chem. Soc. Faraday I*, **70**, 237.
9. Clunie, J. S., Corkill, J. M. and Goodman, J. F. (1966). *Discuss. Faraday Soc.* **42**, 34.
10. Balmbra, R. R., Clunie, J. S., Goodman, J. F. and Ingram, B. T. (1973). *J. Colloid Interface Sci.* **42**, 226.
11. Friberg, S., Linden, St. E. and Saito, H. (1974). *Nature (London)*, **251**, 494.
12. Jones, M. N., Mysels, K. J. and Scholten, P. C. (1966). *Trans. Faraday Soc.* **62**, 1336.
13. Ibbotson, G. and Jones, M. N. (1969). *Trans. Faraday Soc.* **65**, 1146.
14. Clunie, J. S., Corkill, J. M., Goodman, J. F. and Ingram, B. T. (1970). *Special Discuss. Faraday Soc.* **1**, 30.
15. Ingram, B. T. (1972). *J. Chem. Soc. Faraday I*, **68**, 2230.
16. Verwey, E. J. W. and Overbeek, J. Th. G. (1948). "Theory of the Stability of Lyophobic Colloids". Elsevier, Amsterdam.
17. Clunie, J. S., Corkill, J. M., Goodman, J. F. and Ogden, C. P. (1967). *Trans. Faraday Soc.* **63**, 505.
18. Clunie, J. S., Goodman, J. F. and Tate, J. R. (1968). *Trans. Faraday Soc.* **64**, 1965.
19. Platikanov, D., Rangelova, N. and Sheludko, A. (1966). *Ann. Univ. Sofia, Fac. Chem.* **60**, 293.
20. Clint, J. H., Clunie, J. S., Goodman, J. F. and Tate, J. R. (1969). *Nature (London)*, **223**, 291.
21. Prins, A. (1969). *J. Colloid Interface Sci.* **29**, 177.
22. Mysels, K. J., Huisman, H. F. and Razouk, R. I. (1966). *J. Phys. Chem.* **70**, 1339.
23. Kolarov, T., Sheludko, A. and Exerowa, D. (1968). *Trans. Faraday Soc.* **64**, 2864.
24. Princen, H. M. (1968). *J. Phys. Chem.* **72**, 3342.
25. Rusanov, A. I. (1967). *Colloid J. (USSR)*, **29**, 118.
26. De Feijter, J. A. (1973). Thesis, Utrecht.
27. Huisman, H. F. and Mysels, K. J. (1969). *J. Phys. Chem.* **73**, 489.
28. Rijnbout, J. B., Donners, W. A. B. and Vrij, A. (1974). *Nature (London)*, **249**, 242.
29. Clunie, J. S., Corkill, J. M. and Goodman, J. F. (1965). *Proc. Roy. Soc.* **A285**, 520.
30. Clunie, J. S., Corkill, J. M., Goodman, J. F., Symons, P. C. and Tate, J. R. (1967). *Trans. Faraday Soc.* **63**, 2839.

31. Princen, H. M. and Frankel, S. (1971). *J. Colloid Interface Sci.* **35**, 386.
32. Ninham, B. W. and Parsegian, V. A. (1970). *J. Chem. Phys.* **53**, 3398.
33. Barker, J. A. and Watts, R. O. (1969). *Chem. Phys. Lett.* **3**, 144.
34. Ninham, B. W. and Parsegian, V. A. (1970). *J. Chem. Phys.* **52**, 4578.

Discussion

Smith The thermodynamic treatment of Hall and Everett indicates that, under equilibrium conditions, if an increase of concentration in a solution component leads to enhanced attraction between surfaces then the adsorption of that component must increase as the surfaces approach.

Would you like to comment on the application of this principle to your results in the case where polymer is present and also to comment on the analogy between your results and the familiar phenomenon of enhanced attraction between solid surfaces with adsorbed polymer layers, sometimes referred to as polymer bridging.

Ingram I mentioned in my paper that equation (1) showed that an increase in $\Delta\sigma$ (i.e. film attraction) as the polymer concentration increased could signify increasing adsorption of polymer in the film, and I discussed this point in the text.

The mechanism by which polymer enhanced the film attraction may indeed have involved some form of polymer bridging. It is worth noting, however, that this would not necessarily rule out the possibility that the increase in $|\Delta\sigma|$ was caused by changes in the van der Waals energy.

Friberg (1) You put some emphasis on an opinion that the structure of the water in the Newton black film is more ordered than bulk water. Do you have any *experimental* evidence of the enhanced order?
(2) In your lecture you referred to the thin films being *in equilibrium* with the lamellar liquid crystalline phase. Did you really arrange for an equilibrium?

Ingram (1) We have no direct experimental evidence of structure differences between water in Newton black films and bulk water, although there is the indirect evidence to which I referred in my paper. However, because the aqueous layer in these films is so thin, most of the water molecules are very close to charged or polar surfactant head groups and unlikely to have the same orientation as in bulk water.
(2) Some of the films were formed from solutions containing neat phase and there was neat phase in the Plateau borders, so it is reasonable to assume that the films were in equilibrium with the liquid crystals.

Vrij You mentioned in your paper that polymer could be trapped by the rapidly thinning film. This would be a nonequilibrium situation. Can you give us an indication in how far the polymer in the film is in equilibrium with polymer in the bulk of the solution?

Ingram We have no indication that the polymer in the film was not in equilibrium with polymer in the bulk solution. The film tension was constant within experimental error until the film was ruptured (i.e. for periods of up to an hour) and the tension was reproducible for each solution.

Nevertheless, the diffusion of polymer from these very thin films might be extremely slow so that it is difficult to be certain that the films were in true equilibrium with the bulk solution.

6

Equilibrium properties of foam films formed from nonionic surface-active agents

R. BUSCALL*, R. B. DONALDSON†, R. H. OTTEWILL and D. SEGAL

School of Chemistry, University of Bristol,
Bristol BS8 1TS, England

Summary

The equilibrium thicknesses of foam films formed from aqueous solutions of the nonionic surface-active agents. decylmethyl sulphoxide and decyldimethylphosphine oxide, were studied in the presence of sodium chloride and potassium thiocyanate. Stable films were formed and it was found that the film thickness depended on the concentration and type of anion in the bulk solution from which the films were drawn. Examination of the electrophoretic mobility of dodecane droplets stabilized with the surface active agents enabled a comparison to be made between the mobility of the drop and the film thickness. The behaviour of foam films formed from solutions of dodecyldimethylamine oxide were also examined. This material behaves as a cationic surface-active agent below pH 7 and as a nonionic one above this pH. This behaviour was reflected in the film thickness measurements as a function of pH. Measurements made with dodecyldimethylamine oxide at pH 9·2 in the presence of potassium iodide showed that, as a nonionic surface-active agent, this material behaved in a similar manner to declymethyl sulphoxide and decyldimethylphosphine oxide. The data indicated that the common black foam films formed from the three nonionic surface-active agents were charge stabilized owing to the adsorption of anions.

1 Introduction

In a macroscopic foam, the walls of the individual bubble cells are composed of lamellae. These gradually become thinner with time until a stable situation

---

* Present address: Department of Pharmacy, University of Aston in Birmingham. Gosta Green, Birmingham B4 7ET, England.

† Present address: Unilever Research Laboratory, Port Sunlight, Wirral, Merseyside, L62 4XN, England.

is reached when the films appear black, a physical situation encountered when the light waves reflected from the two sides of lamella are in counter-phase. For the purposes of detailed study it is easier to form foam films as isolated single units and a considerable study has already been carried out on films of this type formed from ionic surface-active agents.[1,2] It has been well established now that such films consist of two monolayers of the surface-active molecules separated by an aqueous core.[2]

Only a few studies appear to have been carried out on foam films formed using aqueous solutions of nonionic surface-active agents. Films stabilized by decylmethyl sulphoxide (DMS) have been investigated by Clunie et al.,[3] by Ingram[4] and by two of the present authors.[5] In these studies it was found that inorganic electrolytes had a considerable influence on the equilibrium thickness of the film. At low concentrations of 1:1 electrolytes the film thickness (about 5 nm) was found to be almost independent of electrolyte concentration and the films were considered to be in an equilibrium state in a potential energy primary minimum; following the recent IUPAC report[6] such films are termed Newton black films. At a particular concentration of salt, however, the film thickness increased and with further increases in salt concentration it decreased in the manner expected for a charge stabilized film. Both the film thickness studies and complementary electrophoretic measurements[5] indicated that the charge on the film in this region arose as a consequence of the adsorption of anions to sulphoxide groups in the film surface. These films were considered to be stabilized in a "secondary minimum" of potential energy. Following the IUPAC convention[6] these are now termed common black films.

The present paper describes investigations on films formed from three different surface-active agents, decylmethylsulphoxide, decyldimethylphosphine oxide and dodecyldimethylamine oxide. The first two of these materials are nonionic over the pH range 3 to 10 and the latter is cationic below pH 7 and nonionic above this value. The foam films formed from all of these materials showed pronounced changes, in the presence of electrolytes, to both anion type and anion concentration.

2 Experimental

2.1 MATERIALS

Fresh water, doubly distilled from an all-Pyrex apparatus, was always used.

The sodium chloride used was Analar material and was roasted before use to remove surface-active contaminants.

Analar potassium thiocyanate was recrystallized from water. In order to remove any residual surface-active impurities, the recrystallized salt was washed with redistilled diethyl ether.

Potassium iodide was Analar material which was dried before use.

Dodecane was Fluka "Purum" material.

The decylmethylsulphoxide (DMS) and the decyldimethylphosphine (DDPO) oxide were pure materials kindly supplied by Dr T. Walker of Procter & Gamble Limited. The dodecyldimethylamine oxide (DDAO) was also a pure sample which had been prepared by the method of Lake and Hoh.[7] It was kindly provided by the late Dr J. M. Corkill of Procter & Gamble Limited. The surface tension against log concentration curves obtained with these materials showed them to be free of more surface-active materials. The critical micelle concentrations are given in Table 1.

TABLE 1

Critical micelle concentrations (CMC) of film forming surface-active agents in aqueous solution at 22°C

Material	CMC mol dm^{-3}	pH
DDAO	$1 \cdot 1 \times 10^{-3}$	$9 \cdot 0$*
	$2 \cdot 2 \times 10^{-3}$	$3 \cdot 5$†
DMS	$2 \cdot 0 \times 10^{-3}$	
DDPO	$4 \cdot 0 \times 10^{-3}$	

* Adjusted with NaOH.
† Adjusted with "Analar" HCl.

2.2 MEASUREMENT OF FILM THICKNESS

2.2.1 Horizontal films

The films were supported by a stainless steel frame which was made from a rectangular block by boring two concentric countersunk holes, one from each direction so that a thin ring c. 4 mm in diameter was left in the centre. The steel frame was mounted on a hollow Perspex rod; a thermistor was inserted in the hollow centre to monitor the temperature of the frame. A cell was built to house the film holder and this consisted of a glass bulb of approximately 10 cm^3 capacity with an optical glass window at the top. The Perspex rod on the foam film holder was inserted into a neck in the side of the glass cell via plastic sleeve in order to form a water-tight seal. By using this device it was possible to rotate the film frame. The glass cell was mounted in a Perspex box which was double-jacketed in order to allow thermostatic control. Films were formed by dipping the stainless steel frame into a surface-active agent solution and allowing the excess liquid to drain off. The frame was then inserted into the cell. A small pool of surface-active agent solution was kept in the bottom of the cell to maintain the vapour pressure within the system. Unless otherwise stated, all measurements were carried out on horizontal films.

2.2.2 *Vertical films*

For measurements on vertical films the same optical system was used as for measurements on horizontal films. The films, however, were supported on an inverted U-shaped frame, made from stainless steel, and roughened with fine carborundum powder to provide better wettability. In order to form the film, the glass cell (see above) was filled to just above the top of the frame. After allowing a period for temperature equilibration some of the liquid in the cell was withdrawn with a syringe. As the liquid level in the cell fell, a film was left on the frame which was in contact with its own bulk solution and equilibrium vapour.

2.2.3 *Calculation of film thickness*

The film thicknesses were determined from measurement of the intensity of light reflected from the film at near to normal incidence. An incidence light beam of wavelength 509·5 nm was obtained by means of a tungsten–iodine lamp and an interference filter. The procedure of measuring the ratio of the intensity of the reflected light from the first bright fringe and from the black film was adopted.[8] Assuming the film to be a triple layer of hydrocarbon–water–hydrocarbon and applying the treatment of Caballero[9] a full solution was found[13] for the overall thickness of the film d. For $d \leqslant 10$ nm this reduced to,

$$d = A \left[\frac{I_R}{I_{max}} \right]^{\frac{1}{2}} + B \text{ nm} \qquad (1)$$

where I_R = the intensity of reflection from the black film and I_{max} = the reflected intensity from the first bright fringe, both after correction for background intensity. For films formed from DMS and DPPO A was taken as 51·36 and B as 0·19; for films of DDAO A was taken as 58·34 and B as $-6·6$. The total film thickness was taken as

$$d = 2d_1 + d_2 \qquad (2)$$

where d_2 = the thickness of the aqueous core and d_1 = the length of the hydrocarbon chain of the surface-active agent. At least three measurements were made on each solution and the mean value was taken. The reproducibility of the experiments was within ± 5 per cent.

2.3 MICROELECTROPHORESIS

Mobility measurements were made using an apparatus similar to that described by Alexander and Saggers[10] modified to use slit ultramicroscope illumination. The cell used was of the Mattson type[11] and was mounted in a water bath. Refraction effects in the cell were allowed for by applying the Henry correction. All mobility measurements were made at the stationary level in the cell at room temperature, $20 \pm 0.5°C$.

The emulsions for electrophoresis measurements were prepared by adding one drop of dodecane to about 20 cm^3 of a surface-active agent solution and subjecting the mixture to ultrasonic treatment for 30 seconds. The emulsion was then aged overnight and remixed by shaking just prior to use. This procedure produced drops of approximately 1 to 2 μm. Zeta-potentials (ζ) were calculated from the Smoluchowski equation.

$$u = \frac{\varepsilon\zeta}{4\pi\eta}.$$

where ε = relative permittivity of the medium, η = viscosity and u = electrophoretic mobility.

3 Results

3.1 DECYLMETHYLSULPHOXIDE (DMS)

3.1.1 Effect of sodium chloride

The results obtained for the thickness of foam films formed from solutions of DMS containing sodium chloride at various concentrations are given in Fig. 1. At low salt concentrations, below $10^{-2} \text{ mol dm}^{-3}$, films with a thickness of 6 nm were formed (Newton black films); in this region the film thickness was independent of salt concentration. At a sodium chloride concentration of $10^{-2} \text{ mol dm}^{-3}$, two coexisting films were observed with the thicker, common black film having a thickness of 8·5 nm. With a further increase in salt concentration the thickness of the common black film decreased until it reached a thickness of 6 nm at a salt concentration of 0.5 mol dm^{-3}, i.e. it became a Newton black film.

In order to obtain an estimate of the charge at the sulphoxide–aqueous solution interface, the effect of sodium chloride on the electrophoretic mobility of dodecane droplets stabilized with DMS was investigated. The results are also shown in Fig. 1. The droplets were found to be essentially uncharged at low concentrations of sodium chloride, but between 3×10^{-3} and $5 \times 10^{-3} \text{ mol dm}^{-3}$ the mobility increased to a value of $-1.2 \times 10^{-8} \text{ m}^2 \text{ V}^{-1} \text{ s}^{-1}$. The similarity between the two curves presented in Fig. 1 is striking in that the film thickness increases at the same bulk concentration of sodium chloride as the electrophoretic mobility increases abruptly. It immediately suggests that at this concentration the number of chloride ions adsorbed to the sulphoxide head groups becomes sufficient to produce an

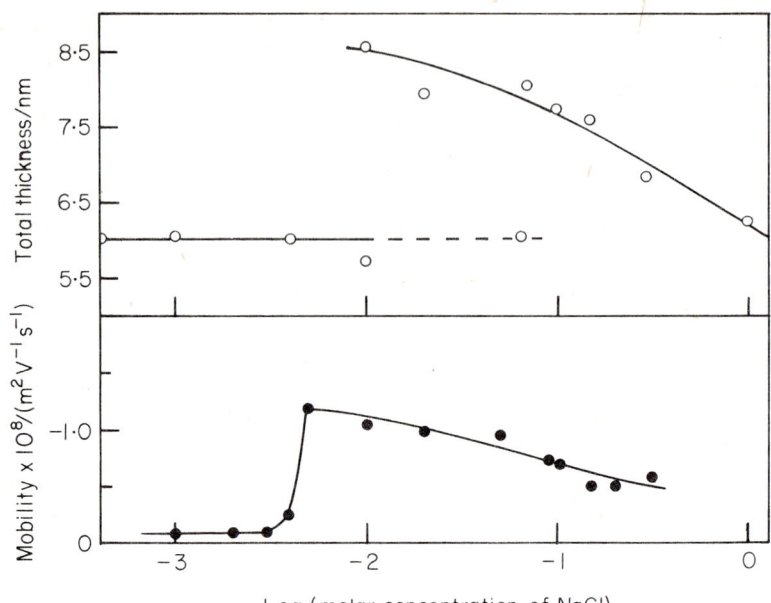

F<small>IG</small>. 1. Total film thickness (DMS) and electrophoretic mobility of DMS-stabilized dodecane droplets as a function of sodium chloride concentration. O. film thickness; ●. electrophoretic mobility.

electrostatic potential which is high enough to provide electrostatic repulsion between the monolayers in the soap film.

3.1.2 *Effect of potassium thiocyanate*

The equilibrium thickness results for DMS foam films formed from solutions containing potassium thiocyanate, at various concentrations, are shown in Fig. 2. In this system at the lowest salt concentration examined, 4×10^{-4} mol dm^{-3}, stable films were formed with a thickness of 97·5 nm. These were clearly common black films. Correspondingly, the thickness decreased with increasing salt concentration and reached a value of 5·9 nm, the dimensions of a Newton black film, at 0·5 mol dm^{-3} potassium thiocyanate. The electrophoretic mobility of dodecane droplets, stabilized with DMS, as a function of potassium thiocyanate concentration are shown in Fig. 2. The droplets had a mobility of $-3\cdot0 \times 10^{-8}$ m^2 V^{-1} s^{-1}, at a salt concentration of 4×10^{-4} mol dm^{-3}, which then decreased. The higher mobility values, extending over a wide salt concentration range, and the greater thickness of the foam films formed suggested that the free energy of adsorption of the thiocyanate ion at the film surface–solution interface was greater than that

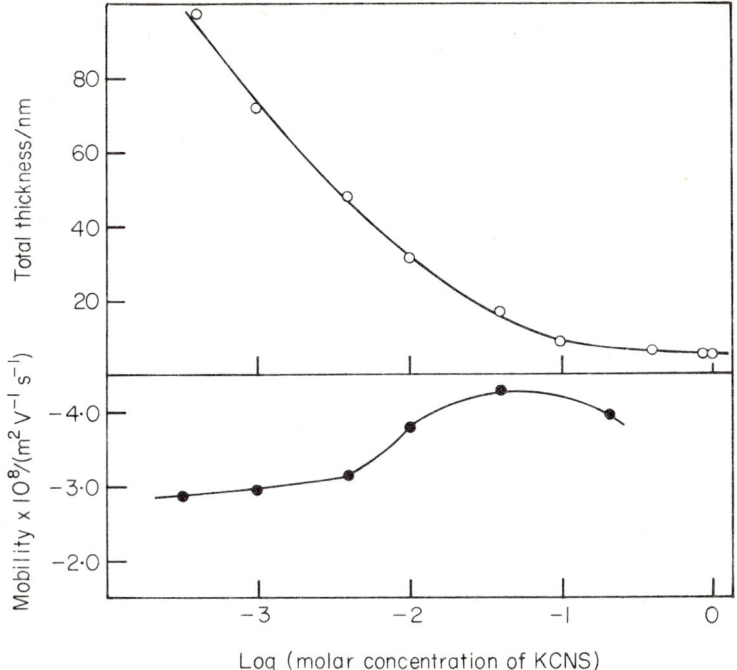

Fig. 2. Total film thickness (DMS) and electrophoretic mobility of DMS-stabilized dodecane droplets as a function of potassium thiocyanate concentration. ○, film thickness; ●, electrophoretic mobility.

of the chloride ion. Hence, it also implied that the extent of adsorption overall was greater for the thiocyanate ion than for the chloride ion leading to thick foam films and high electrophoretic mobilities, even at low salt concentrations.

3.2 DECYLDIMETHYLPHOSPHINE OXIDE (DDPO)

3.2.1 *Effect of sodium chloride*

The film thickness and electrophoretic mobility results for DDPO systems in the presence of sodium chloride are shown in Fig. 3. The film thickness was found to be substantially greater than that obtained for DMS films in $10^{-4} \, \mathrm{mol \, dm^{-3}}$ sodium chloride. Consistent with this a higher electrophoretic mobility value was obtained for the dodecane droplets stabilized by DDPO. The electrophoretic mobility had a value of $-3.0 \times 10^{-8} \, \mathrm{m^2}$ $\mathrm{V^{-1} \, s^{-1}}$ at $10^{-4} \, \mathrm{mol \, dm^{-3}}$ sodium chloride decreasing gradually to $0.5 \times 10^{-8} \, \mathrm{m^2 \, V^{-1} \, s^{-1}}$ at $10^{-1} \, \mathrm{mol \, dm^{-3}}$.

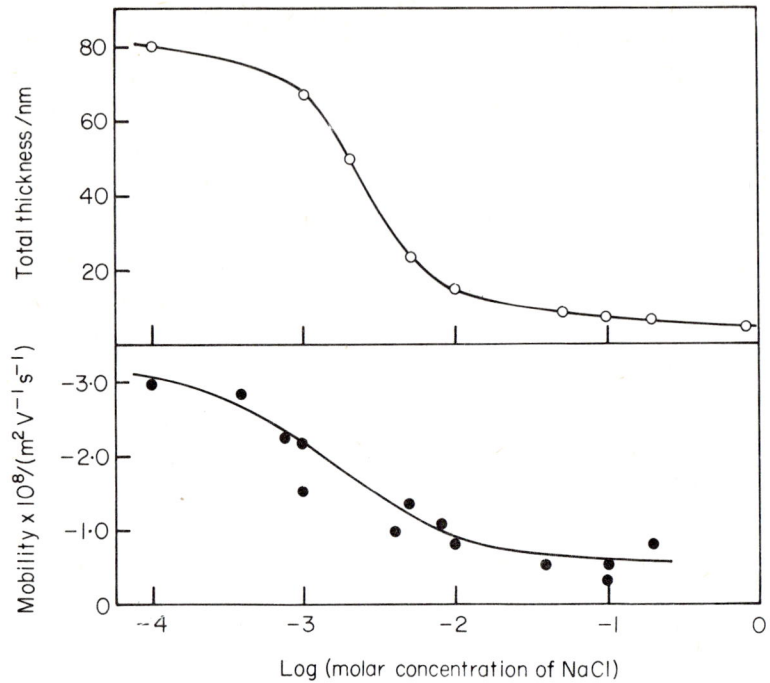

FIG. 3. Total film thickness (DDPO) and electrophoretic mobility of DDPO-stabilized dodecane droplets as a function of sodium chloride concentration. ○, film thickness; ●, electrophoretic mobility.

3.2.2 *Effect of potassium thiocyanate*

The film thickness and electrophoretic results obtained for the DDPO–potassium thiocyanate system are given in Fig. 4. As expected from the DMS results thick films were formed at low salt concentrations. The film thicknesses gradually decreased with increase in salt concentration and stable Newton black films were formed in 1 mol dm^{-3} potassium thiocyanate. The electrophoretic mobility increased from a value of -2.1×10^{-8} m^2 V^{-1}s^{-1} at 3×10^{-4} mol dm^{-3} to about -4.6×10^{-8} m^2 V^{-1}s^{-1} at 3×10^{-2} mol dm^{-3} and then became virtually constant over the higher salt concentration range. A comparison of the results with those obtained with DMS suggests that the initial rate of increase of charge with increase in electrolyte concentration is greater for DDPO than for DMS. The implication would then appear to be that the phosphine oxide head group adsorbs anions more strongly than the sulphoxide. The stronger adsorption of the chloride ion to the phosphine oxide head group is also indicated by the thicker films and larger electrophoretic mobilities obtained with DDPO.

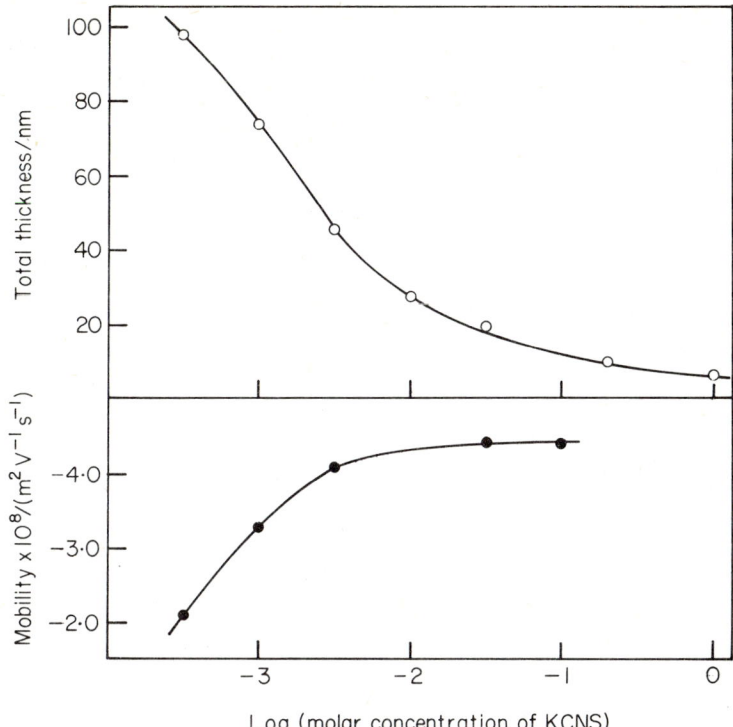

FIG. 4. Total film thickness (DDPO) and electrophoretic mobility of DDPO stabilized dodecane droplets as a function of potassium thiocyanate concentration. ○, film thickness; ●, electrophoretic mobility.

3.3 DODECYLDIMETHYLAMINE OXIDE (DDAO)

3.3.1 *Effect of pH and sodium chloride*

DDAO although a nonionic surface-active agent at high pH values ($> pH$ 7) becomes a cationic surface-active agent below pH 7.[12] For this material both the film thickness and the electrophoretic mobility were studied in the first instance as a function of pH. The thickness measurements were made on small vertical films (*c.* 3 mm × 5 mm) and in each case the thickness was determined on an element at a height of approximately 3 mm above the Plateau border. The thickness of the foam film, as a function of pH at various sodium chloride concentrations, is shown in Fig. 5. All the curves show a maximum in the foam film thickness at acid pH values, the position varying from pH 4 in 10^{-1} mol dm^{-3} electrolyte, to pH 5·5 in 10^{-3} mol dm^{-3} electrolyte. At lower pH values the film thickness decreased as anticipated in view of the compression of the electrical double layer with the increase in ionic strength.

At pH 7 ± 0.3 all the films showed a transition from the common black to the Newton black form. In 10^{-3} mol dm^{-3} salt solution the films were not sufficiently stable to allow the thickness of the Newton black film to be

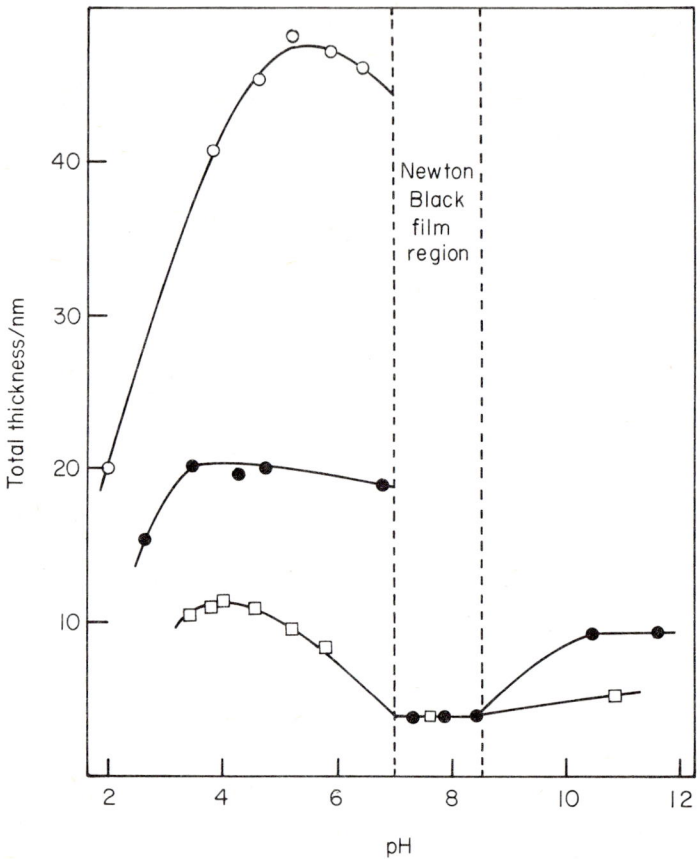

FIG. 5. Film thickness of vertical foam films of DDAO as a function of pH in various concentrations of sodium chloride: ○, 10^{-3} mol dm^{-3}; ●, 10^{-2} mol dm^{-3}; □, 10^{-1} mol dm^{-3}.

observed. With Newton black films, in 10^{-1} and 10^{-2} mol dm^{-3} sodium chloride, it was possible to make satisfactory measurements although the films were not particularly stable.

In the pH range 8.6 to 10.9, small increases in film thickness were observed with variation of electrolyte concentration. In 0.1 mol dm^{-3} sodium chloride the film thickness increased from 3·8 nm to 5·2 nm at pH 10·5 and in 10^{-2} mol dm^{-3} the thickness increased from 3·8 nm at 9·3 nm at pH 10·9.

The mobility results obtained using dodecane droplets stabilized by DDAO are shown in Fig. 6. As anticipated from the properties of DDAO the droplets are positively charged below about pH 8. The positive mobility values were higher in 10^{-3} mol dm^{-3} salt than 10^{-1} mol dm^{-3} salt and maxima occurred in the pH range 3 to 6. The decrease in mobility on the

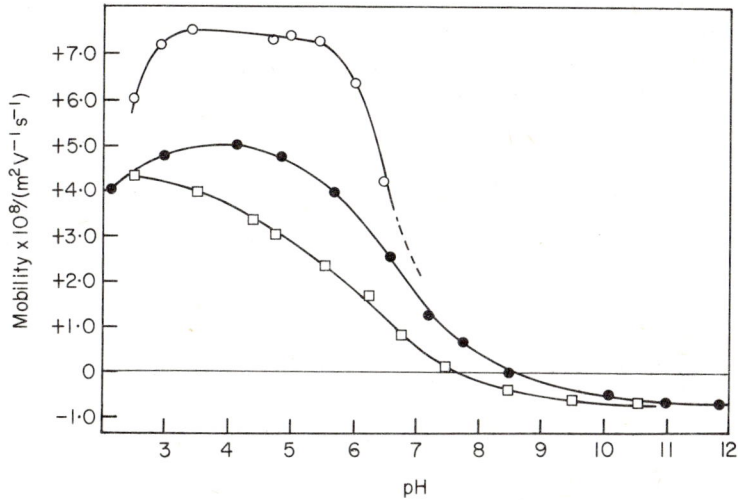

FIG. 6. Electrophoretic mobility of DDAO-stabilized dodecane droplets as a function of pH in various concentrations of sodium chloride: ○, 10^{-3} mol dm^{-3}; ●, 10^{-2} mol dm^{-3}; □, 10^{-1} mol dm^{-3}.

acid side was ascribed to compression of the electrical double layer arising from the charged film. In the neutral pH range the low mobilities were ascribed to the reversion of the cationic head group to the nonionic form. The small negative mobilities observed in the alkaline pH range were consistent with the weak binding of anions to the nonionic amine oxide head group.

3.3.2 Effect of potassium iodide

It was of considerable interest to determine whether films formed from DDAO in the pH range in which it was a nonionic surface-active agent behaved in the same way as foam films formed from DMS solutions. A direct comparison using sodium chloride would have been desirable but, as mentioned above, it was found that chloride ions were not as effective in stabilizing nonionic DDAO films as they were with DMS and DDPO.

However, with potassium iodide, the films formed at all ionic strengths were very stable. These systems were controlled at pH 9·2 by addition of some veronal buffer but in no case did this contribute more than 5 per cent to the total ionic strength. Its effect on the film properties appeared to be negligible. Films with a thickness of c. 7.0 nm were formed at iodide concentrations between 4×10^{-3} and $2 \times 10^{-2}\,\mathrm{mol\,dm^{-3}}$. Between 2×10^{-2} and $4 \times 10^{-2}\,\mathrm{mol\,dm^{-3}}$ iodide the film increased in thickness and then decreased to a value of c. 6.0 nm in 0·5 $\mathrm{mol\,dm^{-3}}$ salt; the latter film was presumed to be a Newton black film. The electrophoretic mobility also showed an increase in magnitude at about the same concentration as the film thickness increased (Fig. 7). The general behaviour of the DDAO–potassium iodide

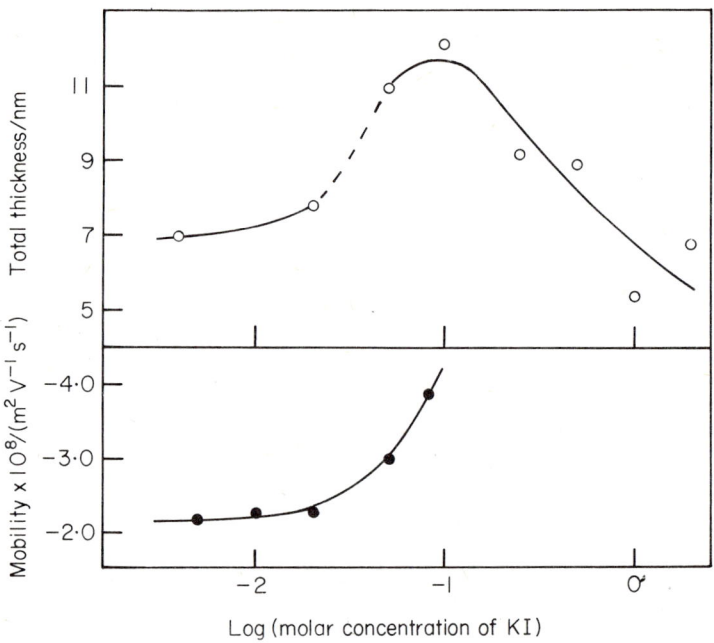

Fig. 7. Total film thickness (DDAO) and electrophoretic mobility of DDAO-stabilized dodecane droplets as a function of potassium iodide concentration at pH 9·2. ○, film thickness; ●, electrophoretic mobility.

system closely resembled that observed with DMS–sodium chloride solutions. However, the transition occurred at a higher salt concentration and it appeared that a higher potential was required in the film to cause the initial transition from Newton to common black. The reasons for this could be that the binding of the anions to the neutral amine oxide head group is weaker than the binding to the sulphoxide head group and that in addition there is an extra van der Waals attraction in the DDAO film owing to the two additional $—CH_2—$ groupings in the alkyl chain.

4 Discussion

The foam films formed from DMS in sodium chloride solutions ranging in concentration from 4×10^{-4} to 10^{-2} mol dm^{-3} had thicknesses of $c.$ 6·0 nm. At 10^{-2} mol dm^{-3} salt, however, two coexisting films were observed with the thicker film having a thickness of 8·5 $\pm$ 0·3 nm and the thinner 6·0 nm. As the concentration of salt was increased the thicker film became the stable species and the thickness decreased with subsequent increases of salt until at 0·5 mol dm^{-3} the thickness again reached a value of 6·0 nm. The form of the results obtained with DMS was very similar to that obtained by Clunie et al.[3] and Ingram.[4] In agreement with these authors we consider that the invariance of the film thickness with salt concentration at low salt concentrations indicates that the 6·0-nm film is a Newton black film.

The sudden increase in film thickness at $c.$ 10^{-2} mol dm^{-3} suggested the onset, in the film core, of electrostatic repulsive forces. Unfortunately it is not an easy matter to measure directly the magnitude of the surface potential in thin foam films. Hence dodecane emulsion droplets stabilized by DMS were used to provide a model of the film–water interface. Although the exact similarity between the film and the drop can be questioned. it would seem reasonable to assume that the sulphoxide head group, since it is in the aqueous phase in both the film and the emulsion, would behave similarly in the two cases. This appeared to be so, since below 3×10^{-3} mol dm^{-3} sodium chloride, the observed mobility was very small (Fig. 1). Just above this concentration, however, an increase in mobility occurred. The negative sign of the mobility suggested that preferential adsorption of the chloride ion to the sulphoxide head group had occurred, and that the charges formed on the film surfaces by this mechanism had caused electrostatic repulsion between them and consequently an increase in thickness of the film to the common black form.

In solutions of potassium thiocyanate the thicker films formed at the low concentrations and the magnitude of the electrophoretic mobility showed that the layer of sulphoxide head groups attained a higher negative potential in the presence of thiocyanate ions than it did in the presence of chloride ions. Consequently, this implied that the double-layer repulsion in the film was also greater. These results, therefore, appear to indicate a higher binding energy for thiocyanate ions than chloride ions to the sulphoxide head groups and confirm the dominance of the anion in determining film thickness. an effect observed by Ingram[4] and by Buscall.[13] Independent investigations of the behaviour of insoluble monolayers of octadecyl sulphoxide have given approximate estimates of the binding energies to sulphoxide head groups as 1–2 kT for chloride ions and 6–7 kT for thiocyanate ions.[5]

With DDPO thick films were formed in 10^{-4} mol dm^{-3} sodium chloride. In agreement with this the electrophoretic mobility at this concentration was *c.* $-3{\cdot}0 \times 10^{-8}$ m^2 V^{-1} s^{-1}. With increasing sodium chloride concentration both the film thickness and the electrophoretic mobility decreased gradually. In fact, the curves above 10^{-4} mol dm^{-3} sodium chloride resembled in form those obtained at higher salt concentrations with DMS films. The results suggested that the binding of chloride ions to phosphine oxide head groups occurred at a lower salt concentration than with sulphoxide, a factor which would indicate a higher binding energy for a chloride ion to the phosphine oxide group. The gradual decrease of electrophoretic mobility with salt concentration (Fig. 3), however, would suggest that the extent of chloride ion binding may not increase substantially beyond 10^{-4} mol dm^{-3} and that only double-layer compression at nearly constant potential could be occurring beyond this point.

The results obtained with DDPO in the presence of potassium thiocyanate suggest a similar behaviour to that observed with DMS films. Thick common black films were formed at low salt concentrations and the electrophoretic mobility of the dodecane droplets was high. With increasing salt concentration the film thickness decreased gradually whereas the mobility increased to a plateau value of *c.* $-4{\cdot}4 \times 10^{-8}$ m^2 V^{-1} s^{-1} between 10^{-2} and 10^{-1} mol dm^{-3} salt. As in the case of DMS quite stable films were formed in 1 mol dm^{-3} salt solutions.

In the case of the DDAO films formed at pH 9·2, i.e. the pH region where the surface-active agent was a nonionic material, the lack of stability of the films formed and the low electrophoretic mobility of the drops in solutions containing chloride suggested that this ion is less strongly bound to an amine oxide head group than to a sulphoxide head and that the head group sequence for the binding of chloride ions is DDPO > DMS > DDAO.

The behaviour of DDAO films, at pH 9·2, in the presence of iodide ions, resembles to some extent that obtained with DMS films in sodium chloride. However, the film thickness at the lower concentrations was somewhat higher than expected for a Newton black film and the high electrophoretic mobility suggested that the film was probably electrostatically stabilized in this region. The increase in film thickness and the parallel increase in electrophoretic mobility which occurred between 2×10^{-2} mol dm^{-3} and 10^{-1} mol dm^{-3} iodide indicated that an increase in iodide ion absorption occurred in this region.

It has been noted by Davies and Rideal[14] that the order of polarization for anions at positively charged surfaces is CNS$^-$ > I$^-$ > Br$^-$ > Cl$^-$. The present results would appear to confirm this order, with the different head groups examined, and this suggests that the anions interact with the electro-positive atom in the head group. Information on the dipole moments of the head groups is sparse but published data indicates a value of c. 4·8 D for an amine oxide group[15] and c. 3·9 D for a sulphoxide group.[16] If these results are applicable to the surface-active agents it would appear that the inter-action does not correlate with the magnitude of the head group dipole. Possibly, steric effects play an important role.

The results presented indicate that the nonionic head groups of surface-active agents can bind ions. The charge, so acquired, can give electrostatic stabilization of the foam film formed. The extent of the ion-binding to the film is not at present known. Similarly the mechanism of the ion-binding process is not completely understood and further studies will be required to elucidate this and to establish whether any part is played by the inorganic cation.

An interesting feature of the foam film studies is the stability of the Newton black films at low and high electrolyte concentrations. The most probably stabilizing factor in these conditions is the so-called "steric stabilization". However, with the small head groups it would seem unlikely that con-figurational entropy terms could be involved and thus the stabilization must arise as a consequence of the solvation of the head groups. If the head groups in the film are staggered, as has been found in insoluble monolayers of octadecylmethylsulphoxide.[5] this would increase the thickness of the solvated layer and hence aid stability.

Acknowledgements

We wish to thank Unilever Limited for support of R.B.D. and the Science Research Council for support of R.B. and D.S., the former on a CAPS studentship in conjunction with Procter & Gamble Limited and the latter on an Advanced Course Studentship. We also wish to thank Drs B. Ingram and T. Walker for a number of useful discussions and for assistance with materials.

References

1. Mysels, K. J., Shinoda, K. and Frankel. S. (1959). "Soap Films". Pergamon Press, London.
2. Clunie, J. S., Goodman, J. F. and Ingram. B. T. (1971). *In* "Surface and Colloid Science" (E. Matijević. Ed.), Vol. 3, p. 167. John Wiley, New York.
3. Clunie, J. S., Corkill, J. M., Goodman, J. F. and Ingram. B. T. (1970). *Spec. Discuss. Faraday Soc.* **1**, 30.
4. Ingram. B. T. (1972). *J. Chem. Soc. Faraday I.* **68**, 2230.

D

5. Buscall, R. and Ottewill, R. H. (1975). *Advan. Chem. Ser.* **144**, 83.
6. Everett, D. H. (1972). *Pure Appl. Chem.* **31**, 613.
7. Lake, D. B. and Hoh, G. L. K. (1963). *J. Amer. Oil Chem. Soc.* **40**, 628.
8. Corkill, J. M., Goodman, J. F. and Ogden, C. P. (1965). *Trans. Faraday Soc.* **61**, 583.
9. Cabellero, D. (1947). *J. Opt. Soc. Amer.* **37**, 176.
10. Alexander, A. E. and Saggers, L. (1948). *J. Sci. Instrum.* **25**, 374.
11. Mattson, S. (1933). *J. Phys. Chem.* **37**, 223.
12. Benjamin, L. (1964). *J. Phys. Chem.* **68**, 3575.
13. Buscall, R. (1973). Ph.D. thesis, University of Bristol.
14. Davies, J. T. and Rideal, E. K. (1961). "Interfacial Phenomena". Academic Press, London and New York.
15. Benjamin, L. (1966). *J. Phys. Chem.* **70**. 3790.
16. McClellan, A. L. (1963). "Tables of Experimental Dipole Moments". W. H. Freeman and Co., San Francisco.

Discussion

Ingram Do you think that the relative sizes of the hydrated cation and anion of the inorganic salt effect the distance of nearest approach of the ions to the surfaces such that the anions are closest and hence the surface is negative if charged?

Laughlin Would you care to speculate as to why preferential adsorption of anions, rather than cations, occurs to those nonionic semipolar surfactants.

Buscall and **co-workers** The relative sizes of the cation and anion appear to affect the properties of the Newton black films but not those of the common black films. For example, we have found that Newton black films formed in the presence of potassium iodide are significantly thinner than those formed in the presence of sodium iodide. However, the common black film thicknesses and droplet zeta potentials for sodium and potassium iodides are identical, suggesting that the relative sizes of cation and anion do not affect the charge. As to the reason why anions are adsorbed in preference to cations, we suspect that interaction between the anion and the large sulphur, nitrogen or phosphorus atom determines adsorption. From a study of insoluble monolayers of octadecylmethylsulphoxide we have found that the SO group dipole is oriented close to the plane of the surface;[1] thus from the solution phase the monolayer appears as an array of large electropositive sulphur atoms and smaller electronegative oxygens. We also find that the adsorption increases both with the size of the anion and with the size of the electropositive atom, thus dispersion interactions, as well as coulombic interactions, appear to be important.

Exerowa In our paper we described the production of Newton black films from

[1] R. Buscall and R. H. Ottewill (1975). Advances in Chemistry Series, **144**, 83.

solutions of DMS with low electrolyte concentration. For the same substance we have obtained microscopic thin films, 90 nm thick, at an electrolyte concentration of 3×10^{-4} mol dm^{-3}. The measured value of the ζ potential was 42 mV. For such a potential the energy barrier for the transition common $\rightarrow$ Newton black film is low. The probable cause of the absence of this transition as described by Buscall *et al.* is due to the use of large films in which the energy barrier is easily crossed and the common black film not observed. Hence there is no contradiction between our results.

For the same material, kindly provided by T. Walker (Proctor & Gamble), we observed a transition common $\rightarrow$ Newton black when ζ was changed by altering pH.

This result, and others, leads us to believe that at the air–solution interface of nonionic surfactant solutions a definite potential is developed although not of great magnitude. This is shown by the results of Huddlestone and Smith for the electrophoretic force on a bubble in a solution of DMS.

Buscall and **co-workers** In the presence of potassium thiocyanate and potassium iodide we also found films with thicknesses of *c.* 90 nm at salt concentrations of *c.* 3×10^{-4} mol dm^{-3}. Exerowa and co-workers do not refer to the temperature at which their experiments were performed; in the case of DMS/NaCl solutions we have found this is to be critical. For example, we found that in the presence of 10^{-3} mol dm^{-3} sodium chloride thick films ($\geqslant 60$ nm) were formed below 24°C and in 10^{-4} mol dm^{-3} films with thicknesses of 94 nm were formed below 13°C. By interpolating, we expect that, in 3×10^{-4} mol dm^{-3} sodium chloride, thick films would be formed below 20°C. We do not, however, know whether this temperature effect is dependent on the size of the film.

7

Interaction forces in soap films

A. VRIJ, J. A. DE FEIJTER, W. G. M. AGTEROF and H. M. FIJNAUT

Van 't Hoff Laboratory for Physical and Colloid Chemistry,
University of Utrecht, Padualaan 8, Utrecht, The Netherlands

Summary

Some properties of foam lamellae are intimately connected with the long-range interaction forces that are operative in those thin liquid structures. These interaction forces include electrical double-layer repulsion and van der Waals attraction forces.

In this paper it will be shown how these forces and also some dynamic properties can be studied by measurements of contact angles and of light-scattering-intensity-fluctuation correlations.

Results will be presented for very thin films containing much electrolyte (contact angles) and for thicker films containing less electrolyte (contact angles and light scattering).

1 Introduction

At this Symposium we would like to report the results of some experiments on single liquid soap films, the structural elements of a foam. In particular we will treat the thinner films with thicknesses between, say, 5 and 100 nm, in which long-range interaction forces are operative, e.g. electrostatic double-layer repulsion forces and van der Waals' attraction forces. The purpose of these studies is twofold. Firstly soap films can serve as model systems to investigate the above mentioned forces that are important for understanding colloids in general and secondly knowledge on these structures is of particular and more direct importance for several kinds of films which are found in practice, e.g. as occur in foams and emulsions. Here we will review some recent results[1] obtained from *contact angle measurements* in very thin films having thicknesses of 4 to 5 nm, of which the full details will be published elsewhere.[2] Some results will also be given of contact angle and light-scattering-intensity-fluctuation spectroscopy measurements of thicker films.

2 Contact angles in soap films

When a glass frame is dipped into a detergent solution and then drawn upwards, a thin, liquid lamellae soap film is formed showing the well-known interference colours in reflected light. Upon draining, the film becomes thinner until it looks gray or black, showing that its thickness is much smaller than the wavelength of light. Upon further draining the film will either break or attain a final thickness in a stationary or equilibrium state. In the last case the transition region (where film and bulk liquid meet at the top of the meniscus) often contains a break with a contact angle θ. This implies that the surface tension of the film surface, σ^f, is smaller than the surface tension of the bulk surface, σ^α, and that both are related as follows:

$$\sigma^f = \sigma^\alpha \cos \theta. \tag{1}$$

This difference between σ^f and σ^α can be ascribed to the excess free energy of the film, ΔF_e, due to the interaction forces mentioned above.[3-6] At equilibrium

$$\Delta F_e = 2\sigma^f - 2\sigma^\alpha. \tag{2}$$

Substituting equation (1) into equation (2) gives

$$\Delta F_e = 2\sigma^\alpha(\cos \theta - 1). \tag{3}$$

In this way θ can be used to obtain the interaction free enegy ΔF_e. In our investigations, θ was measured by an optical diffraction technique as devised by Princen and Frankel.[7]

It is necessary that the films should be enclosed in a vapour-tight vessel to give meaningful results. Also the temperature should be kept constant between narrow limits.

3 Films with high electrolyte content

Contact angles were measured on films drawn from solutions containing sodium dodecylsulphate (NaDDS) and electrolyte at several temperatures. Film thicknesses were calculated from the measured optical reflection coefficients of the (black) films and expressed as a so-called "equivalent water thickness", h_w.[6]

The electrolyte content was rather high (0·2 to 0·5 mol dm^{-3}). In this range h_w was constant and equal to 4·4 nm and σ^α ranged from 33·1 $\rightarrow$ 31·1 mJ m^{-2} at 23°C. The temperature variation of σ^α was small and of minor importance for the calculation of ΔF_e, Measurements at lower electrolyte concentrations are given in the next section.

3.1 RESULTS

In Fig. 1, ΔF_e, as calculated using equation (3), is plotted versus the activity of NaCl in the bulk solution. Some peculiar features will be apparent.

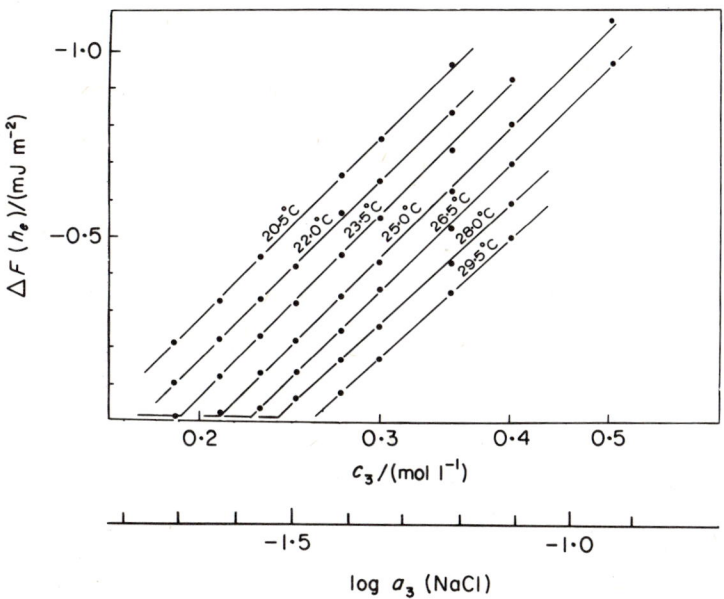

FIG. 1. Interaction free energies, ΔF_e, for NB films drawn from aqueous solutions containing 0·05 per cent Na-dodecylsulphate (NaDDS) and NaCl at several temperatures.

Firstly one observes that below a certain NaCl activity, ΔF_e becomes very small and rather constant as indicated by the horizontal lines. Above this activity, θ, and thus also $-\Delta F_e$, increase steeply. This indicates a kind of "phase-transition" between two types of films called "common black" (CB) films and "Newton black" (NB) films.[8]

Secondly one observes a large temperature dependency which we have measured systematically for the first time, as far as we know. These large variations with electrolyte activity and temperature of the NB films cannot be explained[6] by the rather unspecified van der Waals forces and diffuse electrical double-layer forces alone. That is why we followed a more modest approach to describe the NB films and turned our attention first to a coherent thermodynamic description of those films, in order to investigate what kind of information could be extracted from the slopes of ΔF_e versus log a_3(NaCl) which apparently have only a small temperature dependence.

3.2 GIBBS' ADSORPTION EQUATION FOR FILMS

The Gibbs' adsorption equation for a bulk surface, α, is well known and can be written in the following form

$$d\sigma^\alpha = - \,^s\!s^\alpha dT - \sum_{i=2}^{n} \Gamma_i^\alpha \, d\mu_i. \tag{4}$$

Here σ^α is the surface tension of surface α, T the absolute temperature and μ_i the chemical potential of component i. The total number of components is n. Further $^s\!s^\alpha$ and Γ_i^α are surface excesses. $^s\!s^\alpha$ is the surface excess entropy per unit area and Γ_i^α the surface excess amount of component i per unit area, chosen with respect to a Gibbs dividing plane in which the surface excess of component 1, the solvent, Γ_1^α, is zero.

A similar equation applies for a film. To give a derivation here would be too lengthy, hence only the result is shown:

$$d\gamma = - \, 2^s\!s^f dT + hd\Pi - \sum_{i=2}^{n} 2\Gamma_i^f \, d\mu_i. \tag{5}$$

Equation (5) contains an extra term, $hd\Pi$, because the film has an extra degree of freedom, i.e. thickness. In equation (5), γ is the film tension, Π is the so-called disjoining pressure (pressure difference between vapour and bulk phase at the meniscus top) and h is the film thickness (more exactly, the distance of separation of the two Gibbs's dividing planes). Further, Γ_i^f is the surface excess of component i in the film. The composition of the film is then completely defined with respect to the composition of the surrounding vapour phase and of an inner film reference phase (bulk phase) taken as continuous up to the dividing planes.

There is a close relationship between Π and ΔF_e, because ΔF_e is defined by[3-5]

$$\Delta F_e = -\int_{\infty}^{h_e} \Pi(h)dh, \tag{6}$$

where h_e is the equilibrium film thickness. This expression can be used to obtain the relation between ΔF_e and γ.

Upon integration of equation (5) at constant T and μ_i $(i \geqslant 2)$ it follows that

$$\gamma - 2\sigma^\alpha = \int_{h=\infty}^{h=h_e} hd\Pi = \Pi_e h_e - \int_{h=\infty}^{h_e} \Pi dh = \Pi_e h_e + \Delta F_e. \tag{7}$$

It should be recognized that the state $h = \infty$ is a macroscopic slab in which $\gamma(h = \infty) = 2\sigma^\alpha$ and $\Pi(h = \infty) = 0$. It further follows with equation (2) that

$$\gamma = 2\sigma^f + \Pi_e h_e. \tag{8}$$

This means the film tension, γ, contains two surface components, σ^f, and a pressure component across the film.

A useful expression can now be obtained by subtracting equation (4) from equation (5), giving

$$d(\Delta F_e + \Pi_e h_e) = -2\Delta^s s dT + h d\Pi - 2\sum_{i=2}^{n} \Delta\Gamma_i d\mu_i, \tag{9}$$

where $\Delta^s s = {}^s s^f - {}^s s^\alpha$ and $\Delta\Gamma_i = \Gamma_i^f - \Gamma_i^\alpha$ are excess quantities of the film surface over that of the bulk surface.

In the cases considered here, $i = 1$ is the solvent, $i = 2$ is sodium dodecyl-sulphate and $i = 3$ is added salt.

3.3 INTERPRETATION OF CONTACT ANGLES IN TERMS OF SURFACE EXCESSES OF SALT AND DETERGENT

It follows from equation (9) that $\Delta\Gamma_2$ and $\Delta\Gamma_3$ can be obtained from $\partial\Delta F_e/\partial\mu_2$ and $\partial\Delta F_e/\partial\mu_3$, respectively. From the plots in Fig. 1, values of $\Delta\Gamma_3$ were calculated (see Table 1). (The term $\Pi_e h_e$ is negligible for this case and a slight correction for the change of μ_2 with μ_3 was taken into account.)

TABLE 1

Adsorption of electrolyte in NB soap films drawn from aqueous solutions containing 0·05 per cent NaDDS and NaCl respectively. Na_2SO_4 as calculated from Fig. 1 and equation (9).

$\Delta\Gamma_3 \times 10^7/(\text{equivalent m}^{-2})$						
$t°C$	20·5	22·0	23·5	25·0	26·5	29·5
$\Delta\Gamma_{NaCl}$	1·39	1·35	1·34	1·33	1·29	1·22
$\Delta\Gamma_{Na_2SO_4}$		1·72	1·71			

The value of $\Delta\Gamma_2$ was found to be small ($<0.5 \times 10^{-7}$ mol m^{-2}). At the bulk surface $\Gamma_2^\alpha = 42 \times 10^{-7}$ mol m^{-2}.

From Table 1 it follows that $\Delta\Gamma_3$ is positive and only slightly dependent on T; it is larger for Na_2SO_4 than for NaCl.

To interpret these results let us first consider what the theory of diffuse electrical double layers tells us about $\Delta\Gamma_3$. For a *single* double layer, counter-ions are attracted and co-ions are expelled from the surface. Expressed in electroneutral components this means that neutral electrolyte is expelled from the surface. This is called "negative adsorption". For high surface potentials $\Gamma_3^\alpha \simeq -2c_3\kappa^{-1}$, where κ^{-1} is the "thickness" of the diffuse double layer. Thus, formally, in a surface layer with a thickness of $2\kappa^{-1}$ all electrolyte is expelled. In a film with two *overlapping* diffuse double layers the expulsion of electrolyte is less efficient than for two nonoverlapping double layers. This means that $\Delta\Gamma_3$ is positive which is in accordance with what was found experimentally. The calculation of $\Delta\Gamma_3$ for diffuse, *overlapping* double layers is a complicated matter which involves elliptical integrals.[1]

At the bulk and film surfaces the whole electrical double layer cannot expect to be diffuse because of the high electrolyte concentrations present. We therefore attempted to interpret our results as follows.

Assume that both in the film and at the bulk surfaces a small part of the double layers is still diffuse and that the film is so thin that one expects a nearly total overlap of double layers in the film so that $\Gamma_3^f \sim 0$. Then $\Delta\Gamma_3 \simeq 0 - (\Gamma_3^\alpha) \simeq -\Gamma_3^\alpha$.

This implies that one might (in an indirect manner) obtain information about the double layer at the *bulk* surfaces.

To elaborate further on this one needs the thickness of the diffuse part of the double layer, i.e. the distance, $2d$, between the "Outer Helmholtz" (OH) planes in the film. This distance was estimated as follows. From the measured equivalent water thickness of the films, h_w, and an assessed density and refractive index of the hydrocarbon chains we found a thickness $h_2 \simeq 1.6$ nm for the aqueous core of the film. To assess d, the diameter of the hydrated SO_4^--

TABLE 2

Values of the surface potential, $\psi_0^\infty/(\text{mV})$, of the bulk surface, for the system NaDDS + NaCl at 23°C. The ζ-potentials are from data on NaDDS micelles

$C_3(\text{NaCl})/(\text{mol dm}^{-3})$	0·2	0·3	0·4	0·5
for $d = 0\cdot1$ nm and				
$\psi_0 = -\infty$	-41	-34	-30	-28
$\psi_0 = \psi_0^\infty$	-40	-32	-28	-25
for $d = 0\cdot4$ nm and				
$\psi_0 = -\infty$	-75	-70	-70	-71
$\psi_0 = \psi_0^\infty$	-70	-64	-63	-62
ζ (micelles)	-73	-70	-68	-66

group (say, 0·48 nm) and the radius of the co-ion Cl$^-$, say, 0·24 nm, was subtracted from $\frac{1}{2}h_2$. This gives for d a value in the order of a few tenths of a nm. In further calculations we made two choices, i.e. $d = 0\cdot1$ nm and $d = 0\cdot4$ nm. Secondly one needs the potentials at the (OH) planes in the film, ψ_0, and at the surface of the bulk liquid, ψ_0^∞. We considered two limiting cases

$\psi_0 = \psi_0^\infty$ and $\psi_0 = -\infty$. Then from the measured values of $\Delta\Gamma_3 \simeq \Gamma_3^f -$ $\Gamma_3^\alpha \simeq -\Gamma_3^\alpha$ we assessed ψ_0^∞, which was corrected afterwards by estimating the small value of Γ_3^f using elliptical integrals. The results are given in Table 2.

This Table shows that the choice of ψ_0 is not critical which means that there is indeed much overlap of double layers. The calculated values of ψ_0^∞ (of the bulk surface) look reasonable but are still rather sensitive to the choice of d. The values calculated with $d = 0.4$ nm conform rather well with ζ-potentials of micelles.[9]

For Na_2SO_4 similar calculations were made from which it followed that the ratio $\Delta\Gamma_3(Na_2SO_4)/\Delta\Gamma_3(NaCl)$ should be 1·42. The measured ratio is 1·28.

3.4 TEMPERATURE DEPENDENCE OF ΔF_e, EXCESS FILM ENTROPY

It follows from equation (9) that $\partial\Delta F_e/\partial T$ is related to the excess surface entropy of the film over that of the bulk liquid surfaces. Because the μ_i's will change with T under the usual experimental conditions, it is difficult to calculate Δ^ss. Therefore we introduce the interaction entropy

$$\Delta S_e = - d\Delta F_e/dT \tag{10a}$$

and use this further in our qualitative discussion.

From Fig. 1 it follows that for the NB films $\Delta S_e \simeq -0.066 \pm 0.003$ mJ $m^{-2} K^{-1}$, nearly independent of c_{NaCl} and of T. For the CB films, ΔS_e was found to be at least a factor of 50 smaller.

From ΔF_e and ΔS_e one may calculate the interaction energy

$$\Delta U_e = \Delta F_e + T\Delta S_e. \tag{10b}$$

It follows that $\Delta U_e \simeq -20$ mJ m^{-2}.

This means that the formation of a NB film from a CB film is energetically a favourable but entropically an unfavourable process. Both contributions nearly cancel because $\Delta F_e = U_e - T\Delta S_e = -0.1$ to -1 mJ m^{-2}. The large, negative value of ΔS_e imply that NB films have a more organized structure than CB films. This was also found by Jones and Mysels[10] and by den Engelsen and Frens.[11] The large value of ΔU_e cannot be explained by van der Waals forces alone. We think that it must be connected with specific electrostatic interactions between ions of opposite charge juxtaposed to each other in an organized manner.

However, no quantitative model is available at the moment.

4 Films with low electrolyte content

Films with low electrolyte content (0·001–0·15 mol dm^{-3}) show a rather different behaviour than those with high electrolyte content. The former

films are of the CB type and do not show the large temperature dependence
found in NB films. The contact angles are much smaller and the thickness
varies greatly with electrolyte concentration. Because of the low electrolyte
concentrations the CB films are much more sensitive to disturbances of the
film-vapour equilibrium than NB films. We found it impossible to obtain
reproducible results although our measuring cells were well thermostated
($\pm$ 0·003°C) and vapour-tight. We therefore added to all our solutions
glycerol (1 mol dm^{-3}), which greatly improved the reproducibility.

4.1 RESULTS

In Fig. 2 is given the equivalent water thickness, h_w, of the vertical films as a
function of the height H, measured from the meniscus top, at several NaCl
concentrations. The intercepts with the vertical axis are the h_w at the meniscus

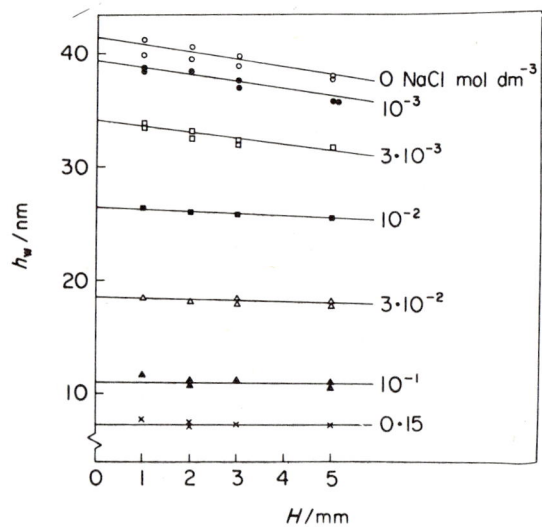

FIG. 2. Equivalent water thickness. h_w, of vertical CB soap films drawn from aqueous solutions
containing 0·03 per cent NaDDS, 1 mol dm^{-3} glycerol and NaCl at 25°C, as a function of height,
H, above the bottom meniscus top. The height of the meniscus top from the bulk surface is
about $z_0 = 3$ mm.

top situated at a height, z_0, above the bulk surface. They are plotted in Fig. 3
as a function of the total active electrolyte concentration in the bulk solution
($c_t = c_{NaCl} + c_{NaDDS}$ when $c_{NaDDS} <$ cmc; $c_t = c_{NaDDS} +$ cmc, when $c_{NaDDS} >$
cmc).

Also some measurements on small horizontal films with a diameter of about

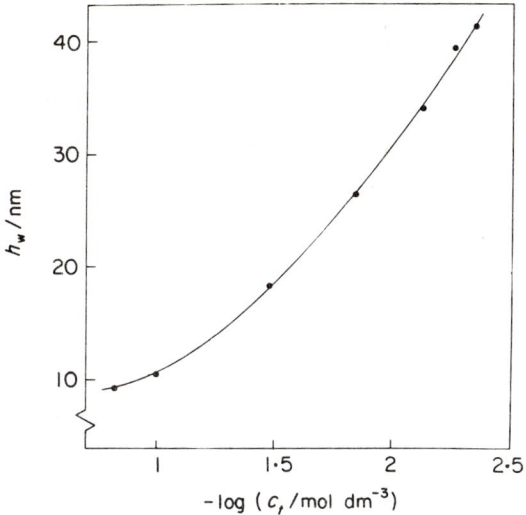

FIG. 3. Equivalent water thickness, h_w, of CB soap films at the meniscus top as a function of the total concentration of counterions, c_t that is active in the bulk solution.

2 mm were made at $c_t = 0.006$ and 0.015 mol dm^{-3} and at various suctions (see further the discussion section below). The results, not shown in Fig. 2, coincide within experimental error with those for vertical films. Therefore we believe that these films are very close to equilibrium. Without glycerol the largest value of h_w was lower than 20 nm in the horizontal films also. In connection herewith we also want to mention some measurements of Bruil[12] on vertical films containing NaDDS plus added salt, with and without the

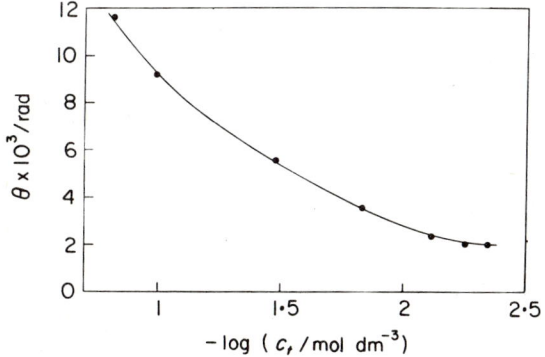

FIG. 4. Contact angles, θ, between CB soap films and bulk solutions at the meniscus top as a function of c_t at 25°C. Composition of the solutions as in Fig. 2.

additive sucrose. At $c_{NaDDS} = 10^{-2}$ mol dm^{-3} and $c_{NaCl} = 10^{-2}$ mol dm^{-3} he found $h_w = 16\cdot0$ nm. With $0\cdot5$ mol dm^{-3} sucrose he found $h_w = 26\cdot3$ nm. These results show a similar trend as we found $h_w = 26\cdot5$ in 10^{-2} mol dm^{-3} NaCl and 1 mol dm^{-3} glycerol. The response of an equilibrium film to external disturbances was studied in particular by Prins and van den Tempel.[13] Mysels and Buchanan,[14] however, reported h_w values of films drawn from solutions *without* glycerol which for comparable electrolyte concentrations were 1 to 3 nm higher than ours.

The measured contact angles, θ, are plotted as a function of the total electrolyte concentration in Fig. 4. Contact angles for similar films as reported by Mysels and Buchanan[14] are appreciably smaller.

4.2 INTERPRETATION

The results we presented here are very recent, so the present discussion is only preliminary. Firstly we wish to draw attention to the variation of h_w with H in Fig. 2. At lower electrolyte concentrations h_w decreases with increasing H. The slope of this line becomes smaller for increasing electrolyte concentration. We interpret this effect as being due to increasing suction on the film due to the increasing hydrostatic head, $dp = -\rho g \, dH$, where ρ is the density difference between the film and the surrounding atmosphere and g the gravitational constant. This suction makes the film thinner and becomes more important when the interaction forces are weak, i.e. in thicker films. We found indeed that the magnitude of the calculated slopes are equal to the measured slopes within experimental error. The same conclusion was arrived at for horizontal films where the hydrostatic head of the film liquid could be changed as well. The measured values of h_w and θ will be analysed with a classical approach[6] as follows.

For the repulsive double force per unit area one writes

$$\Pi_{el} = Bc_t \exp(-\kappa h_2) \tag{11}$$

where $B = 1\cdot59 \times 10^5 \, \gamma^2$ (J mol^{-1} at $t = 25\,^\circ$C) and c_t in mol dm^{-3} and

$$\gamma = \tanh(ze\,\psi_0/4kT) \tag{12}$$

and where ψ_0 is the (diffuse) double-layer potential.

For the van der Waals' attraction force

$$\Pi_w = -(A/6\pi)\,h^{-3} \tag{13}$$

where A is the van der Waals–Hamaker constant.

At equilibrium the sum of these forces plus the hydrostatic suction will cancel, thus

$$Bc_t \exp(-\kappa h_{2e}) - (A/6\pi)\,h_e^{-3} - \rho g z_0 = 0, \tag{14}$$

where z_0 is the height of the meniscus top measured from the bulk surface. From equation (6) one has

$$\Delta F_e = -\int_{\infty}^{h_e} \Pi(h)dh \tag{15}$$

where

$$\Pi(h) = \rho gz = Bc_t \exp(-\kappa h_2) - (A/6\pi) h^{-3}. \tag{16}$$

Substitution of equation (16) into equation (15) and subsequent integration gives

$$\Delta F_e = Bc_t\kappa^{-1} \exp(-\kappa h_{2e}) - (A/12\pi) h_e^{-2} \tag{17}$$

and with equation (14) one finally obtains

$$-(\Delta F_e - \rho gz_0\kappa^{-1}) = (A/6\pi) h_e^{-2} (\kappa h_e - 2)(2\kappa h_e)^{-1}. \tag{18}$$

The variation of $(\kappa h_e - 2)(2\kappa h_e)^{-1}$ is smaller than 5 per cent in our case, so one would expect that $\Delta F_e - \rho gz_0\kappa^{-1}$ is approximately linear with h_e^{-2}.

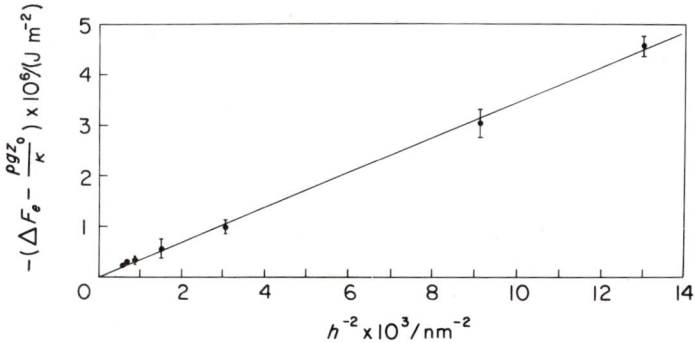

FIG. 5. Interaction free energies, ΔF_e, for CB films drawn from aqueous solutions containing 0.03 per cent NaDDS, 1 mol dm^{-3} glycerol and NaCl, at 25°C, as a function of the reciprocal square of the equilibrium thickness $h = h_e$.

As shown in Fig. 5 this seems to work quite well. From the slope one obtains $A = 1.6 \times 10^{-20}$ J. Substituting this into equation (14) gives

$$\psi_0 \simeq -78 \text{ mV}.$$

Both numbers are not unreasonable. To make a better comparison a more precise expression than equation (13) is required. This will be attempted in the near future.

5 Fluid motions in thin films as observed by light-scattering-fluctuation spectroscopy

Liquid surfaces are not completely smooth but more or less rough. This roughness can be demonstrated by throwing an intense light beam on the surface; most of the incident light is reflected and refracted, but also a very small part of it is scattered (diffusely reflected) in all other directions. The roughness is caused by thermal motion. Surface light scattering observed in a certain direction is, in a similar way to X-ray scattering, intimately connected with a certain Fourier mode of the surface corrugations.[15] In the simple case where the directions of the incident, refracted, reflected and scattered waves are in the same plane perpendicular to the surface, the direction of the corresponding Fourier mode of the surface ripples is in the sample plane. Its wavelength, Λ, or wavenumber, $K = 2\pi/\Lambda$, is connected through a kind of simple "Bragg-law" with the angles of reflection, Θ_0, and of scattering Θ, thus

$$K = (2\pi/\lambda)(\sin \Theta_0 - \sin \Theta), \qquad (19)$$

where λ is the wavelength of light in the gaseous medium.

According to this equation only modes with $\Lambda \gtrsim \lambda$ can be detected. The mean light scattering intensity is proportional to the mean square amplitude of the surface mode, $\langle \zeta_k^2 \rangle$, and to $(n^2 - 1)^2$, where n is the refractive index ratio of the two media separated by the surface. The value of $\langle \zeta_k^2 \rangle$ is obtained by equating the work necessary to create the mode to the thermal energy. In this way it is found that

$$a^2 \langle \zeta_k^2 \rangle = kT/\sigma, \qquad (20)$$

where σ is the surface tension and a^2 is the surface area considered. Thus the light scattering will be proportional to σ^{-1}, which implies that this unconventional method may be of use to measure very small surface tensions.

This is the information that can be extracted when one measures the mean light-scattering intensity. It is possible, however, also to study the *dynamics* of the surface corrugations by the new but now well-known "Light-Scattering-Intensity-Fluctuation-Spectroscopy" technique.[2,16,17] It is not possible to enter here into detail about this method which is very useful in many other areas as well. The principles are roughly as follows.

Instead of measuring only the mean intensity of the scattered light one studies the fluctuations of that intensity, ΔI, in time. In our case ΔI is proportional to $\zeta_k^2 - \langle \zeta_k^2 \rangle$. So doing, it is possible to obtain the mean quantity

$$\langle \Delta I(t_0 + t)\Delta I(t_0) \rangle \qquad (21)$$

by measuring ΔI, at a time t_0, multiplying it by ΔI at $t_0 + t$ and calculating the mean value over many samples.

For a steady-state fluctuation process, as we have here, this quantity does not depend on t_0, but only on t. It will be clear that for large values of t there is no correlation between $\Delta I(t_0)$ and $\Delta I(t_0 + t)$ and their mean product will vanish.

However for smaller values of t, comparable with the relaxation time τ of the modes, this quantity will have a finite value. It can be proved that

$$\Phi_{\Delta I}(t) \equiv \langle \Delta I(t_0 + t)\Delta I(t)\rangle \sim \exp\left(-2t/\tau\right). \tag{22}$$

Generally, τ is a complex number connected with the travelling speed and the damping of the surface mode.

Now let us turn our attention to the surface light scattering of thin, free liquid soap films. In this case we have *two* different types of surface ripple modes: a "stretching mode" in which both film surfaces move parallel to one another, and a "squeezing mode" in which they move antiparallel. The stretching modes have τ's in the order of microseconds and are of no interest here. Only the squeezing modes will be considered further.

Their amplitude is governed not only by the surface tension, but also by the interaction forces considered above, whereas the relaxation time of the mode is in addition determined by liquid and interface properties. For thin liquid films, $h \ll \Lambda$, in the hydrodynamic regime with nonmoving surfaces, we found[18-20] for the relaxation time

$$\tau = 24\,\eta h^{-3}K^{-4}\sigma^{-1}[1 + (2\sigma^{-1}K^{-2}d^2\Delta F/dh^2)]^{-1} \tag{23}$$

where η is the shear viscosity in the liquid film layer having a thickness h and a surface tension σ. The mode is nonrunning and overdamped. The factor h^{-3} is related to Poiseuille's law. Observe that the measurements are not necessarily restricted to equilibrium states as with the contact angle method.

6 Results and discussion

We will show some results of films drawn from aqueous hexadecyltrimethyl-ammoniumbromide (HDTAB) solutions with KBr and glycerol. The light-scattering apparatus was used with an argon-ion laser ($\lambda = 488$ or $514\cdot5$ nm, power about 100 mW) at $\Theta_0 = 45°$ and the "homodyne" correlations were measured either of draining films at constant scattering angle, Θ, or of stationary films ($h \simeq$ constant) with $|\Theta - \Theta_0|$ varying between 5 and 15 degrees (on either side of the reflected beam).

The photocurrent fluctuations were analysed with the aid of a minicomputer.

A resulting correlation function $\phi(t)$ (see equation (22)) given in Fig. 6,

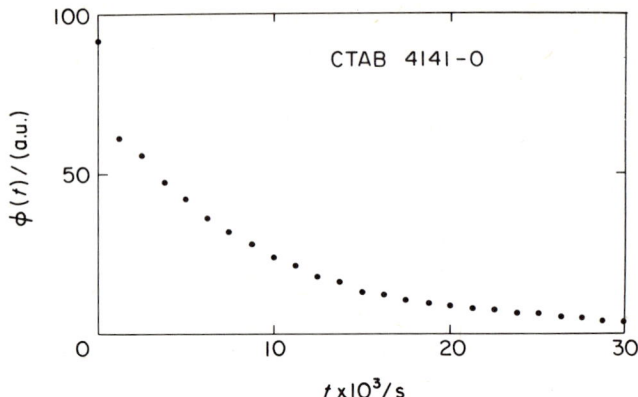

FIG. 6. Example of a correlation function of the intensity fluctuations of light scattered by a soap film (see equation (22)), $h = h_w = 27\cdot2$ nm; $\Theta_0 = 45°$; $\Theta = 51°$.

clearly shows the exponential decay. The relaxation time for this case is $\tau = 19$ ms.

In Fig. 7 values of $(\tau h^3 K^4)^{-1}$ are given for a stationary film with $h_w \simeq 83$ nm for varying Θ (and thus of K). Observe the constancy of τK^4. It means that τ is very angular dependent. Here it changes from 12 ms for $\Theta = 50°$ to

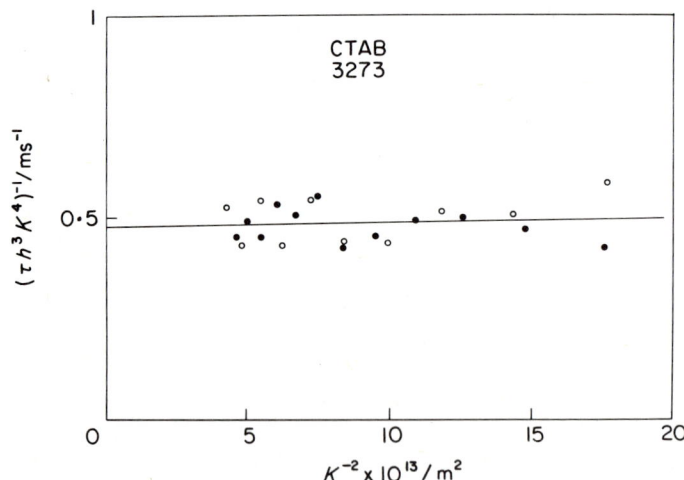

FIG. 7. Values of $(\tau h^3 K^4)^{-1}$ for a soap film as a function of $K^{-2} = (\Lambda/2\pi)^2$, where Λ is the wavelength of the squeezing mode, $h = h_w = 83$ nm. The aqueous solution from which the film is drawn contains $0\cdot293$ g dm^{-3} HDTAB and 85 g dm^{-3} glycerol with $\eta = 1\cdot10 \times 10^{-3}$ J m^{-3} s; $t = 25°$C; $\sigma = 3\cdot82 \times 10^{-2}$ N m^{-1}.

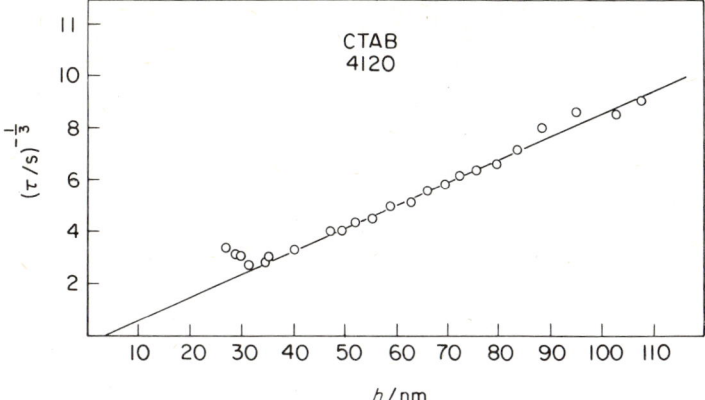

FIG. 8. The cube root of the reciprocal relaxation time of a draining soap film as a function of thickness $h = h_w$, $\Theta_0 = 45°$, $\Theta = 39°$. The aqueous solution from which the film is drawn contains 0.260 g dm^{-3} HDTAB, 10^{-2} mol dm^{-3} KBr and glycerol with $\eta = 1.08 \times 10^{-3}$ J m^{-3} s. $t = 25°C$; $\sigma = 3.37 \times 10^{-2}$ N m^{-1}.

0.8 ms for $\Theta = 55.5°$. This is indeed predicted by equation (23). This equation further predicts that $d^2\Delta F/dh^2$ is apparently too small to measure in these rather thick films.

Figure 8 shows how $\tau^{-\frac{1}{3}}$ changes as a function of film thickness in a draining film at $\Theta = 39°$. This film contains 10^{-2} mol dm^{-3} KBr. A straight line can

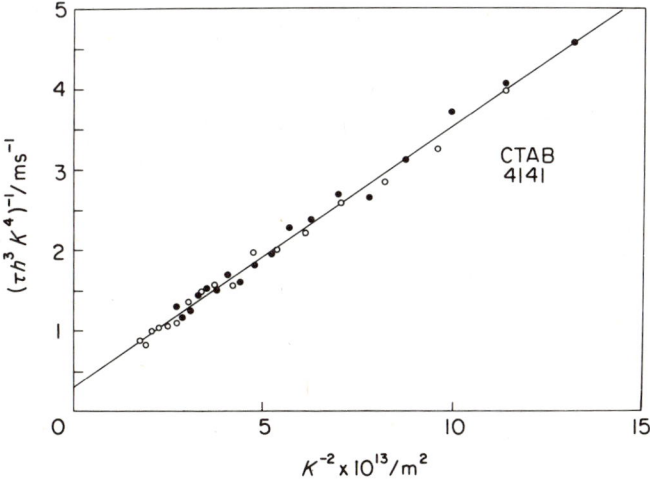

FIG. 9. Values of $(\tau h^3 K^4)^{-1}$ for a soap film as a function of $K^{-2} = (\Lambda/2\pi)^2$, $h = h_w = 27.2$ nm. The composition of the aqueous solution is the same as in Fig. 8.

be drawn through the middle part of the plot as predicted by equation (23). The points around 90 nm are scattered, probably because of the low accuracy of the calculated h_w around the reflection maximum. The points around 30 nm are systematically above the straight line. The value of τ at the smallest thickness would have to be multiplied by a factor of 4 to bring it down on the straight line. Apparently the film is more difficult to deform in this case so that it relaxes faster. This was further tested by doing a similar experiment as shown in Fig. 7 for a film with $h_w = 27.2$ nm. The plot, given in Fig. 9 does indeed show a large slope which, according to equation (23), means that $d^2\Delta F/dh^2$ is large. From the intercepts of the Figs 7 and 9 and from the slope of Fig. 8 it is possible to determine σ/η according to equation (23). We found $\sigma/\eta = 13$, 10 and 20 ms^{-1} respectively.

Taking the bulk values $\sigma = 3.82 \times 10^{-2}$, 3.37×10^{-2} and 3.37×10^{-2} N m^{-1} and $\eta = 1.10 \times 10^{-3}$, 1.08×10^{-3} and 1.08×10^{-3} J m^{-3} s respectively, one obtains $\sigma/\eta = 34.7$, 31.2 and 31.2 ms^{-1}. Thus the ratio σ/η obtained from our light-scattering experiments is 1.5 to 3 times smaller than the bulk values. Our previously reported[21] value of $\sigma/\eta = 24 \times 1.4 = 34$ is in error and should be multiplied by a factor $\frac{1}{2}$, so that it comes into the same range. The discrepancy is puzzling to us. One reason could be an increase of η, e.g. because of electroviscous effects, although an increase of η because of the evaporation of water cannot be entirely ruled out. We are investigating this now more systematically.

References

1. de Feijter, J. A. (1973). Contact angles in soap films. Ph.D. thesis, University of Utrecht.
2. *J. Colloid Interface Sci.* To be published.
3. Princen, H. M. (1965). Shape of fluid drops at fluid–liquid interfaces and permeability of soap films to gases. Ph.D. thesis, University of Utrecht.
4. Princen, H. M. (1965). *J. Colloid Sci.* **20**, 156.
5. Derjaguin, B. V. (1940). *Acta Phys. Chim.* **12**, 181.
6. Huisman, F. and Mysels, K. J. (1969). *J. Phys. Chem.* **73**, 489.
7. Princen, H. M. and Frankel, S. (1971). *J. Colloid Interface Sci.* **35**, 386.
8. Everett, D. H. (1972). *Pure Appl. Chem.* **31** (4), 614.
9. Stigter, D. and Mysels, K. J. (1955). *J. Phys. Chem.* **59**, 45.
10. Jones, M. N., Mysels, K. J. and Scholten, P. C. (1966). *Trans. Faraday Soc.* **62**, 1336.
11. den Engelsen, D. and Frens, G. (1974). *J. Chem. Soc. Faraday Trans. I*, **70**, 193.
12. Bruil, H. G. (1970). *Meded. Landbouwhogeschool Wageningen*, **70–9**, 44. (In English.)
13. Prins, A. and van den Tempel, M. (1970). *Spec. Discuss. Faraday Soc.* **1**, 20.
14. Mysels, K. J. and Buchanan, J. W. (1972). *J. Electroanal. Chem.* **37**, 23.
15. Vrij, A. (1968). *Advan. Colloid Interface Sci.*, **2**, 39.
16. Bouchiat, M. A., Meunier, J. and Brossel, J. (1968). *C.r. hebd. Séanc. Acad. Sci., Paris,* **266B**, 255.

17. Katyl, R. H. and Ingard, U. (1967). *Phys. Rev. Letts*, **19**, 64; (1968). *Phys. Rev. Letts*, **20**, 248.
18. Vrij, A. (1966). *Discuss. Faraday Soc.* **42**, 23.
19. Vrij, A. and Overbeek, J. Th. G. (1968). *J. Amer. Chem. Soc.* **90**, 3074.
20. Vrij, A., Hesselink, F. Th., Lucassen, J. and van den Tempel, M. (1970). *K. Ned. Akad. Wet. Proc. Ser. B.* **73**, 124.
21. Fijnaut, H. M. and Vrij, A. (1973), *Nature Phys. Sci.* **246**, no. 155, 118.

Discussion

Phillips (1) In your paper you consider the surfaces of the thin films to be rigid. Could you please give the values of the rheological parameters that define a rigid film and what levels of adsorption does this imply for common surfactants?

(2) What role do the surface waves you study play in the rupture of the thin films and how do factors such as the dilatational modulus affect film collapse?

Vrij To consider your first question I avoided here the term "rigid" surface, which might lead to confusion, but used the term "nonmoving" surface. Equation (23) to which this condition applies is a special case of the (dispersion) equation of a film in air (see reference 20, equation (5)).

$$\omega^3 - \left(\frac{4i\eta K^2}{\rho}\right)\omega^2 - \frac{K^2}{\rho h}\left(2\varepsilon + \frac{\sigma_{eff}K^2h^2}{2}\right)\omega + \frac{iK^6h^2\varepsilon\sigma_{eff}}{12\eta\rho} = 0 \qquad (a)$$

still under the condition $Kh \ll 1$, i.e. wavelengths large with respect to the film thickness. Here, ω is the (complex) circular frequency, i is the imaginary unit, ρ is the film density, η is the viscosity of the film liquid, σ_{eff} is an effective surface tension, defined by

$$\sigma_{eff} = \sigma + \frac{2}{K^2}\frac{d^2\Delta F}{dh^2}, \qquad (b)$$

where σ is the surface tension of each film surface and $d^2\Delta F/dh^2$ is defined in the text. Further ε is the (real) surface elasticity (dilational modulus) of each surface defined by

$$\varepsilon = d\sigma/d \ln A \qquad (c)$$

where A is the area of a surface element in a single film surface. For sufficiently large ε equation (a) transforms into

$$\omega = \frac{iK^4h^3\sigma_{eff}}{24\eta} \qquad (d)$$

which is equivalent to equation (23) in the text. It is for this situation that we used the term "nonmoving". Sufficiently large ε implies here that it should be large with respect to $K^2h^2\sigma_{eff}/4$.

In our case the last quantity is about 0.2 dyn cm^{-1} or less. This is a rather small number and we expect that ε in our case is large. Influence of ε would also give a deviation of the horizontal line in Fig. 7. Although ε is often not real but complex, because of relaxation due to diffusion, it seems that this is only of importance if ε is already low enough to show an effect. This is under further investigation.

With reference to your second question, equation (23) of the text shows that τ becomes negative, so that spontaneous fluctuations will grow, when $d^2\Delta F/dh^2$ is sufficiently negative. A detailed analysis (see reference 20) shows that in this case ε should be very small, say $< 10^{-5}$ dyn cm^{-1}, to change the regime given by equation (23). It means that this mechanism alone cannot explain the specificity of film collapse due to different surface active agents.

Roberts, K. Do you believe that there are major differences (of orders of magnitude, say) between the relaxation times of thicker films formed from *different surfactants* (19 ms for CTAB in your paper).

Vrij Equation (23) applies only when $K^2h^2 \ll 1$. In this regime we do not expect large differences with different surfactants (see further the answer to Phillips). For $K^2h^2 \gg 1$ the regime is different. Although we did not investigate this case one might expect that the surface ripples below more or less as on a single surface.

Camina (*Paint Research Station, Teddington, England*) It is interesting that the powers of the surface tension, viscosity thickness and wavelength in the first term of equation (23) are the same as that developed for the levelling of paint films, although in the latter case there is no interaction effect. For non-Newtonian flow the equation becomes more complex and I wonder whether the discrepancy between the calculated and experimental values of σ/η could be caused by non-Newtonian flow.

Vrij Although we have no reason to expect non-Newtonian flow in the film liquid, which is a dilute, aqueous surfactant solution containing some glycerol, it could be worth while to investigate this further.

8

Some techniques for the investigation of foam stability

D. EXEROWA, KHR. KHRISTOV and I. PENEV

Bulgarian Academy of Sciences, Institute of Physical Chemistry, Sofia 13, Bulgaria

Summary

The possibility of considering the problems of foam stability on the basis of the properties of the thin liquid films is generalized. The thesis, expressed previously by Scheludko and Exerowa, is that the occurrence of high foam stability is connected with the formation of black spots in the foam films and therefore associated with some peculiarities of the adsorbed layers. This thesis is supported by new results and is further developed. The results show again the great importance of black films on foam stability.

A new technique to study the stability of foam by their forced destruction is presented. Two different methods were used.

The first is destruction of foams and foam films by a dosed disruptive action of α-particles. The time $\tau_{R/2}$ during which the foam disappears at a distance $R/2$ from the α-source is used as a measure of the rate of foam destruction (R is the free path of the α-particles in the air). The values of $\tau_{R/2}$ for several surfactants were determined under standard conditions and it is shown that $\tau_{R/2}$ can be used as a characteristic of the foaming ability of the substances.

The second is destruction of foams through creation of a pressure gradient in the Gibbs channels of the foam and acceleration of the outflow of the solution from the foam. This method can be used for practical destruction of foams, as well as for characterization of foaming agents.

1 Introduction

"It is, perhaps, surprising to find in 1958 that no thoroughly satisfactory explanation has yet been given as to why certain liquids foam strongly, others feebly, and many not at all." This is the beginning of an interesting review by Kitchener and Cooper.[1] Seventeen years later it would be almost true to use the same sentence. This does not mean that foams have not attracted the attention of scientists. Many investigations and several reviews[2,3] have been published recently. It is very difficult to develop a general theory of foam stability : many different—dynamic or static—factors determine it. It is essential, however, to understand the causes of high foam stability. This problem is different from the problem of the stability of lyophobic colloids. Foam bubbles remain in contact with each other and foam stability is closely connected with the properties of the foam films (free liquid films). That is why the investigation of foam films as a model foam is very important and in our opinion is one of the reliable ways of understanding foam stability. This approach is not a new one—many foam film investigations have already been made and some of the results have been reviewed.[1,2,4–7,9,10]

In this paper foam stability is considered on the basis of quantitative results from model experiments on foam films, which permits the identification of the processes occurring within the films. Also. two methods for forced foam destruction are presented. These methods permit the characterization of stable foams and are interesting from a practical point of view.

2 Thin liquid films and foams

Circular microscopic (radius c. 0·01 cm) free films obtained in the middle of a biconcave drop have been used in our laboratory as a basic model of foam films for several years.[6–9,11] This model allows the investigation of all kinds of foam-forming solutions under the same controlled conditions. Many aspects of the influence of adsorption of different surfactants on the behaviour of such films have been studied. First and foremost it is practically impossible to obtain a free liquid film without a surface-active agent. The latter creates conditions for viscous thinning of the film, i.e. outflow of the solution from it, in a finite time.[6,9] In the course of thinning the films reach the so-called state of kinetic instability, characterized by the critical thickness of the film.[7] According to the type and concentration of the surfactant, the film either ruptures or forms thinner stable parts, these appearing black in reflected light—the black spots.[8,9,21,22] Figure 1 shows a microscopic film with radius 0·01 cm showing black spots in it. Both rupture (for unstable films) and formation of black spots (for both unstable and stable films) occur at the same critical thickness, which is about 30 nm for water.[9,17]

Many good foaming surfactants of the "detergent" type[6,7] are able to form black spots in foam films, if added in sufficient concentration to the solution. To this group belong some anionic substances (soaps, alkylsulphates, alkylsulphonates), cationic substances (alkyltrimethylammonium bromides, alkylpiridiniumchlorides) and nonionic substances (alkylpolyglycolethers) etc. Other groups of substances such as lower fatty acids and fatty alcohols do not form black spots in films at any concentration—the films rupture in all cases on reaching the critical thickness. These substances are not good foam stabilizers although they are good surface-active agents. We found a correlation between foam stability and the formation of black

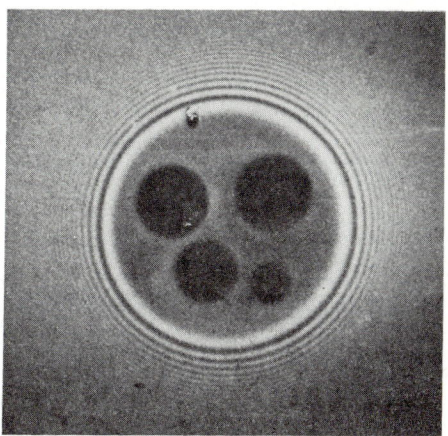

FIG. 1. Black spots in a microscopic foam film from 8×10^{-6} mol dm^{-3} OP 20 + 0·1 mol dm^{-3} KCl aqueous solution.

spots in films.[21] The black spots first appear at that surfactant concentration above which the foam becomes stable with lifetime greater than 10 minutes. At lower concentrations neither black spots are observed, nor are stable foams obtained (the lifetimes being of the order of seconds). This correlation was confirmed for a great number of foaming agents and we propose to use the surfactant concentration C_{bl} at which the black spots first appear in the film as a characteristic of those agents. Of course C_{bl} must always be determined under the same standard conditions.[9] Figure 2 illustrates the relationship between C_{bl} and the surfactant concentration C_s at which the foam becomes very stable.[22] The dependence of the lifetime τ on C_s is shown for several L(EO)$_n$,* τ being measured by the method of Bartsch.[23] The substances were specially prepared for these experiments by Henkel (Düsseldorf) and were kindly sent to us by Professor H. Lange. All precautions were taken to prepare solutions of high purity. The values of C_{bl} for three cases ($n = 4, 9, 15$) are shown in Fig. 2 and it is seen that they are near to the

* The full names of the surface-active substances and their abbreviations are given in a list at the end of the paper.

inflexion points of the curves $\tau(C_s)$. The concentrations, at which $\tau > 10$ min. and the values of C_{bl}, decrease with increasing n, but for $n > 11$ the changes are small. These results are in good agreement with other data about the foaming ability of the same substances.[24, 25] For some other types of foaming agent (for instance the alkylsulphates) the values of C_{bl} are somewhat less

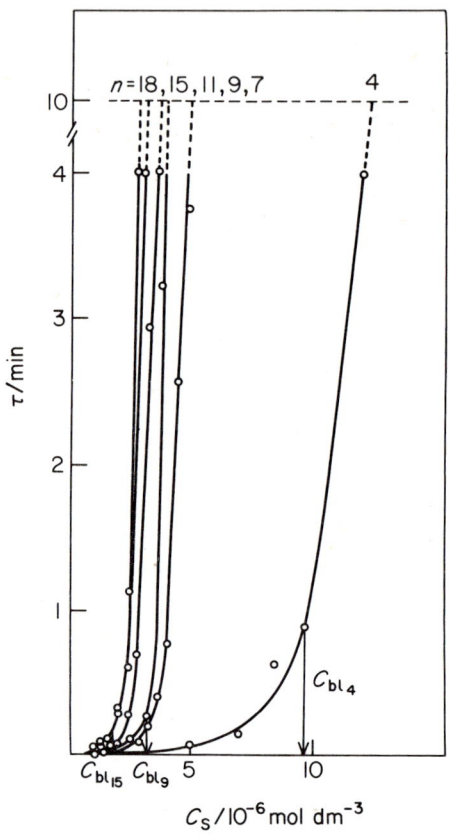

FIG. 2. Dependence of the lifetime τ of foams on the concentration C_s of $L(EO)_n$. The arrows indicate the C_{bl} values for $n = 4.9$ and 15.

than the C_s value which corresponds to the inflexion point of the $\tau(C_s)$ curve.

The action of antifoaming agents can be considered from the same point of view; when an antifoaming agent is added in concentration sufficient to prevent the formation of a stable foam, black spots cease to appear in the microscopic foam films.[28, 29] The limitation to the purification of solutions of surfactants by foaming is also connected with the formation of black spots in the foam films.[31]

In Table 1 the measured C_{bl} values for several surfactants are presented and one can see the regular change of C_{bl} for the members of a homologous series. The data for the anionic alkylsulphates and the nonionic $L(EO)_n$ were obtained very carefully with specially purified reagents.

TABLE 1

Concentrations C_{bl} of the surfactant at which black spots first appear in microscopic foam films for different substances

Chemical pure foam stabilizers	$C_{bl} \times 10^6$ mol dm^{-3}	Chemical pure foam stabilizers	$C_{bl} \times 10^6$ mol dm^{-3}
OP 7	13·6	Sodium undecylsulphate	4·0
OP 9	10·8	Sodium dodecylsulphate	1·6
OP 15	5·2	Sodium tetradecylsulphate	1·4
OP 20	4·6		
		DMS	460·0
$L(EO)_4$	9·7	SLU	35·0
$L(EO)_5$	6·1	Sodium octylsulphonate	650·0
$L(EO)_6$	4·9	Sodium laurate	1·3
$L(EO)_7$	4·8	CPC	14·0
$L(EO)_8$	2·8		
$L(EO)_9$	3·2		
$L(EO)_{10}$	2·6	Technically made	
$L(EO)_{11}$	2·4	stabilizers	$C_{bl} \times 10^6$
$L(EO)_{12}$	2·1		
$L(EO)_{15}$	1·7	Igepon	4
$L(EO)_{18}$	1·3	Nekal	10
		Humektol	10
Sodium octylsulphate	134·0	Saponin	10
Sodium decylsulphate	13·0	Alkaril B	300

The values of C_{bl} are usually 1–2 orders less than the corresponding values of cmc. For instance, for $C_{12}H_{25}(OCH_2CH_2)_4OH$ the cmc $= 5 \times 10^{-5}$ mol dm^{-3} and for $C_{12}H_{25}(OCH_2CH_2)_{18}OH$ the cmc $= 1\cdot3 \times 10^{-4}$ mol dm^{-3}. On the other hand for a homologous series of $L(EO)_n$, C_{bl} decreases, while cmc increases with the number of $-OCH_2CH_2-$ groups. These facts show that high foam stability does not depend on the bulk properties of solutions, such as the micelle formation but that the main effect is on the adsorbed layers. Experimental data on the dependence of surface tension and double-layer potential on surfactant concentration show that at C_{bl} some changes in the adsorbed layer occur.[21,7] These peculiar properties of adsorbed

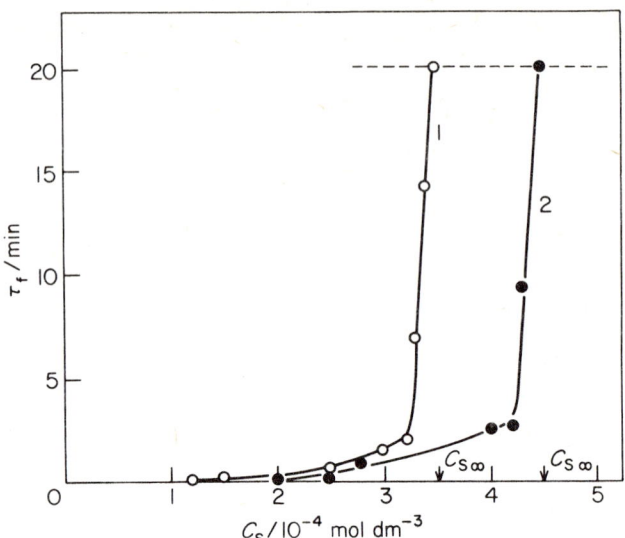

FIG. 3. Dependence of the lifetime τ_f of microscopic films on the concentration of NaLS. Curve 1, common black film ($C_{el} = 0.25$ mol dm^{-3} NaCl); curve 2, Newton black film ($C_{bl} = 0.4$ mol dm^{-3} NaCl).

layers of good foaming agents are closely related with foam stability. Neither the absolute value of the surface tension, nor the double-layer potential ψ_0 (i.e. the electrostatic disjoining pressure) have any correlation with high foam stability.[11,21,27] For instance, the value of ψ_0 is about 43 mV for all members of the L(EO)$_n$ series, whilst they have different C_{bl} values.

TABLE 2

Concentrations $C_{s\infty}$ of the surfactant at which the microscopic black foam films become stable and the concentrations C_{bl} for different substances

Foam stabilizers	$C_{bl} \times 10^6$ mol dm^{-3}	$C_{s\infty} \times 10^6$ mol dm^{-3}
L(EO)$_6$	4.8	50
OP 15	5.5	90
DMS	440.0	1200
CPC	14.0	25
CTAB	8.5	30
NaLS	3.5	450
NaO	40.0	400
SLU	35.0	150

In Fig. 3 the lifetimes τ_f of microscopic common (curve 1) and Newton (curve 2) black films from NaLS solutions are presented as a function of the surfactant concentration C_s. One can see that at concentrations denoted by $C_{s\infty}$ the values of τ_f increase very steeply to values greater than 20 minutes. The curves are similar to the τ (C_s) curves for the foams. The concentrations $C_{s\infty}$ both for the common black film ($3{\cdot}5 \times 10^{-4}$ mol dm^{-3}) and for the Newton black film ($4{\cdot}5 \times 10^{-4}$ mol dm^{-3}) are two orders of magnitude higher than $C_{bl} = 3{\cdot}5 \times 10^{-6}$ mol dm^{-3}.

In Table 2 the values of $C_{s\infty}$ for several surfactants of different type are presented. The data are measured on microscopic films with radii $0{\cdot}025$ cm from solutions with NaCl concentration corresponding to the transition first to second black film.[32, 39] In all cases the $C_{s\infty}$ values are higher than C_{bl}. A quantitative interpretation of these results has not been made yet, but we can assume that at the concentration C_∞ compact adsorbed layers are already formed. Some results for the surface concentration Γ of these adsorbed layers[44] support this assumption. Measurements of foam stability (by the method of Bartsch or the foam column technique) show that in all cases foams at $C_{s\infty}$ are very stable.

3 Investigation of foam stability by forced destruction of foams

Many chemical, mechanical and electrical methods for foam destruction have been described. There are, however, almost no quantitative studies on the destruction of foams with the aim of investigating their stability. Interesting experiments on rupture of foam films by electric sparks, showing the process of development of the holes in the film, were carried out by McEntee and Mysels.[33]

3.1 DESTRUCTION OF FOAMS AND FOAM FILMS BY α-PARTICLES

The disruptive action of α-particles on foams from very concentrated mixed solutions has been observed for a long time,[34, 35, 36] but only qualitatively. The amount of α-radiation can be controlled and its disruptive action defined. Hence this action can be used to characterize the properties of the foams. The effect of α-radiation on a foam from NaLs-solution is shown in Fig. 4. The foam is ruptured in the region of penetration by the α-radiation; a similar picture is obtained in a Wilson camera. We developed experimental methods for such investigations for both foams and foam films.[37, 38]

Our device, in which the foam is produced, irradiated and ruptured is shown schematically in Fig. 5. The solution is put in the vessel and air is blown through the porous plate 3, producing foam. After a definite time (30–60 min) the foam is already fully drained and the α-source 1 is placed

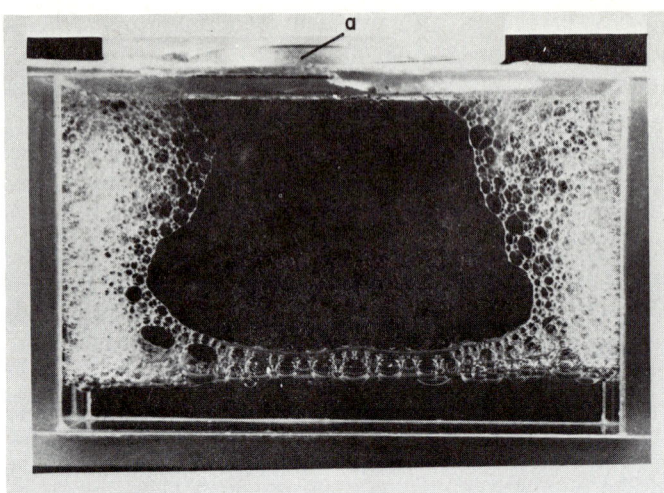

Fig. 4. Destruction of foam in a wide foam column by α-particles from ^{239}Pu source (a).

above it. The destruction of the foam begins immediately with a decreasing rate and ceases when the surface of the foam reaches the distance R from the source, R being the free path of the α particles in air. As a parameter, to characterize foam stability, the time $\tau_{R/2}$ was chosen during which the foam ruptured up to the distance $R/2$ from the α-source.

We investigated the dependence of $\tau_{R/2}$ on the electrolyte concentration C_{el} of the solutions. A curve for a foam from DMS solution, irradiated by

TABLE 3

Order of some foaming agents according to the time $\tau_{R/2}$ for destruction of a foam column by α-particles; α-source is ^{239}Pu. The $\tau_{R/2}$ values are compared with the times τ for spontaneous destruction of the same foams

Foam stabilizers	$\tau_{R/2}$/min	τ/min for subsidence of columns with initial heights	
		20 mm	40 mm
L(EO)$_6$	0	10	40
OP 15	0·5	32	85
DMS	1	47	300
CPC	6	60	110
NaLS	11	130	> 300
NaO	13	200	> 300
SLU	> 300	> 300	> 300

^{241}Am is presented in Fig. 6. One can see that $\tau_{R/2}$ decreases to a minimum with increasing C_{el} and then increases sharply. The minimum corresponds to the electrolyte concentration $C_{el, er} = 0.032 \, mol \, dm^{-3}$ NaCl, at which the transition common black to Newton black film occurs.[39] Similar

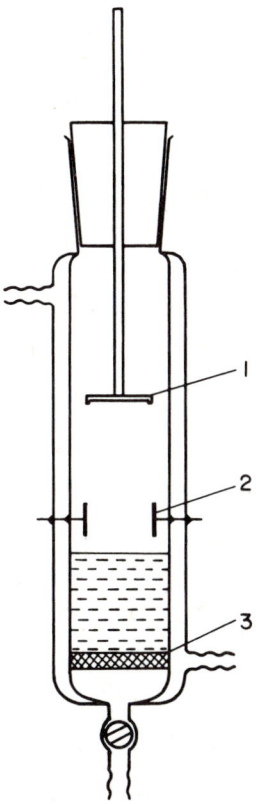

FIG. 5. Drawing of the apparatus for investigating foam destruction by α-particles: 1, α-source; 2, electrodes; 3, porous plate.

results were obtained when single microscopic black films were irradiated with α-rays. A special apparatus was constructed and the lifetimes τ_α of black films under radiation was measured. A curve $\tau_\alpha = f(C_{el})$ for films from the same solution of DMS is shown in Fig. 7. The minimum, corresponding to $C_{el, er}$, is at $0.034 \, mol \, dm^{-3}$ NaCl.

These methods for destruction of foams and films by α-particles allow on one hand the study of the transition common black to Newton black

D. EXEROWA *ET AL.*

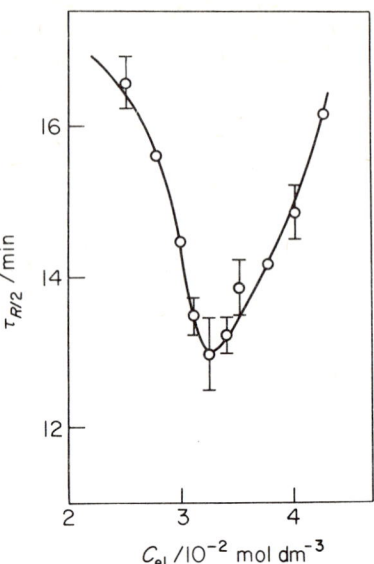

FIG. 6. Lifetime $\tau_{R/2}$ of a foam under α-radiation versus NaCl concentration for $1{\cdot}5 \times 10^{-3}$ mol dm^{-3} DMS aqueous solution. The minimum corresponds to $C_{el, er} = 3{\cdot}2 \times 10^{-2}$ mol dm^{-3}.

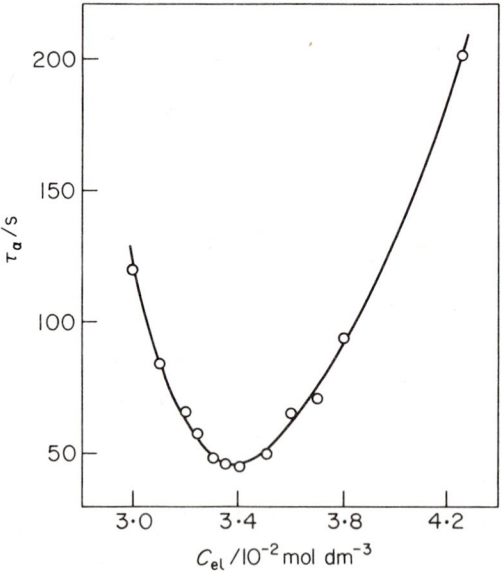

FIG. 7. Lifetime τ_{α} of a black foam film under α-radiation versus NaCl concentration for $1{\cdot}5 \times 10^{-3}$ mol dm^{-3} DMS aqueous solution. The minimum corresponds to $C_{el, er} = 3{\cdot}4 \times 10^{-2}$ mol dm^{-3}.

films and on the other hand the characterization of foams and foam films with respect to their stability. The spontaneous rupture of very stable foams is an extremely slow process. The lifetimes τ (many hours or days) cannot be measured precisely and rapidly and therefore are not convenient for characterization of stability. The times $\tau_{R/2}$ are orders of magnitude less than τ and we propose them as a measure of stability. They can be determined under standard conditions and can be used for comparison and classification of the foam-forming agents. The $\tau_{R/2}$ values, presented in Table 3, are measured under the following conditions: surfactant concentration $1\cdot5 \times 10^{-3}$ mol dm^{-3}, electrolyte concentration $C_{el, er}$, source ^{239}Pu with activity $4\cdot2 \times 10^{4}$ s^{-1}, liquid content of the foam $W = 7 \times 10^{-2}$ vol%. W is determined by measuring the specific conductance of the foam and the solution.

The data in Table 3 show that $\tau_{R/2}$ varies over a wide range from 0 to 3600 s and that different substances can conveniently be compared with respect to their foaming ability.

The $\tau_{R/2}$ values are compared with the time τ for spontaneous rupture of 20 mm of a 40-mm column of the same foam. It is seen that the order of the substances is the same both by $\tau_{R/2}$ and τ. Hence the measurement of $\tau_{R/2}$ is a rapid method for comparison of the lifetime, that is the stability of stable foams.

3.2. DESTRUCTION OF FOAMS THROUGH CREATION OF A PRESSURE GRADIENT IN THE GIBBS CHANNELS OF THE FOAM

The foam, soon after its production, usually contains much liquid (solution) in its films and especially in the Gibbs channels which connect the films. The spontaneous-outflow of this liquid due to gravity (syneresis) is a slow process. In our laboratory, together with P. Kruglyakov and A. Scheludko, we developed a method of accelerating the outflow of solution from the foam.[40] A vessel with a porous plate bottom similar to that shown in Fig. 5 is used. Either by increasing the pressure in the vessel containing the foam, or by decreasing the pressure in the space behind the porous plate, a pressure difference is created between the two parts of the vessel. This difference cannot be greater than the capillary pressure in the pores of the plate which depends on their diameter.

The pressure difference creates a pressure gradient in the Gibbs channels of the foam which accelerates the outflow of liquid. Also the Gibbs channels themselves decrease in size thus decreasing the solution W of the foam. A controlled degree of "drying" of the foam can be achieved by regulating the pressure differential. Figure 8 presents the decrease of W with time for an OP 15 solution spontaneously (curve 1) and due to a pressure differential of 10^{4} N m^{-2} (curve 2). In the second case W decreases faster and after 20 min reaches the value $3\cdot10^{-3}$ vol%.

E

The pressure gradient in the Gibbs channels also causes rapid destruction of the foam, for instance a foam of OP 15 solution ruptures spontaneously in 2 hours and at a pressure differential of $10^4\,\mathrm{N\,m^{-2}}$ in only 17 minutes. The more stable foam from NaLS solution ruptures spontaneously in about 30 hours and at a pressure differential of $5 \times 10^4\,\mathrm{N\,m^{-2}}$ in 2 hours.

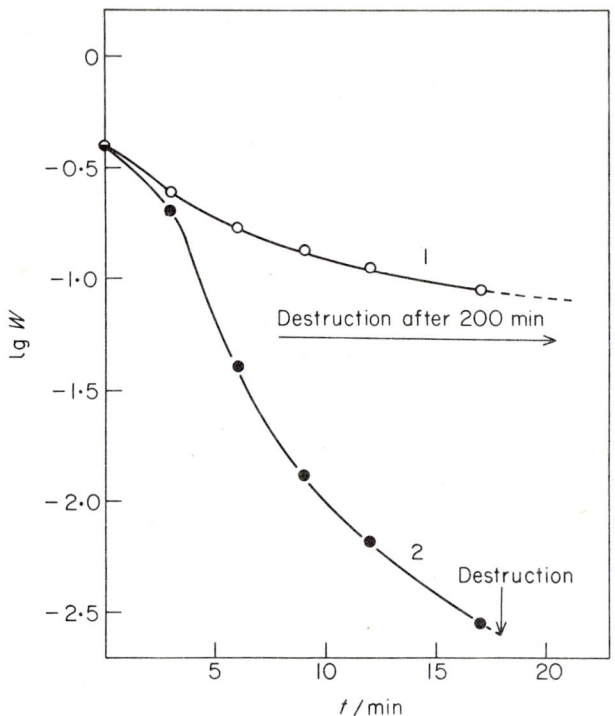

FIG. 8. Water contents of the foam in vol% versus time t for a $1.5 \times 10^{-3}\,\mathrm{mol\,dm^{-3}}$ OP $15 + 6 \times 10^{-2}\,\mathrm{mol\,dm^{-3}}$ KCl aqueous solution. Curve 1, no pressure difference; curve 2, pressure difference $5 \times 10^3\,\mathrm{N\,m^{-2}}$.

This method is very effective for destruction of stable foams and with suitable equipment can be used for practical studies in cases where unwanted foams arise.

The pressure gradient acts disruptively, on the one hand accelerating the outflow of the solution, whilst on the other hand it can be assumed that a critical pressure is reached where it ruptures the films. We are now carrying out experiments with single films to separate these effects.

The times τ_p necessary to rupture a 4-cm foam column under standard

TABLE 4

Order of some foaming agents according to the time for forced destruction of foams: $\tau_{R/2}$ time for destruction by α-particles; τ_P time for destruction by a pressure gradient in the Gibbs channels

Foam stabilizers	$\tau_{R/2}$/min	τ_P/min
OP 15	1	20
CTAB	6·5	42
NaLS	11	46
SLU	> 300	> 300

conditions are compared in Table 4 with the times $\tau_{R/2}$. It is seen that both methods give the same order of the surfactants with respect to their foaming ability. Hence the time τ_p is another characteristic of foaming agents.

4 Conclusion

These experiments confirm again the importance of black foam film for foam stability. They do not give, however, a final answer to many questions. The processes arising in the films at C_{bl} are not yet clear. Many results show that the appearance of black spots is due to peculiarities of the adsorbed layers at the solution surface. According to Scheludko[42] C_{bl} is the surfactant concentration above which a condensed monolayer is formed. Hence all processes in the films above C_{bl} and the high foam stability associated with them may be explained by the tangential forces between the molecules in the adsorbed layer. This hypothesis requires, of course, very precise and reliable investigations of the surface properties of the "detergent type" surfactant solutions. On the other hand the problems of foam stability cannot be solved entirely without understanding the structure and properties of black foam films.

Acknowledgement

The authors are indebted to Professor Dr A. Scheludko for helpful discussions and advice.

Abbreviations and names of surfactants used

OP$_n$ octylphenol polyglycol ether
L(EO)$_n$ dodecyl polyethyleneoxide
DMS   decyl methyl sulphoxide

CPC   cetyl pyridinium chloride
CTAB   cetyl trimethyl ammonium bromide
NaLS   sodium dodecyl sulphate
NaO   sodium oleate
SLU   sucrose dodecyl urethane

References

1. Kitchener, A. J. and Cooper. C. F. (1959). *Quart. Rev.* **13**, 71.
2. Bikerman, J. J. (1973). "Foams". Springer-Verlag, Berlin, Heidelberg, New York.
3. Schick, M. (1967). "Nonionic Surfactants". Marcel Dekker. New York.
4. de Vries, A. J. (1957). "Foam Stability". Rubber Stichting, Delft. Comm. No. 326.
5. Mysels, K. J., Shinoda, K. and Frankel, S. P. (1959). "Soap Films". Pergamon Press, New York.
6. Scheludko, A. (1962). *Proc. Kon. Ned. Akad.* **B65**, 76, Wetenschap.
7. Scheludko, A. (1967). *Advan. Colloid Interface Sci.* **1**. 391.
8. Scheludko, A. (1967/68). *Annu. Univ. Sofia Fac. Chim.* **62**, 47.
9. Exerowa, D. (1969). Thesis, Akad. Bulg. Sci., Sofia.
10. Clunie, J. S., Goodman, J. F. and Ingram, B. T. (1971). "Surface and Colloid Science", Vol. 3, p. 167. John Wiley, London.
11. Exerowa, D. (1969). *Kolloid-Z.* **232**. 703.
12. Derjaguin, B. and Titievskaya, A. (1953). *Kolloid-Z.* **15**, 416; (1957). "Proc. II Int. Congr. Surface Activity", Vol. 1, p. 211.
13. Scheludko, A. and Exerowa, D. (1959). *Kolloid-Z.* **165**, 148.
14. Lyklema, J. and Mysels, K. J. (1965). *J. Amer. Chem. Soc.* **87**, 2539.
15. Lyklema, J., Scholten, P. C. and Mysels, K. J. (1965). *J. Phys. Chem.* **69**, 116.
16. Exerowa, D. and Kolarov, T. (1964/65). *Annu. Univ. Sofia*, **59**, 207.
17. Manev, E., Scheludko, A. and Exerowa, D. (1974). *Colloid Polym. Sci.* **252**, 586.
18. Plateau, J. (1873). "Statique Expérimentale et Théorétique des Liquides Soumis aux Seulles Forces Moléculaires". Gauthier-Villars, Paris.
19. Johonnot, E. S. (1906). *Phyl. Mag.* **11**, 751.
20. Perrin, J. (1918). *Ann. Phys.* **10**, 165.
21. Exerowa, D, and Scheludko, A. (1964). "Proc. 4th Intern. Congr. Surface Activity", Vol. II, p. 1097. Gordon & Breach, London.
22. Exerowa, D., Zacharieva, M. and Radeva, Z. (1974). IV Intern. Tagung über gränzflächenaktive Stoffe, Berlin.
23. Bartsch, O. (1924). *Kolloid-Beih.* **20**, 1.
24. Baldacci, R. (1950). *Ann. Chim.* **40**, 358, 372.
25. Schick, M. and Bayer, E. (1963). *J. Amer. Oil Chem. Soc.* **40**, 66.
26. Dietrich, F., Hager, G., Jehring, H. and Horn, E. (1973). *Tenside*, **10**.
27. Exerowa, D. and Zacharieva, M. (1972). *Res. Surface Forces*, Moskwa, 234.
28. Exerowa, D. and Buleva, M. (1966). *Abh. Deut. Akad. Wiss Berlin*, *Kl. Chem.* **6b**, 938 (1966). *Tenside*, **3**. 210.
29. Krugljakov, P. M. and Kotova, T. T. (1969). *Dok. Akad. Nauk SSSR*, **188**, 865.
30. Krugljakov, P. M. and Koretskj, T. A. (1973). *Izv. Sib. Otd. Akad. Nauk SSSR*, *Chim.* **12**.
31. Derjaguin, B. V. and Gutop, Y. V. (1966). *Res. Surface Forces*, **2**, 36.
32. Exerowa, D. and Platikanov, D. (1970/71). *Annu. Univ. Sofia Fac. Chim.* **65**, 237.

33. McEnke, W. R. and Mysels, K. J. (1969). *J. Phys. Chem.* **73**, 3018.
34. Chaminade, R. (1949). *Compt. rend.* **228**, 480.
35. Ader (1960). *J. Phys. Radium*, **11**, 198.
36. Kato, R. and Kono, T. (1963). *J. Appl. Phys.* **34**, 708.
37. Exerowa, D. and Ivanov, D. (1970). *Compt. Rend. Bulg. Sci.* **23**, 547.
38. Exerowa, D. (1970/71). *Annu. Univ. Sofia Fac. Chim.* **65**, 227.
39. Exerowa, D., Penev, I. and Awgarska, S. (1973). Paper presented at the conference "Modification of Liquid Surfaces", Berlin.
40. Krugljakov, P., Exerowa, D., Christov, Ch. and Scheludko, A. (1974). Authorship Certificate, Bulg. No. 19398.
41. Krugljakov, P., Exerowa, D., Christov, Ch. and Scheludko, A. (1974). Authorship Certificate, Bulg. No 19399.
42. Scheludko, A. "Physical Chemistry of the Surfaces and the Disperse Systems". To be published.
43. Manegold (1953). "Schaum" Heidelberg, Chemie und Technik Verlag.
44. Proust, J. F. and Ter-Minassian-Saraga, L. (1972). *Compt. Rend. Série C*, **274**. 1105.

Discussion

Friberg The experiments using pressure to force liquid out of the Plateau borders should result in a change of the size distribution of the bubbles along the column. Has this phenomenon been explained?

Exerowa On creating a pressure in the Gibbs channels of a foam, a sharp increase in the rate of liquid outflow is obtained, in comparison with spontaneous syneresis (see Fig. 8). This liquid outflow is associated with a marked reduction in the size of the Gibbs channels and a "dry foam" having narrow bubble size distribution. By measuring the foam pressure with a capillary micromanometer it is found that the applied pressure difference in the Gibbs channels is constant over the height of the foam for a column about 8 cm high. This creates conditions for a narrow bubble size distribution in contrast to the situation in spontaneous outflow.

Lee The authors state that in accelerating the drainage of a foam by applying a pressure gradient, the pressure difference across the porous plate cannot be greater that the capillary pressure in the pores. What happens if this pressure difference is exceeded? Do bubbles enter the pores of the plate and does the foam reform on the other side?

Exerowa I would say that the "trick" of the method is that the porous plate only permits passage of the liquid phase, the gas being retained. This occurs when the pressure in the Gibbs channels is less than the capillary pressure in the pores of the porous plate. When the pressure is greater than this value, liquid and gas will be forced through the pores and foam will be reformed on the far side of the porous plate.

Smith In the case of a biliquid foam the pressure applied to the thin films can be varied by centrifugation. When this is done a critical pressure for film breakage and coalescence is found in suitable cases. Do you observe such a critical pressure in your system.

Exerowa When studying the effects of pressure within the Gibbs channels of a foam a critical rupture pressure was found and preliminary experiments indicate that this critical pressure differs with the foam-stabilizing surfactant used. I think that this is an important observation that may lead to an understanding of the mechanism of pressure effects found in quantitative measurements. Model experiments with single foam films are also being carried out to study the same phenomenon.

Laughlin Are experiments carried out at a temperature close to a phase separation boundary in the bulk phase, and could phase separation under the influence of α-particles be responsible for foam instability.

Exerowa Our paper does not consider the mechanism of film rupture by α-particles as this is not yet quantitatively understood. However our experiments show that the composition and temperature of film-forming solutions do not change with radiation. For example the total time for the rupture of Newton black film by interrupted radiation, the periods of radiation and interruption being equal, was the same as those exposed to uninterrupted radiation. It is worth noting that the average lifetimes of irradiated films, which are very sensitive to small changes in film-forming solution composition, are constant for constant radiation levels. Also the critical concentration of the transition common black to Newton black film, as determined by the α-radiation method, agrees with values determined by other techniques. These, and other experiments, prove that a significant overall warming of the film or an energy accumulation does not occur. We consider that we have completely eliminated the possibility of a cumulative mechanism in the rupture of films by α-particles.

I also think that there is no phase separation in films due to the influence of α-radiation, as postulated by Laughlin, and that this is not the mechanism of the observed foam instability.

Barber Could you please explain in rather more detail how the results presented in Table 3 were obtained? In particular, what do you mean by a fully drained foam and were equal heights of foam produced from all the surfactant systems before the collapse started? If this is the case, what was the initial foam height and can you give any idea of the bubble size?

Exerowa The experimental results for spontaneous and α-particle irradiated foam destruction shown in Table 3 were obtained using an 8-cm high foam column produced in the apparatus shown in Fig. 5, the initial bubble size being 1–2 mm. The spontaneous destruction results were obtained by recording the times for the foam column to subside by 20 mm and 40 mm respectively. For foam destruction by α-irradiation the foam column was initially 8 cm and allowed to drain. After 30–60 minutes the foams were found to drain to an equilibrium water content. During this time the foams spontaneously subsided, the more unstable foams, i.e. the first two cases in Table 3, about 35 mm while the more stable foams, the others in the Table, by only about 1–2 mm. I would like to stress that comparison at equilibrium liquid content is not essential. It was shown that the same order of foaming agents was obtained when $\tau_{R/2}$ was measured at equal intervals from foam formation, i.e. at different water contents.

Roberts, K. Why is there a different relative order in the rates of subsidence of 20 mm and 40 mm foams.

Exerowa Roberts comments that the τ-values for a 40 mm foam of DMS and CPC do not increase regularly. It is our view that comparisons of the radiation lifetime $\tau_{R/2}$ must be made with a 20-mm column, as $\tau_{R'2}$ is measured up to this distance, the free path of α-particles being about 40 mm. When spontaneous rupture is studied in columns up to 40 mm high the foam exists for a longer time and may begin to rupture irregularly. "Holes" may appear and the values of τ obtained show considerable scatter. This happens with CPC foams. However, experiments at 20 mm yield reproducible results and as the τ values quoted are based on many measurements, we consider that the order of surfactants obtained is well established.

Prins Can you speculate about the mechanism by which the film on foam collapses when they are subjected to a particle radiation on a pressure gradient?
What is the role of the transition from the common black films to the Newton black film in this mechanism?

Exerowa Foam rupture by α-irradiation and by creating a pressure gradient in the Gibbs channels are different techniques and probably have different mechanisms. The assumption that the action of α-particles is a statistically random rather than cumulative effect (see the reply to Laughlin's question) is based on theoretical and experimental studies of the action of α-particles on Newton black films. The α-particles are thought to augment effects due to external disturbance of the films, the disturbances being due to thermal fluctuations that may lead to spontaneous film rupture, e.g. the cavitation mechanism of film rupture of Derjaguin and Gutop. Hence the action of α-particles is to increase the number of "disturbances" in the film and the probability of rupture and shortened film life. As is shown in Figs 6 and 7, the action of α-particles is dependent on whether the films are common black or Newton black and may provide a technique for identifying the transition between them.
The pressure gradient technique for studying film rupture has only recently been developed and we are not at present in a position to propose a definite mechanism for the phenomena observed. It is clear that a pressure gradient within the Gibbs channels of a foam leads to an accelerated rate of liquid outflow from the foam and that a critical rupture pressure is observed. Preliminary experiments with single films of similar composition at pressures about 1000 times those commonly used also indicate a critical rupture pressure. It is reasonable to postulate that the last sharp branch of the $\pi(h)$ isotherm (π, disjoining pressure; h, film thickness) is a high maximum and that this corresponds to molecular cohesion in the Newton black film.

9

Foams and their clinical implications

J. S. BURTON

Berk Pharmaceuticals Ltd,
Godalming, Surrey, GU4 8HE, England

Summary

The term dyspepsia is usually applied to any form of digestive upset within the stomach or gastrointestinal tract. The two most common complaints for which patients seek medical advice are hyperacidity and flatulence, the latter giving rise to gastric and intestinal distension. Flatulence is caused by a number of factors, the most significant being ingestion of air or aerophagia; a second major cause is the liberation of carbon dioxide following ingestion of food. Both of these effects produce gas bubbles in the stomach or intestine which result in distension.

Clinically the cheapest way to treat dyspepsia is by way of antacids. Where, as is most common, the condition is caused by a combination of both hyperacidity and flatulence, treatment with antacids is frequently not possible since the antacid is prevented from making contact with the excess acid by the gas bubbles.

Thus we have available in the UK a range of pharmaceutical products which contain polydimethylsiloxane as an antifoaming agent in combination with a suitable antacid mixture. Polydimethylsiloxane activated by silica is the antifoaming agent of choice since it spreads very easily to form a surface film which is immiscible with water, although the enhancement of the antifoaming properties of the silicone by silica is less apparent when the oil is formulated into tablets.

Products containing this mixture are of value in the treatment of gastric and intestinal gas by their ability to change the surface tension of mucus-covered gas bubbles allowing small bubbles to coalesce. The free gas formed is eliminated more easily than is the gas held in small, tenacious, mucus-covered bubbles.

These effects can be demonstrated by *in vitro* and *in vivo* studies.

1 Introduction

Whilst not comparing with the total demand for general industrial uses the clinical application of antifoams has become of major significance. Obviously purity is of a much greater consideration for this application and companies supplying materials for these purposes have had to prepare them to a much more rigid specification.

The clinical significance of foams has received wide publicity since it was reported in 1949 by Quin *et al.*[1] that bloat in ruminants could be treated by the use of antifoaming agents.

Previous to that, the clinical indications were grouped under the general heading of dyspepsia and treated by normal antacid therapy, although the discomfort attributable to the presence of excess gas has always been well established.

Even now the term dyspepsia is frequently applied to any form of indigestion, discomfort, pressure, fullness or mild pain within the stomach and gastrointestinal tract. The direct causes of dyspepsia may include a wide range of indications; thus hyperacidity, hypermotility, aerophagia (air swallowing), inflammation of gastric mucosa (gastritis), peptic ulceration and other disorders of the gastrointestinal tract such as hiatus hernia.

It is, however, well established that the two most common complaints occurring in the upper gastrointenstinal tract for which patients seek medical advice are hyperacidity and gas. These two indications frequently occur together. When excess gas is present within the gastrointestinal tract, the condition known clinically as flatulence arises and this is responsible for gastric and/or intestinal distension.

There are a variety of reasons for the formation of flatulence and these will be dealt with later.

The materials used in clinical practice vary and before assessing the effectiveness of a particular formulation, it is necessary to establish exactly what is claimed as the active agent. Having established this, produced a material whose specification is as rigid as the criteria of the Medicines Commission demand and satisfied oneself that the type of antifoam selected is the one best suited to the purpose for which it is intended, the next requirement is to prepare a formulation which presents this active to the foam in the best fashion. This is not a straightforward matter and requires a detailed consideration of the various other ingredients and excipients that are to be used. The development of more efficient formulations is quite apparent from the proven superiority of the modern formulations over what were used in the original trials.

It is desirable also to develop formulations that can be demonstrated to be effective. Quantification of antifoaming potential has developed to an extent that it is possible to demonstrate activity *in vivo*. Finally, it is imperative to prove freedom from toxicity. Although the introduction of antifoams in clinical practice goes back to before the existence of the old Dunlop Committee and later the Committee on Safety of Medicines, a vast amount of work was carried out to prove that the antifoams were safe to use in the doses prescribed.

2 Materials used in clinical practice

Quin and his colleagues in 1949[1] reported on their novel treatment of bloat in cattle using a preparation containing a highly polymerized methyl silicone.

The product was either injected directly into the dorsal vault of the rumen or diluted with water and given perorally as a drench or by stomach tube. Predictably the recovery rate was higher (95 per cent) when an injectable ready to use suspension was employed than when the silicone was administered orally in tablet form (80 per cent). This highly successful trial proved the way for further veterinary and medical treatments of conditions that were attributed to the presence of gastrointestinal foam.

Unfortunately no details are given of the specification for the oil employed, nor of the composition of the formulations in which they were used. However, the work stimulated further veterinary and medical investigations and Rider and Moeller[2] reported in 1960 on their studies on the use of a methylpoly-siloxane used in tablet form. The tablets used contained from 10 to 40 mg of this active, although no further formulation data were supplied. The clinical studies reported by these workers originated in 1953.

With the increasing use of the polydimethylsiloxanes in the treatment of flatulence, their specifications appeared in the British and US Veterinary and Medical Compendia. Also in 1958 a monograph on the "Pharmacy of Silicones" covering all of the pharmaceutical and medical applications of the silicones was published by Levin.[3]

Thus in the 1959 supplement to the 1953 British Veterinary Codex[4(a)] a "Suspension of Silica in Dimethicone" was described as a treatment for frothy bloat in ruminants. In the 1965 British Veterinary Codex[4(b)] this preparation was described as "Silica in Dimethicone Suspension". The preparation was described as containing 6–8 per cent finely divided silica in Dimethicone 1000, a polydimethylsiloxane having a viscosity of around 1000 centistokes.

The first appearance of a monograph on Dimethicone in the British Pharmaceutical Codex was in 1959[5(a)] which covered only the low viscosity (20 centistoke material). At that stage it was not recommended for use in antifoaming preparations. In the next edition of this Codex (1963)[5(b)] a full description of a range of Dimethicones was given. The various grades listed were distinguished by the numbers appended after the name and the monograph listed Dimethicones 20, 200, 350, 500 and 1000. At this stage the monographs mentioned the use as antifrothing agents. It is of course possible to prepare Dimethicones of a much higher viscosity than 1000 centistokes.

The use of silica in the Silica in Dimethicone Suspension B. Vet. C. is of considerable significance. Although not referred to in the Quin[1] or Rider[2] papers it is almost certain that the silicone oil used contained silica. The emergence of a new description for silicone materials containing this active occurred in the USA. In 1963 the United States Adopted Name list (USAN) included the reference to Simethicone.[6a] The specification for this product was as follows: "Simethicone is a mixture of liquid dimethylpolysiloxanes containing 4–4·5 per cent silica . . . ." Note that no specification was given for the polydimethylsiloxane. A later edition (1968)[6(b)] of the same publication also referred to Simethicone but gave no indication of the percentage silica to be used. That there is some confusion with the use of this expression was shown by further publications describing this material such as the Merck Index (8th Edition, 1968)[6(c)] which omitted any reference to silica and the United States Dispensatory and Physicians Pharmacology (1967)[6(d)] which described the material as activated but omitted to refer to the silica content. Although not an approved name in the UK, general usage now would suggest that this name be given to the mixture of Dimethicone 1000 with 4–8 per cent added silica (as described in Martindale's Extra Pharmacopoeia).[6(e)]

The significance of the addition of the silica is obviously of importance, although the relative inertness of both the silica and the Dimethicone are sufficient to render their admixture satisfactory.

In a simple *in vitro* test[7] it can be readily demonstrated that the grade of polydimethylsiloxane used in pharmaceutical formulations, i.e. Dimethicone 1000 has a poor antifoaming property, rather large quantities having to be used and even this causing only minimal defoaming due to the formation of a second layer. Klein and Maluzi in 1967[8] described a process for activating antifoaming liquids by the addition of a small percentage (2–8 per cent by weight) of a hydrophobic silica. The resulting mixture in which the finely divided silica is suspended in the silicone fluid was shown to be much more active as an antifoaming agent and indeed in most non-medical uses it is used in this form.

For medical applications, however, the form in which it is used is critical not only to the assessment of toxicity, but also as regards the description of the active. The term Dimethicone applies to the basic polydimethylsiloxane oil. The addition of the silica results in a mixture with, as stated substantially enhanced antifoaming properties but which *may*, depending on the method of mixing, result in a material which no longer complies with the BPC monograph. If the mixing is carried out at room temperature it can be demonstrated (Buist *et al.*)[9] that chemical interaction does not occur and that the product exists simply as a physical mixture albeit with strong hydrogen bonding between the polydimethylsiloxane and the silica. Thus

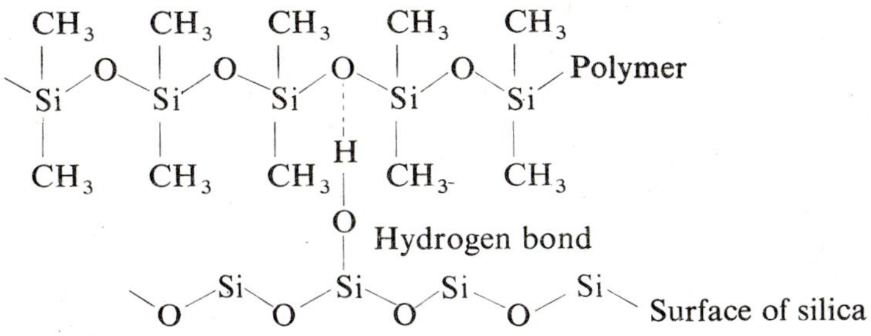

$$
\begin{array}{ccccccccc}
CH_3 & & CH_3 & & CH_3 & & CH_3 & & CH_3 \\
| & O & | & O & | & O & | & O & | \\
Si & & Si & & Si & & Si & & Si \\
| & & | & & | & & | & H & | \\
CH_3 & & CH_3 & & CH_3 & & CH_{3^-} & & CH_3
\end{array}
$$

O   Hydrogen bond

—O—Si—O—Si—O—Si—O—Si—   Surface of silica

At elevated temperatures however (*c.* 300°C) it is possible to form chemical bonding, thus

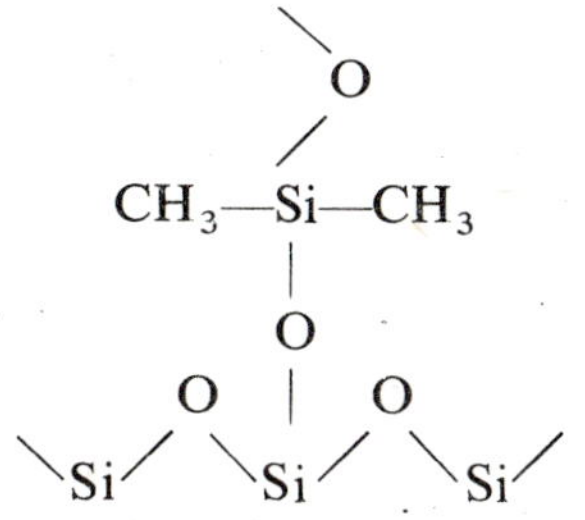

CH_3—Si—CH_3

In this case it is clear that the material will contain a third component, i.e. a silica with either multipoint attachment of a single polydimethylsiloxane chain, or of single-point attachment of many chains or a combustion of both.

This material will obviously not conform to the BPC specification for Dimethicone and more importantly the active is not simply the Dimethicone activated by an inert material (silica) to form a mixture but a combination of two or possibly more components since it is clear that the silica-poly-dimethylsiloxane compounds would not only enhance antifoaming but also have some antifoaming activity.

The work carried out by Buist *et al.*[9] demonstrated that there is multiple hydrogen bonding of the silica commonly used in these preparations (Aerosil 200—Degussa) to the polydimethylsiloxane. Because of this multiple surface bonding and because the silicone fits the surface well, the total physical interaction is very strong.

Further evidence of the physical nature of the bonding has been shown in studies carried out by Chahal and St. Pierre[10, 11] who measured heats of adsorption of polysiloxanes onto silicas. In the absence of a radiation field it is possible to show purely physical adsorption, but with radiation,[11] the polysiloxane becomes chemisorbed to the silica surface.

3 Toxicology

For products designed essentially for oral use as antifoaming agents it is essential to demonstrate freedom from toxicological and pharmacological effects in the body. It is also necessary to demonstrate that absorption into body tissues either as the substance or as a breakdown product does not occur. The mode of activity of Dimethicone is essentially topical and the body may therefore be regarded simply as the vessel in which the foam depression is carried out. The toxicity is obviously dependent therefore on the content of the mixture.

Most of the toxicology on these materials has been done on the simple polydimethylsiloxane–silica mixture conforming to the current (Martindale)[6e] Simethicone formulation. The studies indicate an extremely low rate of toxicity both in chronic and acute dosage in several animal species. More importantly it has been demonstrated that the silicone fluid is excreted unchanged and there is no absorption.

Much of the early work on the toxicity of the silicone fluids was carried out on materials with poorly defined properties. As indicated earlier the terms Simethicone, polydimethylsiloxane, etc., had become widely used for a whole range of products and as the detailed specifications were not given in these early papers, the relevance of the results to materials in current use is dubious. Of particular importance is the level of the low polymers (molecular weight < 800) in the fluid since they have been shown to be responsible for toxicity in certain instances.

However an important series of experiments were commissioned, during the years 1967–1970, by the Dow Corning Corporation on their Antifoam M compound. This material is similar to their earlier Antifoam A, which was derived from material used for general industrial purposes, with the exception that the upper limit for the concentration of low molecular weight species was placed at 0·2 per cent.

The earlier studies include:

i. A 2-year feeding test in rats (1950)[12] of Antifoam A. In this study no significant toxic effects were observed.

ii. Child *et al.* (1951)[13] observed no toxic effects in dogs receiving 3 g kg^{-1} daily of Antifoam A for 6 months.

More recent studies on Antifoam M are:

iii. A short-term feeding test[14] in human subjects each receiving 5 mg kg^{-1}

body weight for 10 days showed no increase in urinary silicate levels and no evidence for the absorption of the silicone. Studies carried out simultaneously on rabbits and rats confirm that the silicone is not absorbed.

iv. Further studies[15] by the same investigators confirmed the freedom from toxicity and the nonabsorption of the silicone in dogs and rats.

v. An additional study using ^{14}C-labelled silicone in rhesus monkeys also proved negligible absorption.[16]

vi. A full lifespan study in mice was carried out by Cutler et al.[17] and this showed no toxic effects due to the silicone whether by oral or injection route, whereas animals injected with liquid paraffin did have an increased incidence of subcutaneous fibromas.

4 Pharmaceutical formulations

As indicated earlier it is possible to dose the silicone fluid directly. As it is unpleasant to take, the appropriate dose of the fluid (usually 0·1 g) is sealed in a soft gelatin capsule. Currently there is no preparation on the market in the UK in this form although the preparation MINIFORM has been shown to be effective in Scandinavia.[18]

One very valuable use of such a preparation is in gastroscopy where the

TABLE 1

Defoaming time of antacid-silicone tablets

| Antacid material | Days on stability at 45°C | | | |
| | 0 | 2 | 7 | 14 |
	s	s	s	s
Calcium carbonate	15	30	30	30
Sodium bicarbonate	15	15	15	15
Sodium citrate	30	15	15	15
Aluminium hydroxide	< 120	< 120	< 120	< 120
Magnesium trisilicate	15	15	90	75
Magnesium hydroxide	15	30	30	30
Magnesium carbonate	< 120	< 120	< 120	< 120
Glycine	15	15	15	15
Magnesium carbonate–aluminium hydroxide co-ppt	90	< 120	< 120	< 120
Magnesium perioxide	40	40	70	90
Bismuth subcarbonate	< 120	< 120	< 120	< 120

Composition and manufacture. The following ingredients were used (mg per tablet): antacid, 400; cornstarch, 30; mannitol, 100; simethicone, 25; polyoxyl, 40; stearate, 3; lactose, 50; PVP, 80; microcrystalline cellulose, 35; magnesium stearate, 7. Total 730 mg.

physician seeks to take an X-ray picture of the stomach. Frequently foam is present and the conventional antacid–polydimethylsiloxane formulations, whilst being effective in breaking up the foam, are of little use since the tablet components obscure the picture.

In normal therapeutic use there are two main types of formulations used which include the polydimethylsiloxane–silica mixture, i.e. tablet and suspension. The suspension may or may not be thickened to form a gel.

The earlier formulations all contained from 25 to 40 mg of the silicone mixture blended with suitable antacids and excipients and compressed to form a tablet. More improved formulation techniques have enabled formulators to prepare tablets containing as much as 250 mg of the mixture. Additionally aqueous suspensions and gels are prepared to contain 125 mg per 5 ml of the mixture. Suitable paediatric preparations are also formulated.

Apart from the soft gelatin capsule presentation referred to and the paediatric formulations, the silicones are always formulated with suitable antacids, since the indications frequently require the treatment of flatulence and hyperacidity. The selection of the antacid can be critical since it has been demonstrated by Rezak[19] that certain antacids adsorb the silicone and reduce its antifoaming potential. This may readily be shown in Table 1.

From the Table it may be seen that aluminium hydroxide is one of the worst offenders and yet somewhat surprisingly it is probably the most important and commonly used. However, it is of interest to note that the most successful products on the market in the UK contain 250 mg of Dimethicone BPC or "activated polydimethylsiloxane" or "Simethicone", whereas the tablets used in the Rezak study contained only 25 mg of Simethicone, i.e. polydimethylsiloxane–silica mixture.

As has been already stated it can be readily demonstrated that the silica greatly enhances the antifoaming properties of the polydimethylsiloxane. As will be shown this is of interest only in the wide industrial and commercial applications and the single case in the medical field where the silicone fluid is used *per se.* The antifoaming activity of polydimethylsiloxane is limited by its high viscosity and its hydrophobic properties. The action of the silica is to optimize the spreading properties and hence the antifoaming properties of the polydimethylsiloxane by the creation of a much enlarged interface between the silicone and the foam. That this effect of the silica is purely that of an activator has been demonstrated by work carried out at Rhône Poulenc[20] which demonstrated that the efficiency of the polydimethylsiloxane–silica mixture as an antifoaming compound was not dependent on the percentage of the silica present. The standard Silica in Dimethicone Suspension B. Vet. C. referred to earlier contains much more fluid than required to form a monomolecular layer which theoretically would be necessary to maximize the antifoaming effects.

When one moves to the tablet or suspensions however, it immediately becomes apparent that the silica has a much less important part to play. When the broken-up tablet (antifoaming tablets are intended to be chewed before swallowing) or the suspension containing the antifoaming system is presented to the gastric foam, there is an abundance of suspended solid materials present which, although not necessarily capable of the same intimate hydrogen bonding to the polydimethylsiloxane as silica (see above) are, however, present in sufficient quantity to provide the spreading of the poly-dimethylsiloxane necessary to optimize the antifoaming capacity of the silicone. Despite this the vast majority of commercial tablet and suspension formulations used in gastroenterology are prepared from a preformulated mixture of the polydimethylsiloxane and the silica. The value of the activated mixture whether or not silica is present, is in its ability to change the surface tension of the mucus-covered gas bubbles causing the bubbles to coalesce.

Two series of experiments, static and dynamic, were described by Carless et al.[21] which clearly demonstrated in vitro that in a formulated tablet the presence of the silica is much less critical than in the unformulated oil. The dynamic froth test showed that removal of Dimethicone BPC by ether or

TABLE 2

Dynamic froth test

Froth	Preparation	Froth reduction Time s	Extent
Cetomacrogol–0·1 N HCl	Asilone tablet (as formulated)	17	complete
	Asilone tablet (ether extracted)	60	80%
	Asilone tablet (CHCl$_3$ extracted)	60	80%
Cetomacrogol–saturated NaHCO$_3$	Asilone tablet (as formulated)	20	complete
	Asilone tablet (ether extracted)		10–0%
Cetomacrogol–water	Calcium carbonate BP		0

chloroform extraction of the powdered tablet markedly reduced the anti-foaming properties of the residual powder, despite the retention of a small percentage of the polydimethylsiloxane dispersed on the silica and aluminium hydroxide content of the tablet. The experiment was conducted on the basis of the time taken for the powdered tablet to remove foam prepared by aeration of cetomacrogol. Because some Dimethicone is retained after ether extraction, this test provides a direct comparison between the tablet as formulated and tablets from which the bulk of the Dimethicone has been removed. Table 2 shows the results obtained in the dynamic froth test.

From this it will be noted that there is a marked difference in antifoaming properties in acid solution shown by the powdered tablets before and after ether extraction and this demonstrates that the unbound extractable Dimethicone is primarily responsible for the antifoaming properties of the tablets. Conversely the adsorbed nonether or chloroform extractable Dimethicone whether bound onto the surface of the silica, aluminium hydroxide or both plays only a minor role in establishing the antifoaming characteristics, since the solvent extracted powder is much slower to take effect and is much less effective in that it is unable to completely destroy the foam. The results described here would also to a certain extent explain the results obtained by Rezak[19] discussed earlier.

The static test was carried out on a number of different tablet granules which were compared to the standard tablet formulation, i.e. in which the product was formulated using the normal polydimethylsiloxane–silica mixture. A number of the formulations were prepared omitting the silica; some were prepared without Dimethicone and others were prepared in which the silica was added, not as a premix with the polydimethylsiloxane, but as a normal tablet ingredient (silica is commonly used in tablet formulations as a disintegrant). The antifoaming properties of the various granules were determined by the time taken to break down a froth produced from a solution of sodium lauryl sulphate. The results are shown in Fig. 1. The formulations Dd. Ee. Ff can be seen to have negligible defoaming action. Each formulation contained no Dimethicone. The granules corresponding

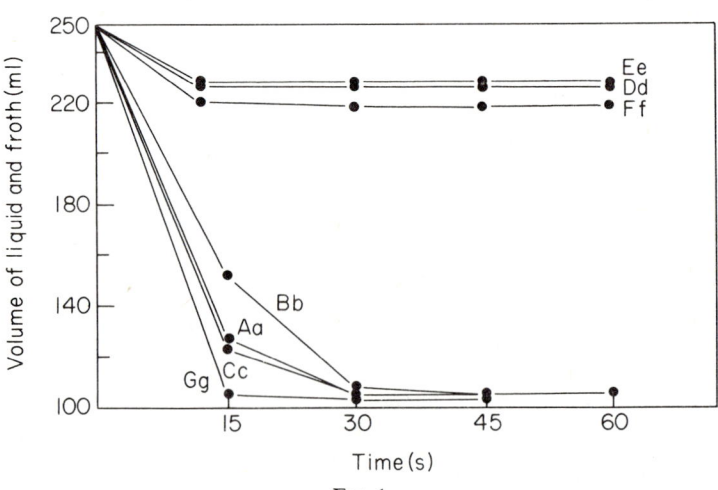

Fig. 1.

to the standard tablet (Gg) were marginally more efficient after 15 s, but from 30 s onwards there is no significant difference between this and granule Aa in which silica was totally absent.

4.1 DRUG ABSORPTION

One possible complication in the use of these antifoam formulations is in the possible interference of the silicone with absorption of other drug substances. With the normal concomitant antacid therapy, this is not a problem since both actives act topically. There are, however, a number of preparations in which the tablet containing the silicone–antacid mixture also contains a drug which is active systematically, i.e. by absorption. Examples of such drugs contained in these formulations are propantheline bromide, dicyclomine hydrochloride and metoclopramide hydrochloride. This subject does not seem to have been studied thoroughly although it has been shown[22] that a silica–antacid mixture containing metoclopramide produces equivalent urinary levels of the metoclopramide metabolite to that of a formulation containing no silicone.

5 *In vivo* experiments

Apart from a brief description of the clinical applications of the various formulations no discussion has been given on the establishment of antifoam activity of formulations *in vivo*. As indicated earlier it has been established that the polydimethylsiloxane is not absorbed and is excreted unchanged. When used clinically as a foam suppressant therefore we may regard the human or animal model as operating solely as equivalent to a vessel containing the foam. Whilst it is possible to demonstrate the antifoaming effect of formulations in humans by the use of X-ray techniques, it is almost impossible to design a controlled method for quantifying the results.

Birtley *et al.*[23] devised a simple animal model for such an experiment. As with the *in vitro* studies described above these experiments were designed to show the effect of silica added to polydimethylsiloxane in the free state.

Rats were chosen as the experimental animals for this study and foam was artificially produced in the stomach by gastric intubation of a freshly prepared solution of citric acid containing sodium iodide followed by sodium carbonate containing saponin to artificially produce foam. Two minutes later the rat was X-rayed whilst in the upright position, and the plates were subsequently examined. Sodium iodide is radio-opaque and water soluble and thus three distinct layers were visible on the image of the stomach: a lower white layer (solution), an overlying grey area (foam) and a darker upper layer of free gas. The height of the foam layer in millimetres was

measured before and 2 minutes after treatment of each rat with polydimethylsiloxane, a 6 per cent w/v aqueous silica suspension and the
polydimethylsiloxane–silica mixture. The results are shown in Table 3.

These studies indicate that both polydimethylsiloxane and the aqueous
silica suspension possess relatively weak antifoaming properties when used
separately. The ability of the silica to defoam was attributed to the fine

TABLE 3

Mean percentage reduction in foam height following oral administration of polydimethylsiloxane (PMS). 6 per cent w/v aqueous silica suspension or PMS containing
6 per cent w/v silica

Treatment (by mouth)	volume ml	Dose (mg per rat) PMS	Silica	Mean percentage reduction of foam height ($\pm$s.e.)
PMS	0·25	250	—	19 $\pm$ 4·0*
	0·50	500	—	29 $\pm$ 6·1*
	1·00	1000	—	56 $\pm$ 6·2†
	2·00	2000	—	84 $\pm$ 3·4†
6 per cent aqueous silica suspension	0·05	—	30	28 $\pm$ 4·7
	1·00	—	60	45 $\pm$ 5·2†
	2·00	—	120	59 $\pm$ 3·6†
	4·00	—	240	75 $\pm$ 4·4†
PMS containing 6 per cent w/v silica	0·005	4·7	0·3	45 $\pm$ 6·3†
	0·010	9·4	0·6	58 $\pm$ 4·7†
	0·020	18·9	1·1	61 $\pm$ 4·7†
	0·040	37·7	2·3	87 $\pm$ 2·3†

Significance of difference between mean foam height before and after dosing:
* $P < 0.05$. † $P < 0.001$. A foam layer was produced in the stomach of each rat by oral
administration of 0·5 ml of 0·17 mol dm^{-3} citric acid containing 4 per cent w/v sodium iodide
immediately followed by 0·5 ml of 0·25 mol dm^{-3} sodium carbonate containing 1 per cent w/v
saponin.

particles causing the small bubbles to coalesce. When the two components
were administered as a mixture substantial defoaming occurred.

Further work[24] was carried out by the same investigators to compare the
effect of tablet formulations prepared with Dimethicone or in admixture
with silica as before. These results are shown in Table 4.

The results from this experiment showed that the tablet containing the
polydimethylsiloxane–silica mixture was approximately twice as efficient as
a defoamer as that containing polydimethylsiloxane alone. Certain anomalies
showed up in this study which did not agree with the results obtained in vitro.

TABLE 4

Mean percentage reduction of foam height* in the stomach of the rat following administration by gavage of polymethylsiloxane (PMS); silica; PMS containing 6 per cent silica; aqueous suspension of tablet base alone or containing 0·6 per cent silica; 10 per cent PMS or 10 per cent PMS and 0·6 per cent silica

Group	Number of rats per group	Treatment	Dose volume ml	Dose of PMS, silica or PMS containing 6 per cent silica mg	Group mean % reduction of foam height
1	5	PMS	1·00	1000	43
2	5	6 per cent aqueous suspension of silica	2·00	120	53
3	5	PMS containing 6 per cent silica	0·03	30	84
4	5	Aqueous tablet base suspension containing 10 per cent PMS	0·75	75	57
5	5	Aqueous tablet base suspension 10 per cent PMS and 0·6 per cent silica	0·30	30	47
6	3	Aqueous tablet base suspension (1 g ml^{-1})	0·30	—	46
7	3	Aqueous tablet base suspension containing 0·6 per cent silica	0·30	1·8	55
8	3	Aqueous tablet base suspension containing 10 per cent PMS	0·80	80	67
9	3	Aqueous tablet base suspension containing 10 per cent PMS and 0·6 per cent silica	0·20	20	60

* A foam layer was produced in the stomach of each rat by administration by gavage of 0·5 ml 0·25 N citric acid followed by 0·5 ml 0·25 N sodium carbonote containing 1 per cent saponin.

For example evidence was found that the tablet base, i.e. that containing no polydimethylsiloxane at all had antifoaming properties.

6 Clinical results

As indicated earlier when a physician is called upon to prescribe treatment for a condition characterized by an increase in gastrointestinal gas, he has to first determine the cause of the condition. By far the largest number of complaints of intestinal gas and bloating are functional in origin. Seventy per cent of these cases are due to aerophagia or air swallowing. Nervous tension may not only produce aerophagia but may cause a spasm of the small bowel or colon so that intestinal gas is trapped and not expelled per rectum, thus aggravating the symptoms. Ingested food may also cause similar conditions. In this case carbon dioxide is liberated when the acid gastric juice comes into contact with alkaline food in the stomach.

The most uncomfortable symptom complex results from gas in the colon which forms during the fermentation or decomposition of undigested food such as raw carbohydrate.[25] Numerous foods are high in cellulose content. Since humans cannot digest cellulose, cellulose-protected carbohydrate particles ferment in the colon as a result of bacterial action, thus causing distension and flatulence.

As indicated, functional causes are by far the greatest factor producing flatulence. Organic causes are more serious and result from a deficiency or malfunctioning of the body mechanism. Thus improper preparation of food for digestion may result from improper or inefficient mastication. Other main organic causes producing flatulence are: a deficiency of bile in the undigested food and consequent fermentation in the colon; and where normal physiology is altered, for example where food travels through the stomach so fast that it is undigested (hypermotility). Severe and chronic constipation will also give rise to intestinal gas and bloating.

Obviously for flatulence deriving from organic causes the physician must determine the cause for the disturbance and treat this with the appropriate medication. However, the flatulence arising from the intestinal foaming will also have to be treated. Hence there is a need for the concomitant therapy with other drugs outlined above.

X-ray studies have shown clearly the presence of foam in the upper gastrointestinal tract in humans. As with the studies in rats it is possible to demonstrate visually that the preparation containing the polydimethyl-siloxane–silica mixture is effective. Since these studies a wide range of other clinical trials have been reported.[28]

A further application in clinical usage is in the treatment of postoperative gas pains. Postoperative gas pains and abdominal distension represent the most frequent complication of surgical procedures; yet until comparatively recently their importance did not receive the attention such complications warranted. These relatively minor complications are thought to be the result of an adynamic ileus following intestinal manipulation, although similar complaints have been observed following extraperitoneal procedures such as hernioplasties and mastectomies, etc.[26] Two factors are probably responsible for the distension of the abdomen and gas pains following surgery. They are (1) sympathetic overactivity affecting the gastrointestinal tract, and (2) accumulation of an abnormal quantity of air. The polydimethylsiloxane–silica mixture dosed in a suitable tablet vehicle was found to be very useful for treatment of these conditions.

For lower intestinal distension it is frequently desirable to present the antifoaming agent direct to the site of discomfort. Suitable enteric coatings for a hard gelatin capsule containing the antifoaming agent can be devised[26] so that it is passed unchanged through the stomach. This is obviously the preferable course, since a chewable tablet or suspension will simply be broken down in the stomach and the active polydimethylsiloxane diluted to an extent that it will not longer be active in the intestine.

For organic disorders leading to undigested food in the intestine, a combination of the silicone fluid with a suitable pancreatic enzyme has been tried.[28] The two materials were combined in tablet form with a suitable enteric coating. The release of the silicone fluid from the tablet was reported as being rather poor, although this could have been due to the poor formulation. Further work on combinations of this nature has been reported by Ferraz and Wafae.[29]

In all of the above clinical applications we must assume that the proven effectiveness of the silicone fluid is based on subjective assessment by the

TABLE 5

Protection afforded by polydimethylsiloxane (PMS) against aspirin-induced gastric irritation

Treatment (oral)	Severity of gastric irritation in each rat (0–6 scale)	Mean score (±s.e.)	Per cent reduction compared with aspirin-treated group
1 per cent cmc (1 ml)	0 0 0 0 0 0 0 0 0 0	0	—
Aspirin (45 mg)	5 5 4 3 4 5 4 1 4 2	3·7 ± 0·42	—
Aspirin (45 mg) + PMS (0·25 ml)	2 2 3 0 0 4 2 4 3 1	2·1 ± 0·46*	43
Aspirin (45 mg) + PMS (0·5 ml)	0 4 2 4 3 2 2 0 4 3	2·3 ± 0·50	37
Aspirin (45 mg) + PMS (1·0 ml)	1 2 3 2 0 1 3 2 2 2	1·8 ± 0·29†	51
Aspirin (45 mg) + PMS (2·0 ml)	2 0 1 0 2 0 1 2 3 0	1·1 ± 0·35‡	70
PMS (2·0 ml)	0 0 0 0 0 0 0 0 0 0	0	—

Significance of difference from groups receiving aspirin only: * P 0–05. † P 0–01. ‡ P 0–001. PMS was given 10 min before administration of aspirin; animals were killed 2 h later. There were 10 rats in each group.

patient and doctor. Gastroscopic experiments can show quite readily that the silicones are effective in reducing foam levels in the stomach, etc. Ultimately, however, the best test of any medicament is the relief experienced by the patient. With the many years of practical experience behind us there can now be no doubt that the Dimethicone dosed in the various forms does provide an extremely useful way of treating gastric discomfort and therefore does not suffer too much from the influence of gastrointestinal foam.

One of the benefits obtained by using some drugs is the appearance of activity in applications other than those for which it was originally intended.

Although the polydimethylsiloxane–silica mixture was originally introduced as a medical and veterinary aid to foam suppression, clinicians noted also the ability of formulations containing it to provide relief against gastric irritants, such as aspirin or tetracycline. Some of the products containing this preparation are therefore marketed as mucosal protectants as well as antiflatulent agents.

The protectant effect can be readily shown in rats as will be seen from Table 5.

This work was reported by Birtley et al.[23] and provides adequate evidence of the protection afforded by polydimethylsiloxane.

The work described in this review has therefore clearly shown that the effectiveness of the silicones combined with their chemical inertness and their freedom from toxicological side effects has won them an important place in modern therapy.

References

1. Quin, A. H., Austin, J. A. and Ratcliff, K. (1949). J. Amer. Vet. Med. Ass. **114**, 313.
2. Rider, J. A. and Moeller, H. C. (1960). J. Amer. Med. Ass. **174**, 2052.
3. Levin, R. (1958). "Pharmacy of Silicones". Published at offices of Chemist and Druggist.
4. (a) British Veterinary Codex, 1959 supplement to 1953 Edition, 54;
   (b) British Veterinary Codex, 2nd Codex (1965), 343.
5. (a) British Pharmaceutical Codex, 1959, 261;
   (b) British Pharmaceutical Codex, 1963, 268.
6. (a) (1963). J. Amer. Med. Ass. **184**, 226;
   (b) (1968). J. Amer. Med. Ass. **203**, 761;
   (c) (1968). Merck Index, 8th Edition, 372;
   (d) United States Dispensatory and Physicians Pharmacology, 26 Edition, 1031;
   (e) Extra Pharmacopoeia (Martindale) (1967), 25th Edition, 1318.
7. Burton, J. S. Unpublished work.
8. Klein, K. and Maluzi, J. (1967). British Patent No. 1,204,383.
9. Buist, G. J., Burton, J. S. and Elvidge, J. A. (1973). J. Pharm. Pharmacol. **25**, 854.
10. Chahal, R. S. and St. Pierre, L. E. (1968). Macromolecules, **1**, 152.
11. Chahal, R. S. and St. Pierre, L. E. (1969). Can. J. Chem. **47**, 2311.
12. Rowe, V. K., Spencer, H. C. and Bass, S. L. (1950). Arch. Ind. Hyg. **1**, 539.

13. Child, G. P., Paquin, H. O. and Deichmann, W. B. (1951). *Arch. Ind. Hyg.* **3**, 479.
14. Unpublished report, *Department of Medical and Pharmacology* (1967). University of Birmingham.
15. Unpublished report. *Department of Medical Biochemistry and Pharmacology* (1967, 1968). University of Birmingham.
16. Unpublished report, Dow Corning Corporation.
17. Cutler, M. G., Collings, A. J., Kiss, I. S. and Sharratt, M. (1974). *Food Cosmet. Toxicol.* **12**, 443.
18. Monograph on MINIFORM (1971). *Fass*, 1, 325.
19. Rezak, M. J. (1966). *J. Pharm. Sci.* **55**, 538.
20. Rhone Poulenc (1972). Unpublished work.
21. Carless, J. E., Stenlake, J. B. and Williams, W. D. (1973). *J. Pharm. Pharmacol.* **25**, 849.
22. Burton, J. S. Unpublished work.
23. Birtley, R. D. N., Burton, J. S., Kellett, D. N., Oswald, B. J. and Pennington, J. C. (1973). *J. Pharm. Pharmacol.* **25**, 859.
24. Birtley, R. D. N., Burton, J. S., Kellett, D. N., Oswald, B. J., Pennington, J. C. (1973). Unpublished work.
25. Rider, J. A. (1960). *Amer. Practit.* **11**, 1960.
26. Roberts, M., Settel, E., Arlen, M. and Friedman, H. P. (1963). *J. Amer. Med. Ass.* **183**, 595.
27. Jones, B. E. (1970). *Manuf. Chemist*, **41**, (May).
28. Rider, J. A. (1965). *Modern Treatment General Practice*, **2**, 976.
29. Ferraz, W. and Wafae, N. (1965). *Hospital Rio de J.* **68**, 1385.

Discussion

O'Neill (*Laboratory of the Government Chemist, London, England*) Referring to Fig. 1 (p. 136), what do we take to be T_0, i.e. the zero time for the x-axis?

The T_0 values of 250 ml are connected to the 15-second values by straight lines (Fig. 1), yet presumably these lines do not represent the actual foam behaviour. Why are the first readings taken only at 15 seconds when the more effective antifoam materials have almost completed foam collapse?

If it is not possible to measure foam height before 15 seconds, then the experimental method does not serve to differentiate good form excellent antifoam substances (despite what is stated in the text) and only differentiates antifoam substances from others.

Burton It is physically impossible to measure the foam height before the 15 seconds have elapsed. The experiments discussed were not attempting to demonstrate the difference between good and excellent antifoam substances. The object was to demonstrate that formulations containing these mixtures will produce a comparable *in vitro* effect and therefore would be expected to be approximately equivalent on clinical usage.

Neustadter (*BP Research Centre, Sudbury-on-Thames, England*) Have you any ideas as to how fresh the silica was that was used in the experiments of Kitchener who showed that the age of silica surfaces can greatly affect their ability to adsorb polymer, e.g. polyacrylamides.

Burton I do not have any idea how old the silica was as used in the experiments of Kitchener but it can be demonstrated that heat will cause hydrophyllic silica to become hydrophobic and thereby reduce its ability to adsorb polymeric materials. Assuming that the silica used by Kitchener was hydrophyllic this may have produced a situation in which the silica has lost its available hydrogen with time. If hydrophobic silica, and in this case I am talking about the pure silica from which the water adsorbed on the surface is removed, then this should not have any effect.

Thompson (*ICI Agricultural Division, Billingham, England*) Have methyl hydrogen polysiloxanes been tested under clinical conditions. Such compounds react readily with silica at room temperature to produce a hydrophobic silica with true chemical bonding.

Burton Methyl hydrogen polysiloxanes have not to my knowledge been tested under clinical conditions. This would be a hazardous operation since not only can the materials react with materials such as silica but they could also react with body chemicals and thereby produce potentially toxic substances. The value of polydimethylsiloxanes clinically is their inertness.

Roberts, K. What is the correlation between model systems used in your *in vivo* studies and real animals or people eating food.

Burton Our studies represent the first attempts to demonstrate that formulations used clinically and shown to be successful clinically do in fact act as antifoaming agents. Because we were using the animals as pure models they were fasted before the experiments started. In real animal or human situations you have to take into account that undigested food may be present which can either cause foam or even suppress foam depending on the nature of the food present. The only way to demonstrate the efficiency of a preparation is to carry out full clinical studies. These have been carried out both in animals and humans with preparations containing the mixture described.

Ingram Do you know how the size of the silica or antacid suspension particles affects the antifoaming properties?

Burton The tableted preparations containing the polydimethylsiloxane/silica are intended to be chewed so that the particle size of the material presented to the stomach contents is variable. Empirically one would feel that the smaller the size of the particle the greater would be the surface area coverage and therefore the greater the antifoaming effect.

James (*Welsh School of Pharmacy, Cardiff, Wales*) Volatile oils are commonly used for the treatment of flatulence and are very effective in this respect. There is no simple chemical relationship between the various volatile oils employed, but their constituents are physically similar in that they are mainly hydrophobic and also have a substantial hydrophilic component. This structure suggests that they may concentrate at gas–liquid interaces. Could the speaker comment on the possibility that the carminative activity of volatile oils is due to foam breaking?

Burton I would suggest that the carminative activity of volatile oils may well be due in part to their ability to break foams. As with the silicone mixtures described in my paper, it is necessary for them to be adsorbed on to a suitable surface and it is quite possible that the combination with an adsorbent produces a suitable antifoaming mixture, thereby reducing flatulence.

As with the silicones it is important to ensure that the formulation is presented in a palatable form.

Vrij Is anything known about the mechanism by which the silica enhances the foam destabilization.

Burton The mechanism of silica enhancement of foam destabilization has been dealt with by a number of experts in surface chemistry. Ross and Ottewill, both of whom are contributors to the Symposium, are two leading workers in this field of physical adsorption and their work should be consulted for details of the mechanism.

10

Bubble coalescence in aqueous solutions of n-alcohols*

N. H. SAGERT, M. J. QUINN, S. C. CRIBBS and E. L. J. ROSINGER

Research Chemistry Branch, Whiteshell Nuclear Research Establishment, Atomic Energy of Canada Limited, Pinawa, Manitoba ROE 1LO, Canada

Summary

High-speed cinematography was used to measure coalescence times of two nitrogen bubbles formed on adjacent nozzles in dilute aqueous solutions of C_2 to C_6 n-alcohols. Coalescence times from 0·5 to 500 ms could be measured. They were proportional to alcohol concentration for ethanol and propanol and were proportional to almost the second power of alcohol concentration for n-pentyl and n-hexyl alcohols.

A model for bubble coalescence was developed. By supposing that the dynamic rise in surface tension as the film between the bubbles stretched is the principal force stabilizing the film, a time t_s, was calculated for stretching to any given thickness. However, this calculation does not predict the thickness at which the film actually breaks. To estimate that thickness, the breaking time, t_B, for films of constant thickness was calculated by Ruckenstein's method. The actual breaking times were then taken as the minimum in the plot of $(t_s + t_b)$ against film thickness. Agreement between these calculated breaking times and the experimentally observed breaking times were satisfactory, even though there were no adjustable parameters in the model. The model gives much closer agreement with the observations than previous models based on diffusion into the film.

1 Introduction

Gas–liquid contacting is very important industrially and, in fact, forms the basis of many production processes. Amongst such processes is the Girdler–Sulphide (GS) process for producing heavy water, in which liquid water and gaseous hydrogen sulphide are contacted over sieve trays.[1] In most of these

* Issued as AECL No. 5066.

processes, an adequate interfacial area must be maintained for good mass transport, but foaming may be excessive, which in turn may lead to instabilities, especially in equipment such as sieve trays.

The behaviour of a short-lived foam such as that formed in the GS process is complex. However the foaming tendency may be related to the rate at which liquid films thin and break since, when two bubbles come together, the film between them thins and breaks if the time in contact is sufficient.[2]

Bubble coalescence phenomena have been studied traditionally in two ways, firstly in relatively large columns of foam,[3,4] and secondly in agitated tanks,[5] and empirical correlations may be made. Such studies provide very useful information, but generally give little insight into the ways in which the liquid film between any two bubbles thins and breaks. On the other hand, there is a body of very detailed knowledge about the draining and thinning of very stable films formed by highly surface-active materials.[6,7]

Recently Marrucci and his colleagues have begun to study bubble coalescence by an intermediate approach.[2,8,9] The coalescence times of two bubbles grown on adjacent nozzles were studied on the millisecond to 1 second time scale. Although the actual film thickness as a function of time cannot yet be measured, nevertheless the coalescence time is a quantitative parameter which can be obtained for the thin liquid film between two individual bubbles. Marrucci et al.[9] studied coalescence of bubbles stabilized with electrolytes, and developed a theory based on rapid thinning of the liquid film to a quasi-equilibrium film thickness, followed by diffusion of surface-active material into this quasi-equilibrium film which necessarily has a deficit of the more surface-active material.

We have applied the Marrucci technique to aqueous solutions of some typical nonelectrolytes, the normal alcohols, to see if the previous theories of Marrucci were adequate to explain film breaking properties of films stabilized with very small concentrations of nonelectrolytes.

2 Experimental methods

Coalescence times of two bubbles growing on adjacent nozzles were measured using an apparatus shown in Fig. 1, and essentially similar to that used by Marrucci et al.[2,8] The external vessel was made from 60 mm square Pyrex tubing. The nozzles were made by drilling two holes of 2 mm diameter in a Pyrex block, the edges of the holes being 1·1 mm apart (see insert to Fig. 1). Two independent gas delivery systems were used, one for each nozzle. Non-contaminating regulators were used and the gas was pre-saturated with the vapour from the solution being examined. Flow rates were measured with a soap film flowmeter.

All solutions were prepared from triple distilled water. The alcohols were reagent grade. They were purified by distillation using a spinning band column and their purity was checked by gas chromatography. The nitrogen was Linde high-purity grade and was passed over activated molecular sieve 5A. Reagent grade KCl was used for a few runs with added electrolyte. It was heated to 800 K for 12 h to remove organic impurities.

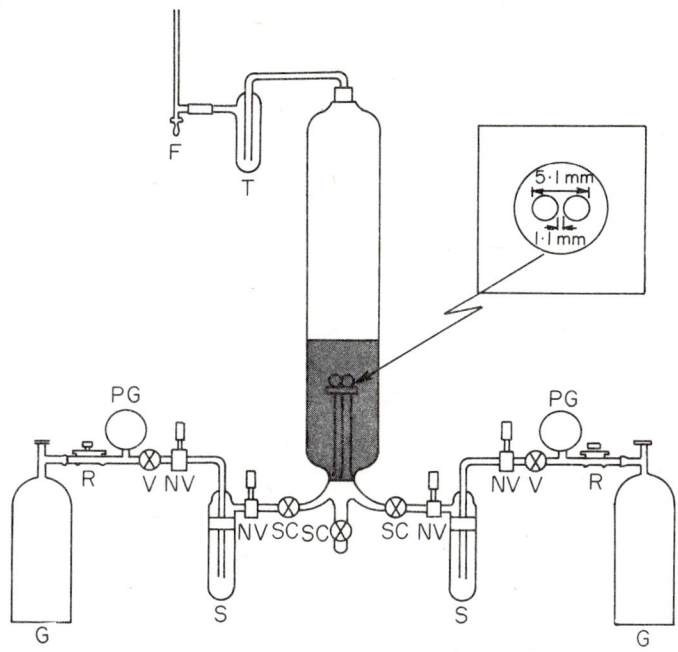

FIG. 1. Schematic diagram of the apparatus. SC, greaseless stopcocks; NV, needle valves; V, valves; PG, pressure gauges; R. regulators; G, nitrogen cylinders; and S, saturators. T is a trap and F a flowmeter.

Before each run the Pyrex parts of the system, consisting of the saturators, the nozzle assembly and the outer container, were cleaned with chromic acid or potassium permanganate sulphuric acid cleaning solution, washed with triple distilled water and dried in an oven at·420 K. All the glassware used for making up solutions was cleaned in a similar way. Glassware was stored in the oven at 420 K until just before use.

To begin a run, the saturator was filled with the solution to be examined, and the container was filled with solution to a level about 50 mm above the nozzles. This depth was found to be the optimum depth for suppressing bubble oscillations. The nitrogen flow was started through one nozzle and adjusted to a flow rate of (usually) 33 mm³ s⁻¹. Then the flow from the

FIG. 2. A portion of a typical film. The bubbles are coalescing in a solution containing 2·5 mg kg^{-1} *n*-butanol. The film is running at 4700 frames per second and the frames shown are those at final film rupture.

other nozzle was started and the total flow adjusted to twice the previous value.

Coalescence times were measured by high-speed cinematography using a Hycam motion picture camera operating at 500 to 5000 frames per second. Several frames from a typical sequence are shown in Fig. 2 at the time when the bubbles are coalescing. Film speeds were determined from a calibrated timing light. Coalescence times are defined as the time between first touching of the bubbles as evidenced from the disappearance of an uninterrupted light

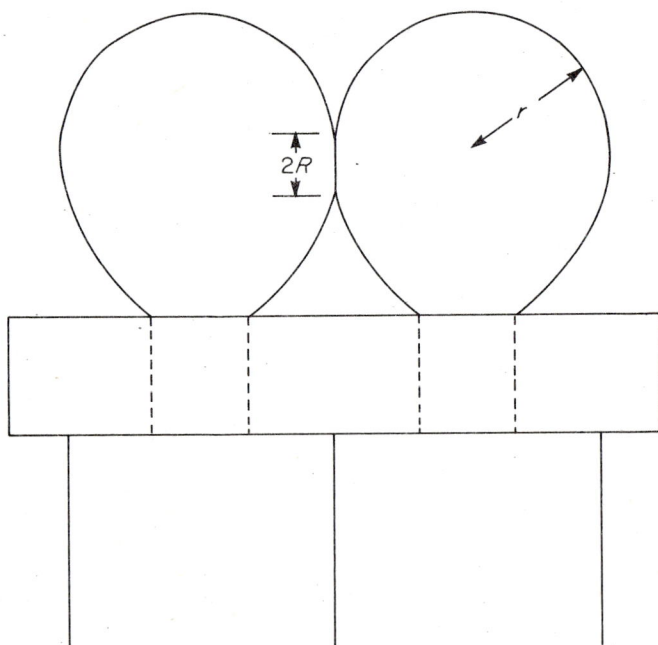

FIG. 3. An idealized drawing of the bubbles on the nozzles showing the definition of *r* and *R*.

zone between the two bubbles, and the final rupture of the film between them. The point at which the films finally ruptured was very distinct (Fig. 2) and never took more than one frame even at 5000 frames per second. However, the time at which the bubbles first touch was less distinct, and often involved an uncertainty of two or three frames. Usually ten to fifteen sequences were photographed and the average coalescence time was taken. The individual values were usually within ± 20 per cent of the mean value. The effective bubble radius, r, was measured from the films as was the radius, R, of the contact area (Fig. 3).

A value for the decrease in surface tension (γ) as a function of concentration, c, of solute at very low concentrations was required. Surface tensions were measured by the Wilhelmy plate method[10] using an electrobalance. The sensor was a platinum foil 6 μm thick and 10 mm wide which was cleaned with nitric acid and washed with distilled liquid before use. After initial wetting, liquid was withdrawn from the reservoir until the bottom of the sensor was just even with the liquid level. Measurements were made at $298 \cdot 16 \pm 0 \cdot 02$ K in a container which was isolated from the atmosphere as completely as possible.[10]

3 Results

The measured value of $d\gamma/dc$, where c is the concentration of solute in $mol\,kg^{-1}$, and γ is the surface tension in $mM\,m^{-1}$, are shown in Table 1.

TABLE 1

Surface tension measurements for aqueous alcohol solutions

Alcohol	Maximum concentration $c/mol\,kg^{-1}$	Surface tension change $(d\gamma/dc)/mN\,kg\,mol^{-1}\,m^{-1}$
C_2H_5OH	0·174	22·4
n-C_3H_7OH	0·0267	94·8
n-C_4H_9OH	0·0108	222·4
n-$C_5H_{11}OH$	$1 \cdot 85 \times 10^{-3}$	643·7
n-$C_6H_{13}OH$	$3 \cdot 54 \times 10^{-3}$	1538·0

The concentration ranges in which γ was measured are from zero to the indicated maximum. At the small concentrations used, the surface tension depressions were small, and generally less than 5 $mN\,m^{-1}$, $d\gamma/dc$ is estimated to be accurate to ± 10 per cent. Nicodemo et al.[9] have shown that radius, R, of the circular contact area between the bubbles (Fig. 3) should vary with

F

the square root of the time for short contact times.

$$R = C t^{\frac{1}{2}}, \tag{1}$$

where C is a constant defined by the flow rates and the bubble volumes just before contact. In our case for a bubble radius, r, equal to 1·4 mm and a

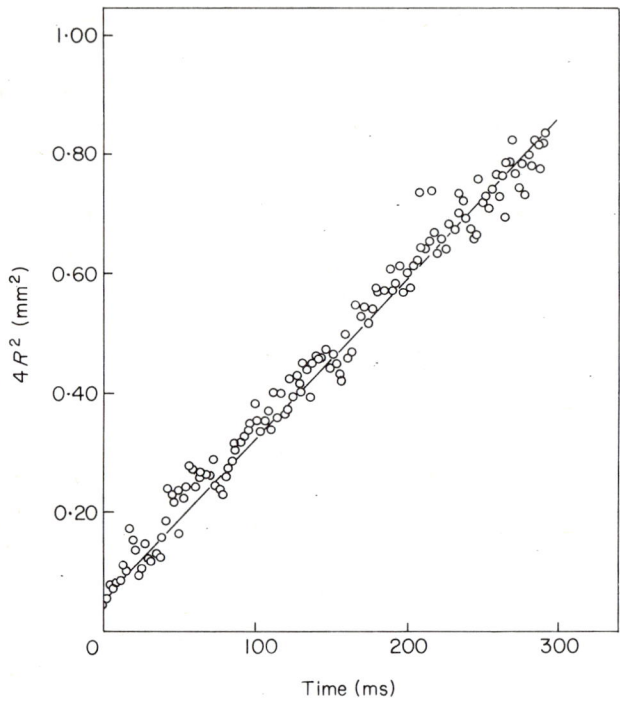

FIG. 4. A plot of film diameter ($2R$) squared against time for a solution containing 40 mg kg^{-1} n-amyl alcohol.

flow rate per nozzle of 33 mm^3 s^{-1}, the theoretical expression for R at times equivalent to those actually employed is

$$R = 6·6 \times 10^{-4} t^{\frac{1}{2}} m. \tag{2}$$

Figure 4 shows an actual plot of $4R^2$ against t for a solution of n-amyl alcohol containing 40 mg kg^{-1} n-amyl alcohol. The scatter is large since there are bubble oscillations superimposed on the steady growth of R, but the plot is quite linear. In this case $C = 8·1 \times 10^{-4}$ m s$^{-\frac{1}{2}}$, which is only fair agreement with the theoretical prediction. The difference may arise because the bubbles are not spherical. Nevertheless the agreement with regard to the

functional dependence and the general order of magnitude of agreement in
C provides some assurance that the system is behaving as expected.

Experimentally determined coalescence times are shown in Fig. 5 as a
function of alcohol concentration. For the lower alcohols, coalescence times
were proportional to the alcohol concentration, but for n-hexyl alcohol,
coalescence times were proportional to the square of the alcohol concentration. N-amyl alcohol behaved in an intermediate manner.

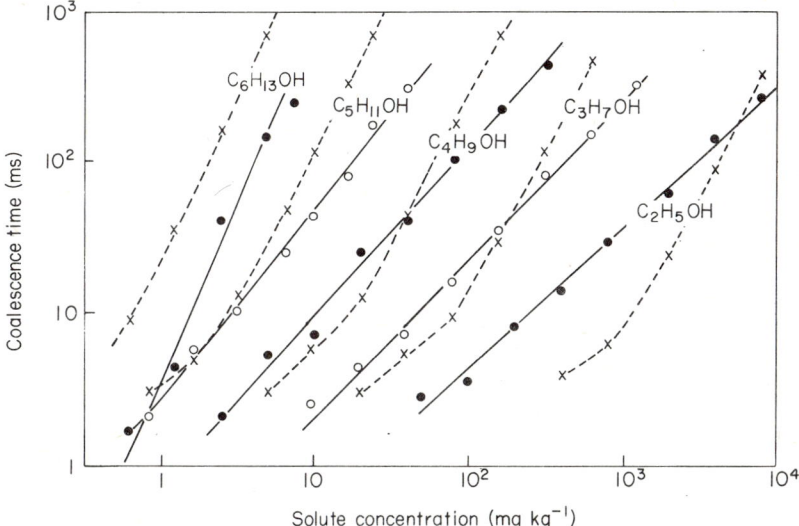

FIG. 5. Double logarithmic plots of both experimental and calculated coalescence times against
solute concentration for the five alcohols investigated. ●, Experimental; ×, theoretical.

Coalescence times of less than 1 ms could not be measured accurately by
our techniques and, as well shall see, such times were not of great interest.
There was also an upper limit to measurement of coalescence times because,
at the smallest flow rate that we could conveniently use, the individual
bubbles remained on the nozzles for at most 0·5 s before breaking away on
their own accord. Thus, if coalescence did not occur within 0·5 s, we could
not observe it.

Some experiments were performed with 101 mg kg^{-1} of n-butanol varying
the flow rate by a factor of $\sim$ 10, i.e. about a factor of 3 above that normally
used and a factor of 3 below it. Coalescence times remained constant to
$\pm$ 20 per cent, indicating that gas flow rate had little effect on the coalescence
time. Also, some experiments were done with KCl in 40 mg kg^{-1} n-butanol
to check if there were any noticeable effects of electrical double layers.

Adding 10^{-3} mol kg^{-1} KCl, which should be sufficient to affect these double layers,[11] resulted in only a 10 per cent increase in coalescence time. Hence such effects were small.

4 Theoretical models

It is apparent from Fig. 5 that very small concentrations of solute can have very large effects on the breaking times of thin liquid films, even though the films are not stable in the classical sense[6, 12] and break in less than one second. Three models were considered to explain this stability. The first is the original model of Marrucci which we shall term a liquid phase diffusion model. The second is a gas-phase diffusion model as suggested by Nicodemo *et al.*,[9] and the third is a nonequilibrium model.

4.1 LIQUID PHASE DIFFUSION MODEL

Marrucci[8] has proposed a theory of coalescence which occurs in two stages. First a rapid thinning takes place to a quasi-equilibrium thickness which balances the surface forces on the film. During this thinning the solute surface is considered in equilibrium with the liquid in the film, and absolute concentration is assumed to be in equilibrium throughout the film. If the solute is positively adsorbed in the surface, the creation of new surface results in a lower concentration of solute in both the bulk and the surface and hence a higher surface tension. This process will halt when the increase in surface tension is balanced by the internal pressure in the film. The film is then in quasi-equilibrium. However, its bulk concentration is lower than that in the rest of the solution, and Marrucci considered the case where the film thinned down and eventually broke as solute diffused into the film (for a positively surface-active solute) through the outer edge of the film. This picture gave reasonable diffusion parameters and a good functional fit to experimental data[9] for coalescence of bubbles stabilized by electrolyte solution. However, as Nicodemo *et al.*[9] point out themselves, very sharp diffusion gradients are required to fit data from alcohol stabilized films. The characteristic diffusion distance (penetration model) would be about 10^{-11} m which is unrealistically small. Thus, this model is considered inadequate.

4.2 GAS PHASE DIFFUSION MODEL

Nicodemo *et al.*[9] suggested that, for volatile solutes, the same quasi-equilibrium state could be set up, but that the concentration difference between the bulk and the film could be equalized by diffusion through the gas phase into the film. We have analysed such a model, following the procedure used by Marrucci.[8] It was assumed that the gas in the bubble had the same partial pressure of alcohol as in equilibrium over the bulk liquid. If the mole fraction of the active component is small, activities can be taken as equal to concentrations. The basis mass balance then becomes

$$\frac{d}{dt}(\Delta c) = -\frac{2}{h}\frac{D}{x}\Delta c + \frac{2c}{R_G T}\left(\frac{d\gamma}{dc}\right)\frac{d}{dt}\left(\frac{1}{h}\right). \tag{3}$$

Here h is the thickness of the film, x is the diffusion distance, c is the concentration of solute in $mol\,kg^{-1}$ and Δc is the concentration difference. Following the procedure of Marrucci,[8] we deduce that, at concentrations actually used,

$$\frac{Dt}{x} = \frac{2\pi}{3\sqrt{3}}\frac{r}{R_G T\gamma}\left(\frac{12\pi\gamma}{A_0 r}\right)^{\frac{1}{4}}\left(\frac{d\gamma}{dc}\right)^2 c. \tag{4}$$

Here r is the bubble radius, as before, and A_0 is the Hamaker–London constant. For ethanol–water, this formula leads to a diffusion distance $x = 5\cdot8\,\mu m$ and for n-amyl alcohol–water it leads to a value of $x = 8\cdot4\,\mu m$. Now it is obvious that these distances should be small with respect to the thickness of the film (say 10 to 20 per cent) for the model to be realistic. However, the model also allows calculations of the initial film thickness when the quasi-equilibrium state is set up, and also of the final thickness at rupture. These are 900 nm and 15 nm respectively, both considerably smaller than the estimated diffusion distance. Since both diffusion models seem inadequate an entirely different approach was tried.

4.3 NONEQUILIBRIUM MODEL

We have started from the assumption that film stretching is fast enough that the solute concentration is never in equilibrium throughout the thickness, h, of the film and the surface excess (Γ_2) of solute at the film surface is in equilibrium only with the solute concentration just below the surface. The procedure will be to consider the retardation in motion caused by not having the solute in the film in equilibrium throughout the film and to calculate from this the time required to stretch the film from essentially infinite thickness to a specified final thickness h_f. This will be done first of all neglecting inertial terms, and then an attempt will be made to correct for inertial effects where they are important. Unlike the two diffusion models considered above, which predict a thickness where $dh/dt = -\infty$, i.e. breakage, there is no inherent prediction of a final breaking thickness in this nonequilibrium film stretching model. Hence a separate film breaking model is considered and used with the thinning model to predict the final film thickness and coalescence time.

4.3.1 *Thinning time without inertial corrections*

If the boundaries of an element of the interface are expanding, any excess of surface-active material will be spread over a larger area and the surface concentration will fall. The interfacial tension in the element will rise and oppose the expansion.[13] It has been suggested that even for a pure liquid there will be a rise in surface tension if the film is quickly expanded[14] but this effect will not be considered here. A specific surface expansion rate $S = (1/A) (dA/dt)$ may be defined where A is the surface area of the element under consideration.[8, 15] Lee and Hodgson[15] have shown from simple kinematics that if x and y are rectangular coordinates in the plane of the surface and v_{s_x} and v_{s_y} are the velocities in these directions, then

$$\frac{1}{A}\left(\frac{dA}{dt}\right) = \frac{\partial}{\partial x}(v_{s_x}) + \frac{\partial}{\partial y}(v_{s_y}). \tag{5}$$

Andrew[16] has considered the case of a steady uniform expansion at the surface of a liquid of infinite depth, where the stretching causes a decrease in the surface concentration because solute cannot be transported to the surface fast enough. This leads to an increase in surface tension, $\Delta\gamma$. Mass transfer to the surface takes place by convective diffusion normal to the surface. Assuming a pseudo-stationary state (no accumulation or depletion in the element of volume with time). Andrew derived equation (6) for dilute solutions:

$$\Delta\gamma = \frac{c\,(d\gamma/dc)^2}{R_G T\left(\dfrac{2D}{\pi S}\right)^{\frac{1}{2}} - \left(\dfrac{d\gamma}{dc}\right) - \left(\dfrac{d^2\gamma}{dc^2}\right)c}. \tag{6}$$

Over the range of solute concentrations used, the third item in the denominator may be neglected. This relation will be assumed to apply to each surface of the film between two bubbles under the conditions of interest.

Following Marrucci,[8] if a segment of film is considered. the criterion for mechanical stability is

$$h\,\Delta P = 2\Delta\gamma \tag{7}$$

where ΔP is the excess internal pressure in the film. This follows from considering a film segment of thickness h and an arbitrary length, l.

The excess film pressure consists of at least three terms. One term is the excess pressure in the bubble because of bubble curvature. This is given by $2\gamma/r$. As the film thins below 100 nm the London forces exert an additional positive attractive pressure equal to $A_0/6\pi h^3$ where A_0 is the Hamaker–London constant for water, taken as $3\cdot8 \times 10^{-20}$ J.[17] If there is electrolyte present, a double-layer repulsion pressure will be set up, given by $\Delta P =$

$64 R_G T c_{el} \sigma^2 \exp(-\kappa h)$ where c_{el} is the electrolyte concentration, $\sigma = \tanh$ $(zF\phi_0/4R_G T) z$ being the ion valency, ϕ_0 the double-layer potential at the phase boundary, and κ the Debye reciprocal length.[6] σ and κ are constants not involving h and can be calculated if the surface potential is known. The surface potential of water was taken as 30 mV, and with this potential, if no electrolyte was added the electrostatic repulsion term was small.

Combining all terms,

$$\Delta P = 2\gamma/r + A_0/6\pi h^3 - 64 R_G T c_{el} \sigma^2 \exp(-\kappa h). \tag{8}$$

If the film is taken to stretch elastically, i.e, volume in the film is conserved, then[8]

$$S = (1/A)\, dA/dt = -(1/h)\, dh/dt. \tag{9}$$

The expression for S in terms of h from equation (9) is substituted into equation (6) and the resulting expression for $\Delta\gamma$ is substituted into equation (7).

The time t_s' for stretching a film from infinite thickness to a final thickness h_f is then given by the equation

$$t_s' = -\frac{\pi}{2\,DR_G^2 T^2} \int_{\alpha}^{h_f} \frac{[2c(d\gamma/dc)^2 + h\,\Delta P(d\gamma/dc)]^2}{h^3\,(\Delta P)^2}\, dh. \tag{10}$$

When the expression for ΔP, equation (8), is substituted into equation (10), the complete expression for t_s' is obtained. The right-hand side could not be integrated to give an analytic expression for t_s', so t_s' was evaluated numerically using PDP10 computer to get t_s' as a function of h_f.

4.3.2 Inertial corrections

In the derivation of t_s', inertial influences were completely neglected. This frequently produces a reasonably good approximation, but at the shortest times inertial effects became important. By using Bernoulli's equation along the edge of the film, Marrucci et al.[2] and more recently Kirkpatrick and Lockett[18] have shown that the velocity, V, with which liquid is moved from between two bubbles as they approach, is given by

$$V = 2\,(\gamma/\rho r)^{\frac{1}{2}}. \tag{11}$$

where ρ is the density of the fluid. Then the inertial time, t_{in}, for movement of the film of liquid is

$$t_{in} = R/2V \ln(h_i/h_f). \tag{12}$$

where h_i and h_f are the initial and final thicknesses of the film. R can be measured approximately from the film sequences and h_f can be calculated, as shown in the next section. The values of h_f are in the range 5 to 10 nm. The initial thickness is harder to estimate. From the photographs it is certainly less than 0·1 mm. Marrucci[2] estimated it to be of the order of 10 μm. If this value of 10 μm is used in equation (12) then an estimate of t_{in} can be made which will be correct to within a factor of 2·3 if we have estimated h_i to within one order of magnitude. Typical values are in the 1 to 3 ms range.

Although t_{in} and t'_s should be coupled, we have simply added t'_s and t_{in} to give t_s, the time for stretching from infinity (or 10 μm) to a final thickness h_f. However, there is nothing in this model to tell us at what thickness the film actually breaks. Thinning can, in principle, occur down to below molecular dimensions.

4.3.3 *Final film rupture*

To estimate the thickness at which the film actually breaks, we have used equations developed by Ruckenstein and Jain[19] for the lifetime of a liquid film of constant thickness. These authors considered a surface perturbation of variable wavelength and derived equations for breaking times as a function of wavelength using hydrodynamic linear stability theory. The films eventually break because a condition can be reached where the wave amplitudes grow without increasing the total free energy of the system. The free energy is augmented as the area increases, but at the same time, it is diminished because of London attractions between opposite interfaces. It is assumed that a complete spectrum of wavelengths is available, and the time for breaking is governed by the fastest growing perturbation only. Thus, it is possible to calculate the breaking time, t_B, of a film of given thickness. The equations developed by Ruckenstein and Jain[19] are implicit functions of the breaking time so breaking times were calculated on the PDP10 computer using a Fibonacci search technique.

4.3.4 *Final estimation of coalescence time*

From the above sections we now have t_s, the time for stretching a film to a given final thickness. These values are calculated for a range of final thicknesses to give t_s as a function of the final thickness. As the final thickness decreases, t_s increases. Similarily, from the Ruckenstein and Jain relations, we can calculate t_B for a number of constant film thicknesses, and find that t_B rises very sharply as h increases. In principle the equations governing t_B should be re-derived for the case where the film thickness is changing with time. This would be very difficult as pointed out by Vrij[20] in his study of the draining and breaking of more stable films. Thus we have used Vrij's[20] approach of taking the actual breaking time as the minimum time in a plot of $(t_s + t_B)$ against thickness. Calculated values are shown in Fig. 5 as the theoretical points.

5 Discussion

The theoretical lines shown in Fig. 5 are very gratifying in that they predict the correct order of magnitude for the breaking times and predict the correct order of the alcohols in stabilizing films (i.e. n-hexyl > ... > ethyl), from relations without any adjustable parameters. However, they are not a totally satisfactory prediction of the experimental results. One source of error which affects the absolute magnitude of the predicted value is the error in $d\gamma/dc$. As equation (10) shows, the coalescence time increases as $(d\gamma/dc)$.[4] If the error in $d\gamma/dc$ is ± 10 per cent, then the predicted value will be in error by ± 46 per cent. Since this is an experimental error, any theory that predicts coalescence times to better than ± 50 per cent must be considered satisfactory.

Another source of uncertainty is that t'_s, t_{in} and t_B have been calculated separately and the actual breaking time taken as the minimum in $(t'_s + t_{in} + t_B)$ as a function of h. In principle a more precise mathematical model of the system would treat the growth of waves in thinning films taking inertial effects into account.

We have used Andrew's[16] expression for the increase in γ as a function of film stretching. In this model a steady uniform expansion is assumed and the liquid is assumed to have infinite depth. Clearly neither of these restrictions are fulfilled rigorously. Experimentally (Fig. 4) we observe that while the film is stretching roughly as $r \propto t^{\frac{1}{2}}$, oscillations are superimposed on this, and R grows very much faster in the initial few milliseconds. Thus any more exact model will have to take into account a more detailed picture of R as a function of h (or t) rather than assuming a steady uniform expansion. The film has a thickness h, so the maximum depth which should be considered is $h/2$. Furthermore, the surface shear stress due to the surface gradient of the interfacial tension is taken essentially as zero. This cannot be rigorously true.[15] If there is any appreciable coefficient of surface dilational viscosity this will effect the surface shear stress and thus the coalescence time, although for the dilute solutions considered here, it is unlikely that surface dilational viscosity will be measureable.[21]

Finally, we have not considered any diffusion of volatile solute to the interface through the gas phase. Although we have shown that this is not the major process controlling coalescence, it may well be a significant secondary consideration.[22]

In summary, we have measured coalescence times of dilute aqueous solutions of n-alcohols and have predicted the results theoretically using a model where thinning of the film is inhibited by the rate with which solute can be transported to the surface. The film finally breaks when it becomes thin enough for surface waves to grow spontaneously within the time-scale involved. This model predicts the observed coalescence times reasonably well, and much better than theories based on maintaining the solute concentration uniform throughout the film thickness and in equilibrium with the surface. However, the correspondence between theory and experiment is not perfect and needs further improvement.

Acknowledgements

We would like to thank our colleagues, G. G. Strathdee, D. R. Prowse, A. Sawatsky, T. R. Heidrick, W. G. Mathers, and K. A. Burrill for useful advice and discussion.

References

1. Becker, E. W. (1962). "Heavy Water Production", pp. 33–43. International Atomic Energy Agency, Vienna.
2. Marrucci, G., Nicodemo, L. and Acierno, D. (1969). *In* "Cocurrent Gas–Liquid Flow" (E. Rhodes and D. S. Scott, Eds), p. 95. Plenum Press, New York.
3. Zieminski, S. A. and Whittemore, R. C. (1971). *Chem. Eng. Sci.* **26**, 509.
4. Bowonder, B. and Kumar, R. (1970). *Chem. Eng. Sci.* **25**, 25.
5. Lee, J. C. and Meyrick, D. L. (1970). *Trans. Inst. Chem. Eng.* **48**, T37.
6. Sheludko, A. (1967). *Advan. Colloid Interface Sci.* **1**, 391.
7. Woods, D. R. and Burrill, K. A. (1972). *J. Electroanal. Chem.* **37**, 191.
8. Marrucci, G. (1969). *Chem. Eng. Sci.* **24**, 975.
9. Nicodemo, L., Marrucci, G. and Acierno, D. (1972). *Quad. Ing. Chim. Ital.* **8**, 1.
10. Vochten, R. and Petre, G. (1973). *J. Colloid Interface Sci.* **42**, 320.
11. Clunie, J. S., Corkill, J. M., Goodman, J. F. and Ingram, B. T. (1970). *Special Discuss. Faraday Soc.* **1**, 30.
12. Lyklema, J. and Mysels, K. J. (1965). *J. Amer. Chem. Soc.* **87**, 2539.
13. Kitchener, J. A. (1964). *In* "Recent Progress in Surface Science" (Ed. J. F. Danielli), Vol. 1, p. 51. Academic Press, New York and London.
14. Kochurova, N. N., Shvechenkov, Yu. A. and Russanov, A. I. (1974). *Kolloid. Zh.* **36**, 785.
15. Lee, J. C. and Hodgson, T. D. (1968). *Chem. Eng. Sci.* **23**, 1375.
16. Andrew, S. P. S. (1960). *In* "International Symposium on Distillation" (P. A. Rottenburg, Ed.), p. 73. Instn. Chem. Engrs., London.
17. Vincent, B. (1970). *Special Discuss. Faraday Soc.* **1**, 78.
18. Kirkpatrick, R. D. and Lockett, M. J. (1974). *Chem. Eng. Sci.* **29**, 2363.
19. Ruckenstein, E. and Jain, R. K. (1974). *J. Chem. Soc. Faraday II*, **70**, 132.
20. Vrij, A. (1966). *Discuss. Faraday Soc.* **42**, 23.
21. Lucassen-Reynders, E. H. (1973). *J. Colloid Interface Sci.* **42**, 573.
22. Marrucci, G. and Nicodemo, L. (1967). *Chem. Eng. Sci.* **22**, 1257.

Discussion

Vrij You mention in your paper that you calculated the breaking time t_B for films of constant thickness by "Ruckenstein's method". May I call your attention to the fact that the same method was previously treated by us? (See also Hansen's Plenary Lecture and references 10 and 11 therein.)

Sagert We certainly acknowledge that the theory of Vrij and his colleagues is very similar to that of Ruckenstein and Jain and was published some four years earlier. Indeed, the papers mentioned were referenced by Ruckenstein and Jain. However, we did employ the equations developed by Ruckenstein and Jain and did so because they were easier for us to use. We regret any implication that these authors developed the theory of spontaneous rupture in its entirety.

Zichy The mathematics accounting for the coalescence times deals with fast liquid movements in thin liquid layers, but there do not appear to be any terms relating to liquid viscosity. Could you explain the reasons for this?

Sagert In this paper we have largely assumed mechanical stability (equation 7) for simplicity, since we consider that the inertial influences are small at any but the shortest times (section 4.3.2.). Viscosity would certainly enter into any more complete description. Viscosity was also neglected in calculating the inertial correction since it was assumed that the film was stretching elastically (reference 18).

Lee (*University College of Swansea, Wales*) In commenting on the gas phase diffusion model for the relaxation of the surface tension gradient induced by film stretching, the paper states that "it is obvious that the diffusion distances calculated should be small with respect to the thickness of the film". However, as the diffusion distance concerned lies in the gas phase adjacent to the surface of the liquid film and in a direction perpendicular to the plane of the film, there seems to be no reason why it could not be considerably larger than the thickness of the liquid film. Furthermore, a relatively large value of a "diffusion distance" calculated from a given mass flux and particular diffusivity on a quasi-steady state basis implies an easy path for mass transfer. It is possible therefore that the gas phase diffusion mechanism could in fact satisfactorily explain the thinning of the film.

Sagert In the gas phase diffusion model the "diffusion distance" calculated was in the liquid phase, assuming that the greatest resistance to mass transfer would be in the liquid. If the resistance to mass transfer is assumed to be entirely in the gas phase the "diffusion distance" would be larger than the bubble diameter. Thus for both conditions (i.e. gas or liquid phase diffusion controlling) the "diffusion distance" is larger than the physical dimensions of the system.

Lee is certainly correct in that the gas diffusion model predicts very fast breaking. In fact, the model represented by equations (3) and (4) probably represents a lower time limit since the model assumes a surface excess in equilibrium with the solute concentration in the film. If the solute is delivered through the face of the film, the surface excess will be somewhat higher[1]. Because the model predicts breaking times much shorter than those observed, we assumed that other factors must be involved which prevent gas phase diffusion from being the controlling factor. Thus we looked for a different model and found one which was reasonably consistent with our observations. However, no model can ever be proven to be correct. It can only be shown to be consistent with the known facts and therefore to be plausible. A more realistic gas phase diffusion model may well give better agreement with experiment. We are presently examining one such model.

[1] D. R. Prowse. Private communication.

11

Electric charge at the air–solution interface

R. W. HUDDLESTON and A. L. SMITH

Unilever Research,
Port Sunlight Laboratory, Wirral, Merseyside L62 4XN, England

Summary

Two electrokinetic experiments are described, designed to give information on charge separation at the gas–solution interface. In the absence of added surfactants the charge is much more dependent on pH than on the concentration of other simple ions; the interface seems to have an isoelectric point around pH = 2 and is negatively charged at normal pH. Anionic and cationic surfactants modify the charge in the expected sense.

1 Introduction

One of the interactions across sufficiently thin films is that arising from the overlap of the electric double layers which in turn results from charge separation at the air–solution interface. Where the solution contains ionic surfactants the potential presence of this effect will not be doubted; in solutions containing nonionic surfactants or only simple electrolytes the existence of such an effect (at least of significant magnitude) is more doubtful. In the absence of experimental data, discussions of the interaction across the films sometimes take the electric potentials involved as zero and sometimes as infinite (the interaction does not increase much for potential higher than ~ 150 mV). While some attempt can be made to calculate the surface potentials in the case of ionic surfactant solutions from surface tension/adsorption data, this involves assumptions about counterion binding and the dielectric state near the interface and is less than satisfactory.

While there is no convincing evidence that the air–aqueous solution interface involves a significant charge separation when the solution contains only simple electrolytes, there is some circumstantial evidence from repeated observations[1] that, for instance, hydrocarbon droplets or particles behave as if negatively charged in such solutions at ordinary pH value, however much the materials are purified in an attempt to exclude ionic surface-active impurities. A simple electrolyte which increases the surface tension of water must be negatively adsorbed at the solution–air interface. Unequal negative adsorption of oppositely charged ions will then constitute a charge separation though this cannot be expected to give charges of the magnitude observed for the particles quoted and will not of itself explain the particularly strong influence of pH.

The presence of adventitious ionizable groups at the surface of apparently inert solid particles may be suspected as the origin of their mobility in an electric field and while this particular suspicion cannot be justified for the gas–solution interface (e.g. gas bubbles in solution), the corresponding one of adsorption of ionic surface-active impurity has to be admitted. It was mainly for this reason that we performed two types of electrokinetic experiment, in one of which the surface could be repeatedly swept clean; this will be referred to as the plane interface experiment. The other experiment, requiring more complex apparatus but otherwise easier to perform, will be referred to as the spinning cylinder experiment. In both cases the experimental difficulties are matched by difficulties of data interpretation.

2 Spinning cylinder experiment

This technique, originally due to Quinke[2] in 1861, used by Alty[3] and McTaggart[4] and significantly modified by Whybrew et al.[5] uses a cylindrical cell rotating about its axis. The test bubble is held on the axis of the cell and an electrical (or gravitational) field can be applied to move the bubble along the axis of rotation. In our apparatus the glass cylinder, of wall thickness 5 mm and internal diameter 15 mm, was held in air bearings and rotated by an air turbine at speeds up to 10 000 r.p.m.

Experimentally we observed the mobility of single gas (nitrogen) bubbles as a function of applied electrical potential gradient, angular speed of rotation and bubble size. Variation of the latter was achieved by gradual adsorption of the bubble into a (partially) degassed test solution.

Two immediate questions arise in any attempt to interpret the data from the spinning cylinder experiments: (1) the contribution of electro-osmotic effects arising from the cell walls to the velocity of bubbles on the axis; and (2) the influence of the cell rotation on the axial movement of the bubble.

2.1 ELECTRO-OSMOTIC EFFECT OF THE CELL WALLS

In a normal nonrotating cylindrical cell particles on the axis will exhibit, quite apart from an electrophoretic velocity resulting from their own charge, a velocity, which could well be much larger, arising from the charge separation at the cell walls. This familiar situation leads to particle electrophoretic velocities being measured at the so-called stationary level, 0·146 of the diameter from the cell walls for cylindrical cells.

Whybrew et al.[5] recognized this complication and, while not checking its existence in the rotating mode, allowed for it in the same way as for stationary cells. Cichos[6] made a partial correction for it. In this work we looked for the existence of this correction using a hollow glass sphere of diameter ~ 0.3 mm and density ~ 0.8 g cm^{-3}. These have the advantage over gas bubbles that their own electrophoretic mobility can be separately determined (after crushing) and their charge altered at will by adsorption of polymeric materials, as can that of the cell wall. Using spheres and wall of the same and opposite charges, which would lead to large electro-osmotic differences, we were not able, within the experimental reproducibility ($\sim 0.5 \times 10^{-8}$ m^2 s^{-1} V^{-1}), to detect any electro-osmotic effect at the centre even when its value for a stationary cell would have been ten times this value. Furthermore gas bubbles themselves exhibited the same velocity in the same applied field regardless of the charge on the cell walls.

This absence of electro-osmotic effect must be related to the rotation though we still do not have a quantitative treatment.

2.2 FORCE ON AN AXIALLY MOVING BUBBLE

A sphere of diameter d and density ρ moving at velocity v along the axis of a cylindrical cell containing an (imaginary) inviscid fluid of density ρ_0 and rotating at an angular velocity Ω experiences a retarding force F_s given by

$$F_s = (2/3) \, \Omega(\rho_0 - \rho) \, d^3 v. \tag{1}$$

This result, obtained by Stewartson,[7] shows that the rotation of the cell produce retardations which can be an order of magnitude greater than the Stokes's forces in a stationary fluid of viscosity η, viz. $3\pi\eta d$. The effect was ignored by workers before Whybrew et al.[5] and has the result that ordinary electrophoretic mobilities cannot be measured—only the electrophoretic force exerted on the bubble or particle by the applied field.

This electrophoretic force could be obtained by investigating the coupling between the axial force due to spin and the Stokes's drag, so that the required electrophoretic "forward" force could be obtained. More convenient, however, is to calibrate the effect by tilting the cell away from the horizontal by a known small angle so that a known gravitational force can be applied with the electric field switched off. Alternatively the two fields can be applied simultaneously in opposition to keep the sphere stationary.

The required electrophoretic force F_E in a given electric field is then obtained from

$$F_E = (1/6)\pi d^3(\rho_0 - \rho)g \sin \theta, \qquad (2)$$

where θ is the angle of tilt necessary to give the same velocity as the electric field in a horizontal tube or alternatively the angle of tilt necessary to keep the sphere stationary with the electric field still applied.

Figure 1 shows bubble velocities as a function of diameter, angular

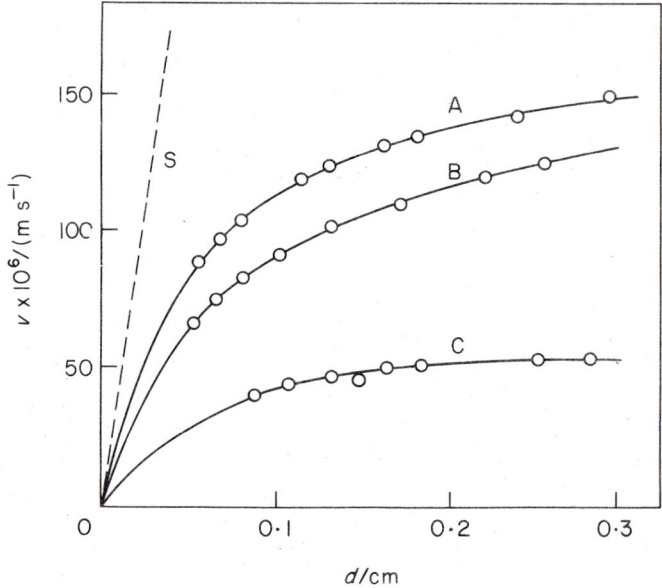

FIG. 1. Bubble velocity v along the axis of a rotating cell as a function of bubble diameter d, angular velocity Ω and inclination to the horizontal θ. Curve A: $\Omega = 632$ s^{-1}, $\sin \theta = 1\cdot292 \times 10^{-2}$; curve B: $\Omega = 886$ s^{-1}, $\sin \theta = 1\cdot260 \times 10^{-2}$. Curve C: $\Omega = 959$ s^{-1}, $\sin \theta = 0\cdot709 \times 10^{-2}$; curve S shows the corresponding Stokes's velocity at $\sin \theta = 1\cdot292 \times 10^{-2}$.

velocity of the cell and angle of tilt. The effect of Stokes's drag alone is shown for comparison.

Larger bubbles depart from spherical shape at high speed of rotation to an extent dependent on the interfacial tension; indeed the spinning cell can be used to measure interfacial tensions. For the purpose of data interpretation below, the surface area of the bubbles at any shape was calculated from microscope observations of radii and the equations of Princen et al.[8]

Direct illumination was sufficient for the (relatively) large bubbles used and the whole apparatus was contained in a thermostatted enclosure, the turbine air supply also being thermostatted to the same temperature.

2.3 RESULTS FOR SPINNING CYLINDER EXPERIMENTS

Figure 2 shows the electrophoretic force F_E at a field strength 100 $V m^{-1}$ and temperature 25°C, obtained by the methods described above, plotted

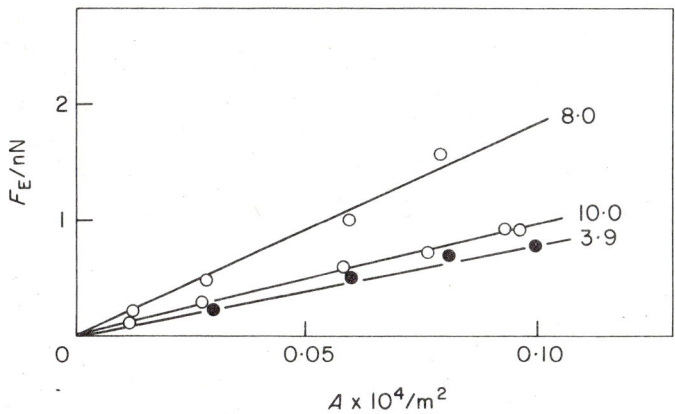

FIG. 2. Electrophoretic force F_E experienced by nitrogen bubbles in 10^{-4} mol dm^{-3} aqueous NaCl solution as a function of bubble surface area A at three pH values: 10·0, 8·0 and 3·9.

against the nitrogen bubble surface area at different pH values in CO_2 free NaCl aqueous solutions (with NaOH/HCl to adjust the pH). The total electrolyte content was kept to $\sim 10^{-4}$ mol dm^{-3} in order to limit heating of the electrolyte by the applied field (~ 100 V cm^{-1}). Motion was in all cases opposite in direction to that of the field (expressed as a negative F_E) as for a negatively charged particle.

Figure 3(a) shows the same plot for sodium dodecyl sulphate SDoS (in this case going up to 10^{-3} mol dm^{-3}), Fig. 3(b) for dodecyl trimethyl ammonium bromide (DoTAB) and Fig. 3(c) for decyl methyl sulphoxide (DMS).

In both NaCl and the surfactant solutions, F_E is proportional to the bubble surface area under constant conditions. The dependence on surface area is slightly closer than that on the square of the long radius (in the presence of some distortion at the higher Ω values) and much closer than dependence on the first power of a radius.

Since the bubbles are large (of order 0·3 mm diameter) the value of κa in 10^{-4} mol dm^{-3} of 1:1 electrolyte is at least 1000, in which range rigid

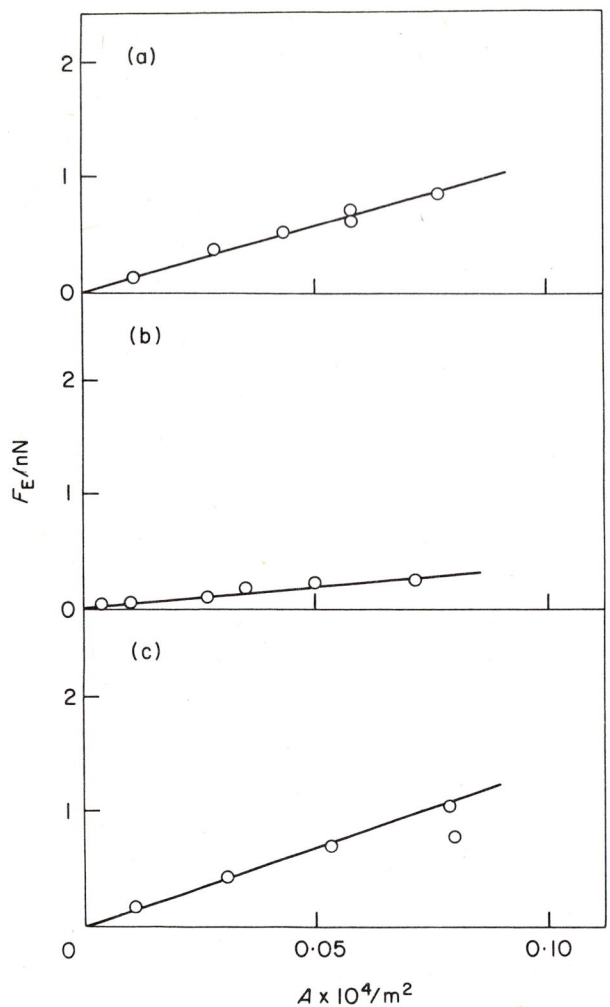

FIG. 3. Electrophoretic force F_E experienced by nitrogen bubbles in 10^{-4} mol dm^{-3} of (a) SDoS, (b) DoTAB, (c) DMS as a function of bubble surface area A.

particles in a stationary cell would obey the Smoluchowski relation

$$v = E\varepsilon\zeta/4\pi\eta \qquad (3)$$

where v is the terminal velocity of particles (independence of shape) under an electric field E in a solution of (unrationalized) permittivity ε and viscosity η. The Stokes's drag for a rigid spherical particle of radius a is $6\pi\eta a$ so that F_E, which just balances this drag, is proportional to the first power of a not the second as for the gas bubbles in the rotating cell. Eliminating η, F_E becomes $1\cdot5\ E\varepsilon a\zeta/v$ which, since v is independent of a in the Smoluchowski region, again has $F_E \propto a^1$.

If F_E were proportional to the particle charge this would give $F_E \propto a^2$; for rigid particles, however, this will only be so in the small κa ($< 0\cdot1$) region or "Henry" region. At the higher κa values the shielding effect of the electrolyte prevents this simple behaviour. At low κa the Henry equation for rigid spherical particles gives

$$v = E\varepsilon\zeta/6\pi\eta. \tag{4}$$

So that F_E, again balancing the Stokes's drag $6\pi\eta a$, is given by

$$F_E = E\varepsilon a\zeta/v \tag{5}$$

but now v is proportional to a^{-1} so that $F_E \propto a^2$. The result follows directly of course if F_E is proportional to a (uniformly distributed) surface charge.

For the gas bubbles of this work κa is very large but $F_E \propto a^2$ (or the surface area) giving an apparent contradiction. Of course the gas bubble is not a rigid particle as described above and it is to be expected that any surface charge can be shifted around the bubble both by the applied field and possibly hydrodynamically by the motion produced. Also, as for the electro-osmotic flow, the cell rotation may be producing effects not allowed for.

The related problem of whether gas bubble obey Stokes's law in a purely gravitational field has been much studied.[9] In this work the force due to the cell rotation is so large that it is not possible to distinguish between the Stokes's drag of $6\pi\eta a$ and the Hadamard[10]–Rybezybski[11] drag of $4\pi\eta a$. The

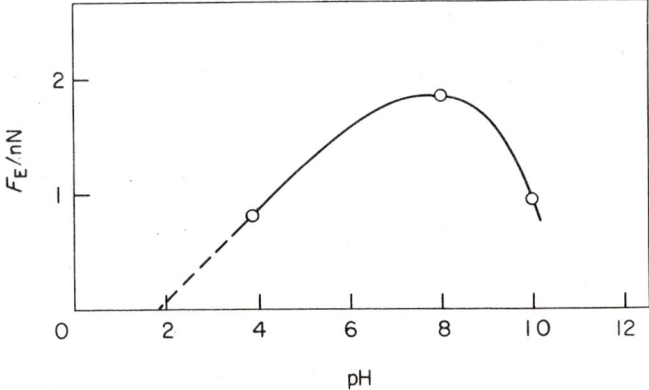

FIG. 4. Electrophoretic force F_E experienced by nitrogen bubbles in 10^{-4} mol dm^{-3} aqueous NaCl solution at bubble surface area 10^{-5} m^2 as a function of pH.

latter should obtain in the case of full momentum transfer across the interface. Booth[12] has shown that in such a case the electrophoretic mobility would be zero at any (uniformally distributed) surface charge.

Since the dependence of F_E on the radius a cannot be explained, it is clearly not admissible to use either the Smoluchowski or Henry equations to get potentials or charges from the data (though some authors have done this[5, 6]). The only nonambiguous situation is that of $F_E = 0$ which presumably corresponds to zero charge. Figure 4 shows F_E at a surface area of 0·1 cm^2 plotted against pH for 10^{-4} mol dm^{-3} aqueous NaCl solution. It can be seen that F_E and the bubble charge are going towards zero at a pH in the region of 1 to 2. The drop at high concentrations of potential-determining ion is a common phenomenon in electrophoresis.[13]

It is not easy to explain why the electrophoretic behaviour is so dependent on pH (not at all on pNa) or why such an isoelectric pH is displaced so far to acid conditions. No reasonable differential negative adsorption based on ion size can explain it and it is pointless to calculate an uninformative "specific adsorption potential" for OH$^-$. One might speculate that water molecules near the interface have a very different dissociation from those in the bulk but we are not able to give a quantitative explanation.

The behaviour in anionic and cationic surfactant solutions is at least qualitatively understandable though here it may be noticed that a bubble in sodium chloride solution at pH = 8 has as large an F_E as one in 10^{-3} mol dm^{-3} SDoS. The relatively large negative F_E for bubbles in DMS solutions may be surprising since this is a nonionic surfactant. However, DMS has a substantial dipole and orientation of this at a surface with a clearly defined potential-determining mechanism can give significant charge separation.[14] Again it is not possible to be quantitative.

3 Plane interface experiment

The spinning cylinder experiment has three main disadvantages:

1. the single bubble is likely to adsorb any surfactant impurity from the solution;

2. the force on the bubble due to the spin (equation (1)) prevents the measurement of a simple electrophoretic mobility and decreases the sensitivity with which the electrophoretic force can be determined;

3. there is an unexplained radius dependence.

For these reasons the cell shown diagrammatically in Fig. 5 was constructed in which the aqueous–gas interface could be cleaned by suction. Particles of well dialysed polystyrene latex or graphitized carbon, whose electrophoretic mobility had been previously determined in a conventional apparatus, were introduced into the solution. With an electric field applied the velocity of these particles was then determined as a function of depth into the solution. Subtraction of the electrophoretic velocity then allowed the solution electro-osmotic velocity to be determined at any level. Extrapolation to the glass–solution interface gives the familiar electro-osmotic velocity at this interface

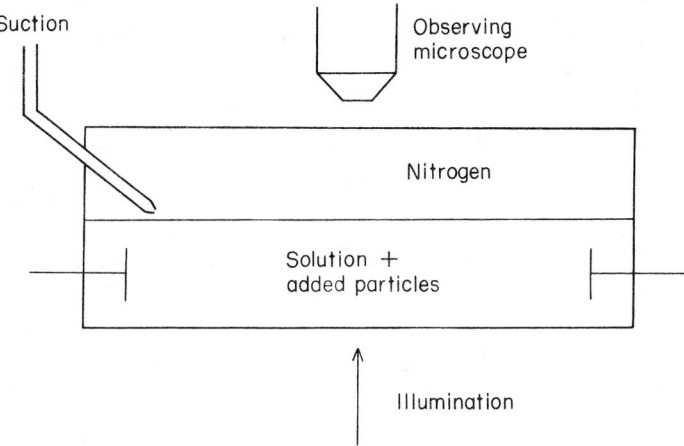

FIG. 5. Diagrammatic representation of the plane interface experiment.

which can be confirmed conventionally as a check on the validity of the calibrations of the cell in Fig. 5.

The cell was constructed of PTFE except where glass was necessary to give transparency. Two factors make the cell much more difficult to use than a flat cell with two rigid walls.

Firstly the meniscus must be taken into account when calibrating the cell with aqueous KCl to get the effective interelectrode spacing 1. The meniscus contributes to the solution cross-sectional area and the former remains constant as the latter is varied; it follows that a correction for the meniscus cross-section is required. Calculation of the meniscus area from contact angle–surface tension measurements allows this correction to be obtained.

Secondly, and more seriously, difficulties arose with respect to the condition of no net solution flow in the cell. This condition is usually assumed (justifiably) for rigid walled cells and can be easily confirmed, at least in the plane of measurement, by determinations of velocity as a function of depth. In our glass–solution–gas cell this condition appeared not always to be obeyed. Without this condition it is difficult to see how observed particle

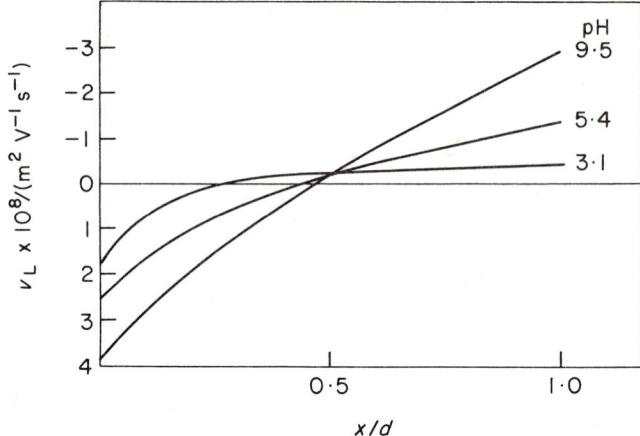

FIG. 6. Liquid velocity v_L at a distance x from the glass wall of the plane interface cell using 10^{-3} mol dm^{-3} aqueous NaCl solution of total depth d. pH values 3·1, 5·4 and 9·5.

velocities can be interpreted, so that some care was necessary to achieve it. The failure to conform to this condition is clearly related not to an actual overall flow (except in the first moments to set up a hydrostatic back pressure) but rather to the presence of lateral return flows, which seemed in practice to be related to the presence of menisci. By using an inset PTFE trough it

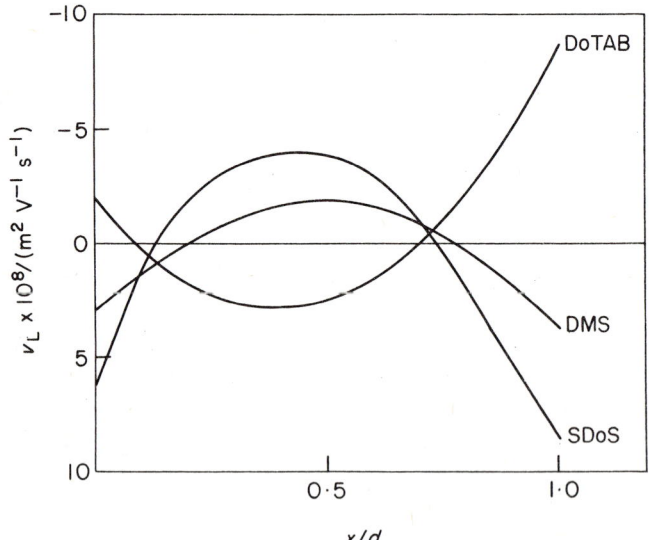

FIG. 7. Liquid velocity profiles in 10^{-3} mol dm^{-3} SDoS, DMS and DoTAB.

was possible to get effective contact angles of 90° at which the difficulty disappeared (at least in the plane of measurement). It was therefore necessary to adopt the extremely tedious procedure of determining the velocity–depth profile at various total solution depths d and interpolating to that which gave the condition of no net flow.

Using this procedure, liquid velocity profiles were determined for aqueous electrolyte (NaCl) as a function of pH (Fig. 6) and for anionic, cationic and nonionic surfactant solutions (Fig. 7). It is immediately apparent that different types of profile are obtained for the surfactant and nonsurfactant systems, in particular the occurrence of a single stationary level for the latter. These observations reflect a basic difference in the hydrodynamic behaviour of a simple air–solution interface and one containing adsorbed surfactant. The air–solution interface can show rigid behaviour in the presence of surfactant if a sufficient interfacial tension gradient becomes established. The similarity of the data for the surfactant case to that observed in glass-walled cells confirms that interfacial rigidity is indeed established in the surfactant experiments here.

3.1 RESULTS FOR SURFACTANT SOLUTIONS

Interfacial rigidity significantly simplifies the interpretation of the surfactant data in that the extrapolated liquid velocity at the interface can be regarded as a conventional electro-osmotic velocity. This allows computation of the potential (ζ) and surface change density (σ_d) associated with it from equation (3) together with

$$\sigma_d = -\frac{\kappa \varepsilon}{4\pi} \frac{2kT}{ze} \sinh \frac{ze\zeta}{2kT} \qquad (6)$$

TABLE 1

Solution	pH	$v_1 \times 10^8/(m^2\ V^{-1}\ s^{-1})$	ζ/mV	$\sigma_d \times 10^4/(\mu C\ m^{-2})$
10^{-3} mol dm^{-3} SDoS	4·7	+8·9	−114	−2·4
10^{-3} mol dm^{-3} DoTAB	5·2	−8·7	+111	+2·4
10^{-3} mol dm^{-3} DMS	5·4	+3·5	−45	−0·5

where κ is the Debye–Hückel reciprocal distance, z the valence of the (assumed symmetrical) electrolyte and k is the Boltzmann constant. Table 1 shows these quantities calculated for solutions of SDoS, DoTAB and DMS from the solution electro-osmotic velocity v_1 at the gas–solution interface, obtained in this rigid surface case by extrapolation of the solution velocity to the interface concerned. ζ and σ_d have the expected sign and ordinary magnitude which gives further confirmation to the assumption of surface rigidity.

3.2 RESULTS FOR SIMPLE ELECTROLYTES

If the surface were rigid then, as in the surfactant case, the electro-osmotic velocity v_1 could be obtained by extrapolating the liquid velocity to the interface. Quite apart from the unlikeliness of this model, any attempt to do so (Fig. 6) would give a negative v_1 implying a positively charged interface, which becomes even more positive as the pH is raised. This need not be proceeded with.

If the surface is completely unrestricted, with $(dv/dx) = 0$ at the interface, then it is readily shown that the liquid velocity v_L (at unit electric field) is is given, at zero overall flow, by

$$v_L = v_2 + (x/d)(v_1 - v_2) - (3/4)(v_1 + v_2)[(2x/d) - (x/d)^2], \qquad (6)$$

where x is the distance into the solution from the glass base at which the electro-osmotic velocity is v_2. The total solution depth is d and the channel width is taken infinite.

Equation (6) has the necessary property $v_L = v_2$ at $x = 0$, but at $x = d$ has v_L not equal to v_1 but rather $(v_1/4) - (3v_2/4)$, so that

$$v_1 = 4v_{L_{(x=d)}} + 3v_2. \qquad (7)$$

However before examining v_1, it is necessary to observe that if the gas–solution interface is completely unrestricted then the whole of the separated charge (if any) at the interface can move so that there should be no observable electrokinetic effect. This can be seen clearly from the balance of electrical and viscous forces

$$E\rho dx = - d[\eta(dv/dx)]. \qquad (8)$$

Where E is the applied field and ρ the charge density. If equation (8) is integrated right across the charge separation layer so that

$$\int \rho dx = 0 \qquad (9)$$

then, at constant η, dv/dx has the same value on both sides of the layer and no observable velocity effect is present. This corresponds to the conclusion of Booth[12] for gas bubbles under similar conditions.

Under these conditions $v_1 = v_2 = v$ so that

$$v_L = v[1 - (3/2)(2f - f^2)] \qquad (10)$$

where $f = (x/d)$. This yields $v_L = v$ at $x = 0$ and $v_L = - (v/2)$ at $x = d$, with a single stationary level at $f = 0.42$. The curves of v_L in Fig. 6 are certainly much nearer to these conditions than those in the presence of surfactant (Fig. 7). Deviations from such conditions can be expressed as $v_1 - v_2$ as in Table 2, taken from the data of Fig. 6.

TABLE 2

pH	$10^8(v^1 - v_2)/(m^2 s^{-1} V^{-1})$	$f(v_L = 0)$
3·1	+ 1·6	0·25
5·4	− 0·2	0·42
9·5	− 3·8	0·47

It may be, with the great experimental difficulties of this apparatus, that $(v_1 - v_2)$ is not differing significantly from zero. If v_1 is taken as a normal electro-osmotic velocity then the sign of $(v_1 - v_2)$ indicates that the gas–solution interface is less negatively charged than the glass–solution interface at high pH but more negatively charged at pH $= 3$. This would not be inconsistent with the bubble experiments which suggested an isoelectric point around pH $= 2$.

4 Complementary observations

AgI dispersions when negatively charged show no signs of surface coagulation, i.e. no "scum" or surface particulate layer is formed. However when the particles are positively charged (pAg < 5) such effects are always present to some extent. If the pH of the dispersion is varied the charge on the AgI particles remains virtually unchanged but the surface coagulation is modified. From pH 2 to pH 4 there is a slightly mobile scum at the surface, and from pH 4 to pH 7 this becomes distinctly more extensive and rigid. This behaviour could well be caused by the migration of positively charged particles into a negatively charged surface region, becoming more marked as the pH rises above 2.

A cloud of gas bubbles can be formed by the sudden decompression of a saturated solution of the gas in water at high pressure. The subsequent behaviour of the bubbles depends somewhat on the pH of the solution at constant total ionic strength. As the pH is raised from 2 to 7 the clearance time is increased by a factor around 2·0. This also is consistent with the bubbles undergoing coalescence at low pH to form larger, and therefore faster rising, bubbles while at high pH the coalescence rate may be reduced by the bubble charge.

5 Conclusions

Both the spinning cell and flat interface experiments present great difficulties. The spinning cell technique obviously requires more sophisticated apparatus but is otherwise easier to handle.

In the presence of a rigid surface layer, the flat interface experiment can yield electrokinetic potentials and associated charges. For the cases studied the values obtained are reasonable. With a free surface, interpretation is doubtful.

Full interpretation of the spinning cell data requires an explanation of the anomalous radius dependence of the electrophoretic force. However the isoelectric point, identified with zero force, is in principle unequivocal though subject to inevitable doubts with respect to the effect of impurities. In these experiments the isoelectric point was at low pH with the bubble behaving as if negatively charged at ordinary pH values.

References

1. Taylor, A. J. and Wood, F. W. (1957), *Trans. Faraday Soc.* **53**, 523.
2. Quincke, G. (1861). *Pogg. Ann.* **CXIII**, 513.
3. Alty, T. (1926). *Proc. Roy. Soc.* **112A**, 235.
4. McTaggart, H. A. (1927). *Trans. Roy. Soc. Can.* **21**, 249.
5. Whybrew, W. E., Kinzer, G. D. and Gunn, R. J. (1952). *J. Geophys. Res.* **57**(4), 459.
6. Cichos, C. (1971). *Neve Bergbautech,* **1** (12), 941.
7. Stewartson, K. (1952). *Proc. Cambridge Phil. Soc.* 168–177.
8. Princen, H. M., Zia, I. Y. Z. and Mason, S. G. (1967). *J. Colloid Interface Sci.* **23**,99.
9. Levich, V. G. (1962). "Physicochemical hydrodynamics", p. 395. Prentice Hall, New York.
10. Hadamard, J. S. (1911). *Compt. Rend.* **152**, 1735.
11. Rybezynski, W. (1911). *Bull. Acad. Sci. (Cracovie)*, **1**, 40.
12. Booth, F. (1951). *J. Chem. Phys.* **19**(11), 1331.
13. Smith, A. L. (1973). *In* "Dispersions of Powders in Liquids", 2nd ed., p. 101. Applied Science Publishers, London.
14. Smith, A. L., McDowell, F. W. and Fairhurst, D. (1972). VI Int. Congr. für grenzflächenaktive stoffe, 679.

Discussion

Jameson (*Imperial College, London, England*) We have recently measured the electrophoretic mobility of small gas bubbles (*c.* 35 μm in diameter) using a Rank Bros. electrophoresis apparatus. A quartz cell of rectangular cross-section (1 mm × 10 mm) was used, and platinum electrodes were sealed into the top and bottom of the cell in such a way that when a potential of 12 V was applied, a small gas bubble was generated electrolytically at the bottom of the cell. Its path as it rose could be tracked by racking the stage holding the cell, and if a horizontal potential gradient was applied in the usual way, its lateral velocity could be measured and a mobility calculated. The electrodes were placed so that bubbles rose up the stationary level of the cell.

Mobilities were measured in connection with flotation experiments. In the presence of CTAB (5×10^{-5} mol dm^{-3}) and sodium sulphate (10^{-3} mol dm^{-3}) at pH = 8·0, the bubbles were found to be positively charged, with mobilities of $4·5 \pm 0·1 \times 10^8$ (m^2 V^{-1} s^{-1}). The technique was not easy to apply, mainly because of the difficulty of keeping a bubble in the centre of the field of vision in the microscope while it was moving upwards. It could undoubtedly be improved by a motorized drive. Nevertheless the technique seems promising.

Exerowa It is recognized that theories of the air–solution electrical double layer are insufficiently understood and are contradictory. I am therefore agreeably surprised with the interesting results obtained concerning the pH dependence of the electrophoretic force of aqueous sodium chloride solutions and the consequent values of the isoelectric point. Using a different method, depending on measurements on equilibrium films and the application of DLVO theory, we have found an analogous dependence of double layer on pH.

Using a 10^{-4} mol dm^{-3} solution of sodium chloride, the same ionic strength as used by Huddlestone and Smith, we found the isoelectric point to be greater than pH 4·5. This is of great interest because it demonstrates that the surface charge of such solutions is largely determined by the adsorption of H$^+$ and OH$^-$ ions.

If for reasons of simplicity we consider that the potential is determined by the adsorption of these ions and that at pH 4·5, $\psi_0 = \psi'_0$ ($\psi'_0 \cong 0$),

$$c_{H+} \exp\left[(E\psi'_0 + \theta_{H+})/kT\right] = c_{OH-} \exp\left[(-F\psi'_0 + \theta_{OH-})/kT\right]$$

$$RTn\left(c_{H+}/c_{OH}\right) = \theta_{OH-} - \theta_{H+}$$

Calculating for pH 4·5 at 25°C gives

$$\theta_{H+} - \theta_{OH-} = 28·6 \text{ kJ mol}^{-1}$$

for the difference in adsorption potentials. This shows that the potential difference is due to the adsorption of OH$^-$.

These approximate calculations only show a trend but do demonstrate the influence of OH$^-$ and allow us to postulate that for a more complex system (e.g. the adsorption of nonionic surfactants) the potential would also be pH dependent.

From examination of the influence of type and concentration of ionic surfactant on the isoelectric point we obtained the following values: DMS, 2·9; DP 20, 3·4; $C_{12}(Eo)_n$ 3·7.

Huddlestone and Smith It is interesting that Exerowa has obtained results similar to ours for the isolelectric point at the gas–water interface, by a different experimental technique.

I doubt, however, whether it is illuminating to calculate an apparent specific adsorption potential without relation to detailed mechanisms, and the obtaining of such a value cannot be taken as a verification of the assumptions made.

12

The use of an aqueous foam as a fibre-suspending medium in quality papermaking

V. W. PUNTON

Wiggins Teape Research and Development,
Butler's Court, Beaconsfield, Buckinghamshire HP9 1RT, England

Summary

An aqueous foam has proved to be an excellent suspending medium for paper fibres. The pseudoplastic nature of the foam allows for the dispersion of relatively long fibres under high shear conditions at consistencies of 0·5 to 1·5 per cent on a weight for weight basis. Under low shear conditions, such as during the transfer of stock to the paper machine, the fibres are in effect "frozen" in their dispersed state. On the paper machine wire the foam–fibre mixture can be drained rapidly.

Excellent fibre dispersion can be achieved in the finished paper even with long fibres.

In an unpressed state paper made by the Radfoam process has the characteristics of high bulk and low strength when compared with a comparable water-laid paper. Pressing raises the strength to the water-laid level whilst correspondingly reducing the bulk. A wider range of bulk–strength relationships is therefore available to the papermaker using this process.

1 Introduction

The papermaker is concerned with the uniformity of the paper he can make from the pulps available. Nonuniformity becomes particularly severe when, to achieve satisfactory properties in the finished paper, long fibres have to be used. Partial solutions to this problem have been achieved by the use of high-dilution processes and by the use of deflocculants.

Another approach is to replace the water used as a suspending medium by foam, making use of the latter's increased viscosity to dampen the relative movement of fibres and so their flocculating tendency.

This is the basis of a new process, the Radfoam process, which is described in this paper.

2 The problem of fibre dispersion

The most critical stage in the papermaking process is the transference of the fibre suspension, from pipelines, onto the papermachine wire. The distribution of fibres within the finished sheet will reflect the distribution of fibres in the suspension immediately prior to the suspension reaching the wire. The behaviour of fibres suspended in water has been studied by many workers, notably Mason. The brief account below[1] should help to explain why this stage in the process continues to challenge the skills of the machine designers.

Elongated particles, such as fibres suspended in a liquid shear field, experience forces that tend to make them rotate. Such forces are set up in fibre suspensions during pipe flow and during the transference of the suspension to the wire. Above a critical concentration $6/A^2$, where A = (fibre length/fibre radius), the fibres will collide and begin to form bundles known as flocs. As the value of A for woodpulp fibres is about 60 to 300, this critical concentration is between 0·16 and 0·007 per cent. If longer fibres are to be used, such as cotton or synthetics, which have fibre lengths up to 20 mm (values of A up to several thousand) the critical concentration would be several orders of magnitude lower.

The papermaking process, in recent years. has caught the attention of would-be manufacturers of nonwoven fabrics who see in the process the ability to make such fabrics at normal papermaking speeds of 300 to 2000 ft min^{-1}. A nonwoven fabric is a material which ideally has the properties of a textile but with a fibre distribution similar to paper. The textile-like properties are obtained by incorporating long synthetic fibres, such as rayon or nylon, into the sheet. Consequently, the desire to make nonwoven fabrics on a paper machine has highlighted the problems of fibre flocculation in fibre suspensions.

There are two solutions to the problem which follow directly from what has been said above:

1. Reduce the fibre concentration to below the critical value.

2. Increase the viscosity of the suspending medium so that the rotational forces, acting on the fibre, are smaller than the viscous forces, thus reducing the likelihood of fibre collisions.

Solution 1 is just feasible with short-fibred woodpulps. Machines, such as the Voith Rotoformer, have been designed to deal with fibre suspensions at concentrations of about 0·01 per cent. Owing to the vast quantities of water that have to be removed to produce a dry sheet, the speed of the machine, and thus its output, is limited. When dealing with long fibres the quantity of water involved would be so enormous that this solution cannot be seriously considered.

Solution 2, in fact, has been used since the earliest days of papermaking. Traditionally, Chinese and Japanese papermakers have included natural gums in the suspension of fibres in water. These gums increase the viscosity of the suspending medium and thus restrict the movement of the fibres. This technique undoubtedly leads to an improvement in the uniformity of distribution of fibres in the finished sheet but the method has one serious disadvantage. The increased viscosity of the suspension means that rapid drainage of the liquid from the network on the wire cannot be achieved and thus the machine speed must be reduced if adequate drainage is to be obtained by the time the network leaves the wire section of the machine. When using long fibres a considerable increase in viscosity would be required to restrict fibre movement sufficiently to prevent flocculation. The viscosities involved would mean that adequate drainage could only be obtained at machine speeds which would be far too slow for economic production.

It is clear, therefore, that although one or other of these two solutions may be employed in the production of short-fibred sheets at a relatively slow rate, neither solution is suitable if one wishes to use long fibres and yet still achieve high production rates.

3 The Radfoam process

A new method of papermaking, which overcomes the problems outlined above, has been developed at the Wiggins Teape Research & Development Laboratories, Beaconsfield.[2] Fibres are delivered to the wire of a conventional Fourdrinier paper machine in suspension in an aqueous foam. The laboratory has called the process the Radfoam process after Mr B. Radvan who headed the development team.

The process originated as a by-product of research which had the objective of increasing the understanding of the phenomena involved in the making of a sheet of paper. The main conclusion of this work was that papers made on a Fourdrinier machine have, necessarily, a layered structure. A non-layered 3-D structure may only be obtained if the paper is made from very high concentrations of fibres in water. Attempts were made to produce a 3-D structure by making a very bulky pad of fibres from an uncollapsed suspension of fibres in froth. The experiments were unsuccessful. It was noticed, however, that the froth sometimes separated out into three layers: top froth, liquid at the bottom, and a middle foam containing fibres. Further investigation showed that good fibre dispersion could be achieved using this thick, creamy, viscous, middle foam. Laboratory experiments showed that long fibres could be dispersed in the foam and that small handmade sheets made from foam–fibre mixtures had an exceptionally uniform fibre distribution compared with sheets made from water–fibre suspensions. Drainage of the foam proved to be no problem. The quantity of water to be drained was actually less than in normal circumstances because about two-thirds of the suspending medium was air. In other words, this foam seemed to have solved the problem of fibre flocculation when dealing with long fibres and appeared to present no insurmountable problems to running machines at normal speeds.

Development continued to scale up the process, firstly to pilot plant scale and then to full-scale production machines. No major problems were encountered.

By the middle of 1970 the process had been successfully demonstrated on a production machine and laboratory work was concentrated on process and product development.

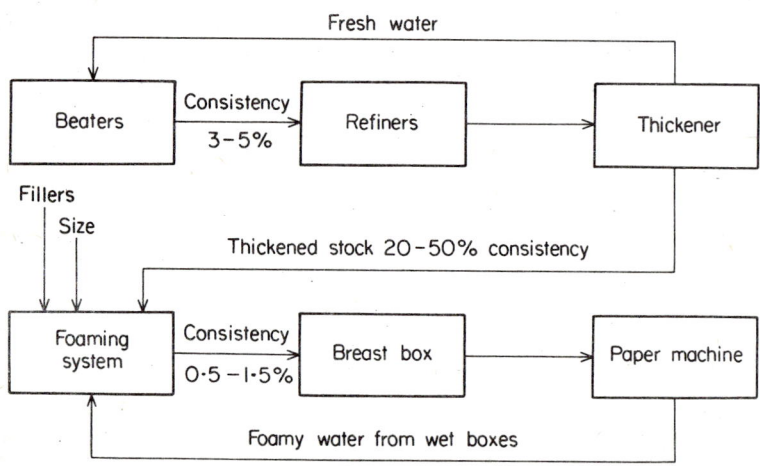

FIG. 1. Typical flow chart of Radfoam process.

A simplified flow chart of the Radfoam process is shown in Fig. 1. Briefly, the system works as follows. The pulp is mechanically treated, as in conventional papermaking, in a beater and a refiner. The amount of treatment given depends on the raw material and on the desired properties of the finished product. A suspension of the treated fibres in water is passed to a thickener which removes some of the water to yield a suspension which contains 20 to 50 per cent fibres. The thickened suspension is then passed into the foaming system. Water removed at the thickener is recycled for use on the next batch of pulp in the beater.

Several systems for producing foam have been investigated but currently an in-line foamer of novel design is used.

Water, added at the foamer, is that water which has been drained out of the paper on the wire of the machine. In other words, the drained water is recirculated. Surfactant solution is also metered into the system at the foamer.

The foam–fibre mixture leaves the foamer containing about 65 v/v air and from 0·5 to 1·5 w/w fibres.

The foam is passed onto the wire of the machine. Water is drained from the fibre web by means of vacuum boxes. Once the water has been drained off, the papermaking process reverts to the normal system.

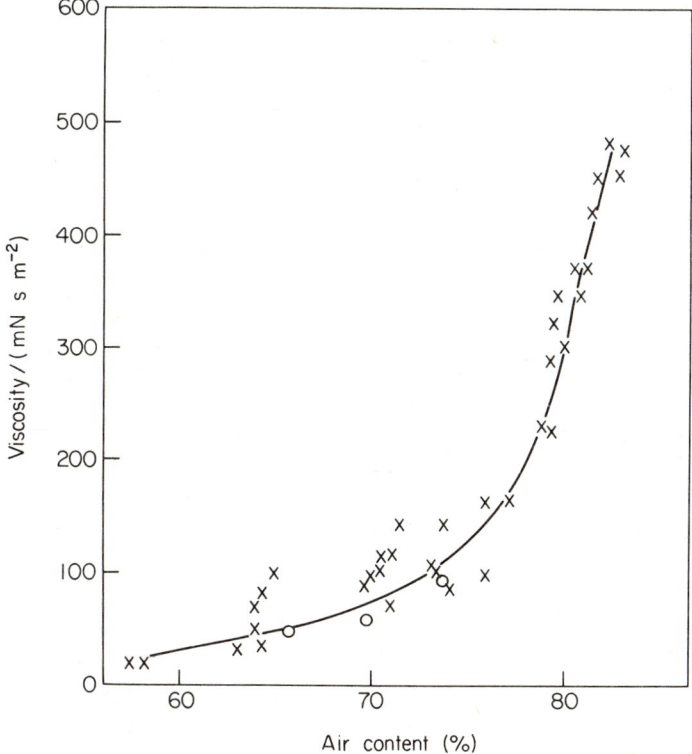

FIG. 2. Relationship between foam viscosity and air content after 10 s drainage.

G

4 Properties of the foam and foam–fibre mixtures

The relationship between air content, the volume percentage of gas in the foam, and viscosity at constant shear rate is shown in Fig. 2. The foam contains 0·1 per cent of an anionic surfactant. Results from measurements,

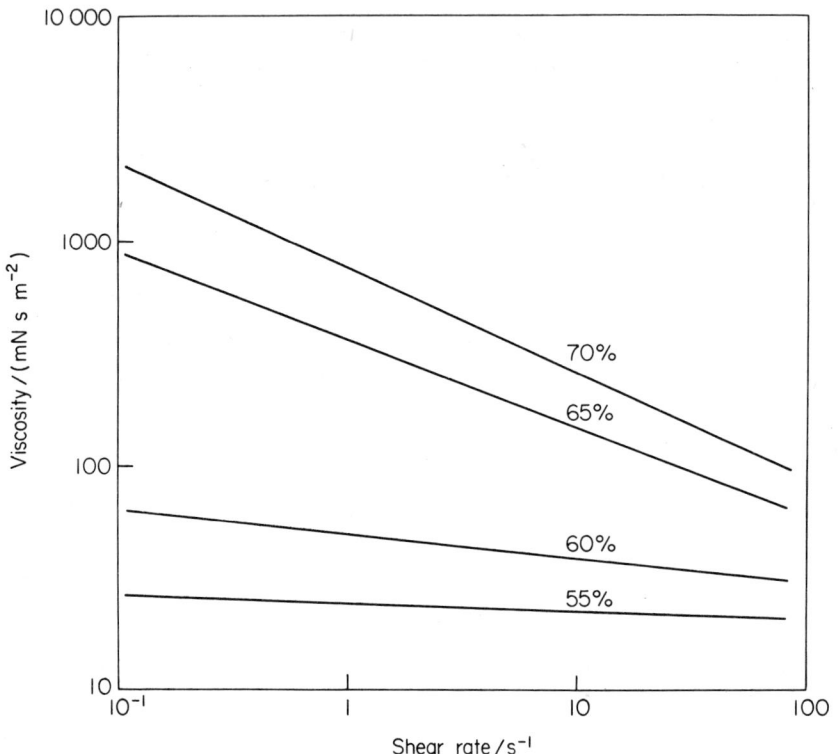

FIG. 3. Logarithmic relationship between viscosity and shear rate for foam containing 0·1 per cent anionic surfactant.

on a similar foam, of viscosity at varying shear rates and over a range of air contents, Fig. 3, indicate that it approximates closely to a "power-law fluid". Figure 4 illustrates the effect of shear rate on viscosity for a foam–fibre mixture plotted on a similar graph to Fig. 3. It can be seen that the suspension retains substantially the straight line relationship characteristic exhibited by the foam. Thus this foam–fibre mixture can be considered to be pseudoplastic in its behaviour.

The range of air contents over which this pseudoplastic behaviour extends appears to be from approximately 55 per cent to 75 per cent. Thus the process in general is operated within these limits. Experience has shown that

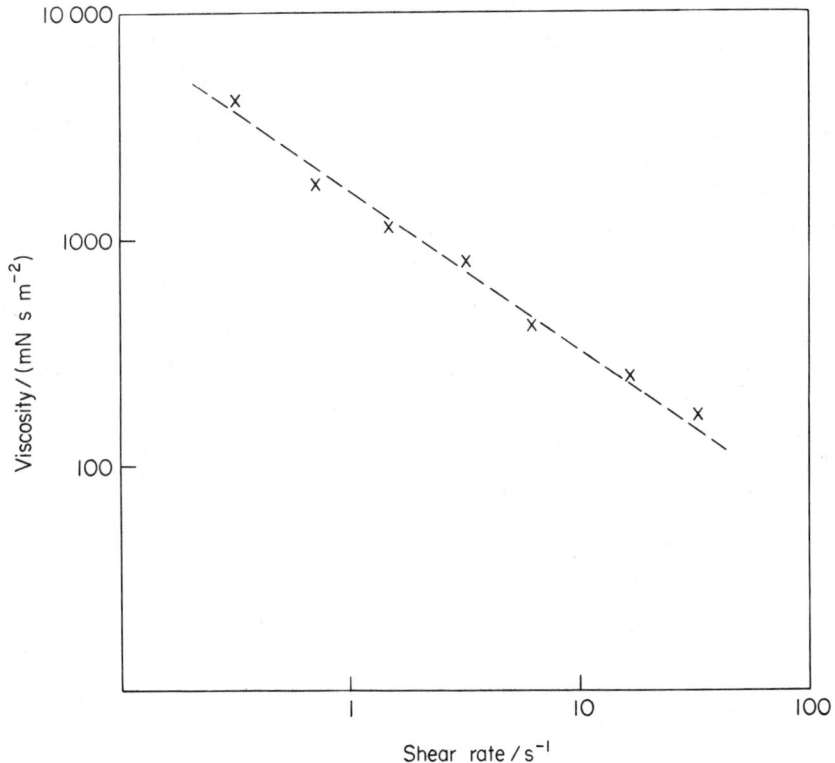

Fig. 4. Logarithmic relationship between viscosity and shear rate for foam containing 0·1 per cent anionic surfactant and hardwood pulp at 1 per cent consistency. Foam at 65 per cent air content.

for most fibres optimum fibre dispersion is achieved when foam air content is approximately 65 per cent.

Figure 5 is a photomicrograph showing bubbles in a typical foam as used in the Radfoam process. The diameter of the bubbles range from 20 to 200 μm with a mean diameter of approximately 50 μm.

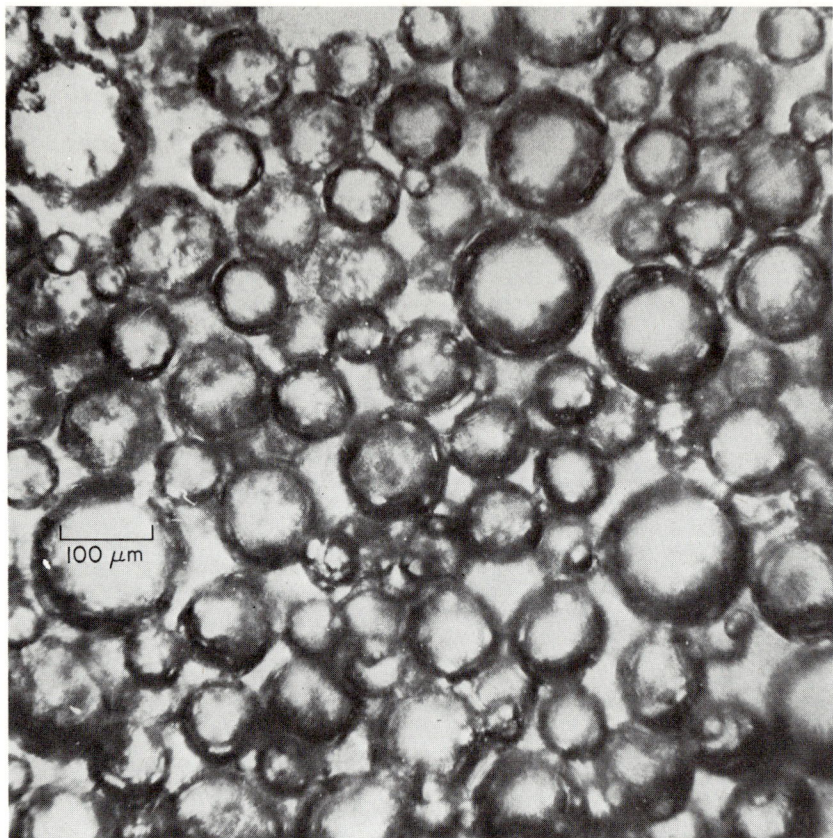

FIG. 5. Photomicrograph of foam bubbles taken from foam used in the Radfoam process.

5 Paper properties

5.1 FIBRE DISTRIBUTION

Just how effective foam is in the production of uniform dispersions of fibres can best be seen by examination of the paper produced by the process. A study of the relationship between the mass density distribution in papers, determined by β-ray transmission, and the size of the inspection area was made by Corte.[3] The technique examines the variation in mass density between paper samples of various areas. For small samples, of the order of 1 mm², the variation is expected to be high because the number of individual fibres in any particular sample will vary considerably. As the sample size increases for conventional papers, this variation remains fairly high because flocculation results in a high level of variability of the number of fibres in any particular sample. It is not until a sample size is reached which is relatively large in comparison with the floc size, at about 5 mm square inspection area, that the variation within samples averages out so that the between-sample variation decreases to lower values. However, it should be noted that the variability does not decrease to zero, and in some cases it remains at a relatively high value.

Figure 6 shows the mass density distribution for a representative selection out of 24 commercial papers examined, with the average fibre lengths of the

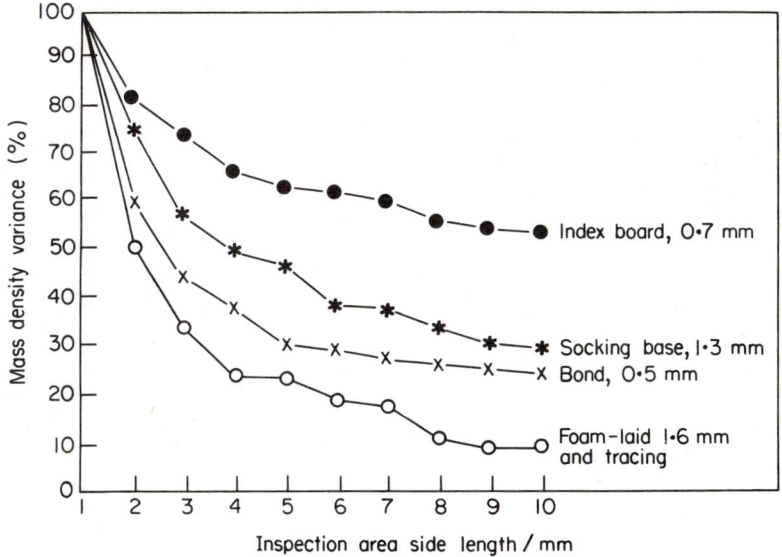

FIG. 6. The distribution of mass density for a range of commercial papers.

furnishes used marked on each curve. The variance at $1 \, \text{mm}^2$ has been normalized to 100 per cent. The higher the curve on this graph, the higher is the variability of the paper at the particular inspection area. The foam paper was made on a production machine converted to foam and running a rag furnish with average fibre length of 1·6 mm. Its variability is the lowest at any inspection area and it is only equalled by a tracing paper, the fibre length of which is very much smaller.

The formation improvement can be appreciated usually by comparing the two micrographs of conventional and foam handsheets of an unbeaten softwood kraft shown in Fig. 7.

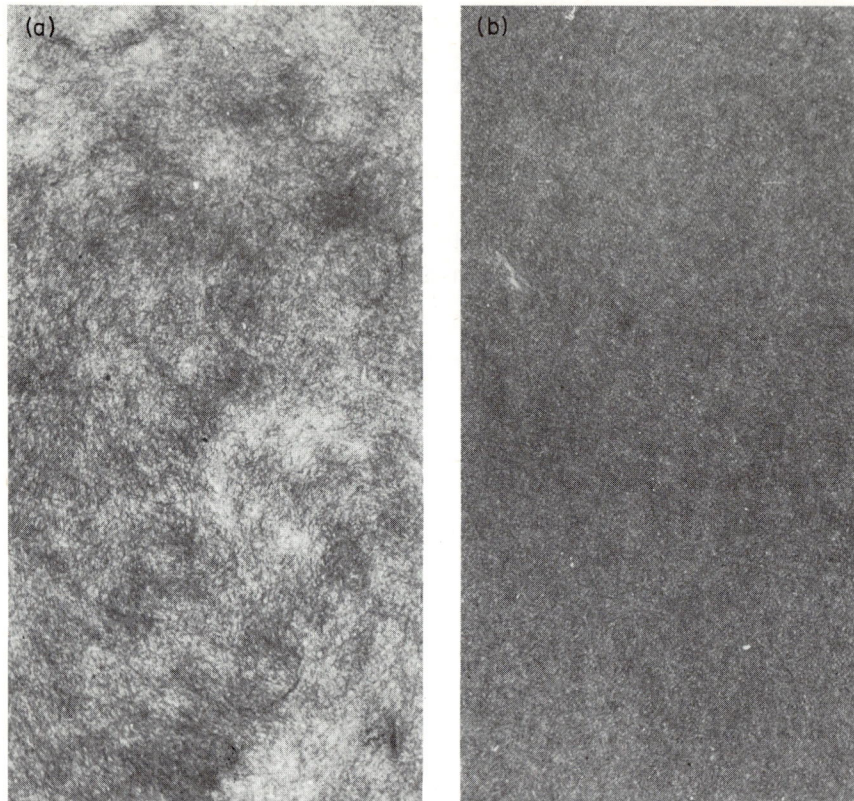

FIG. 7. Softwood Kraft handsheets: (a) water-laid from 0·3 per cent w/w stock; (b) Radfoam-laid from 1·0 per cent stock.

5.2 SURFACE TENSION

Campbell's classical explanation[4] for the bringing together of fibres to sufficiently close proximity to allow inter-fibre bonding relies on the surface tension of the suspending medium. Several authors[5,6] have shown that reducing the surface tension leads to a corresponding decrease in the Campbell forces.

At the surfactant concentration used in the foaming process, the surface tension of the continuous phase is less than one-half that of pure water. As a result, the Campbell forces exerted in foam sheet consolidation will be reduced correspondingly, and there will be a significant reduction in the number of fibres which are drawn sufficiently close to their neighbours for bonding to occur.

The effects of the surface tension changes on sheet properties are shown in Table 1, the first two lines of which show the properties of 60 g m^{-2} handsheets made (a) from a suspension of fibres in water and (b) from surfactant in water solution, without foaming, at the surfactant concentration normally used for the new foaming process.

5.3 PRESENCE OF BUBBLES

When a sheet is formed from foam, it contains many bubbles which can survive couching and the early phases of drying. Figure 8 is a photomicrograph of two handsheets which were removed from the wire of a handsheet machine immediately after formation. The water-laid sheet (a) is saturated

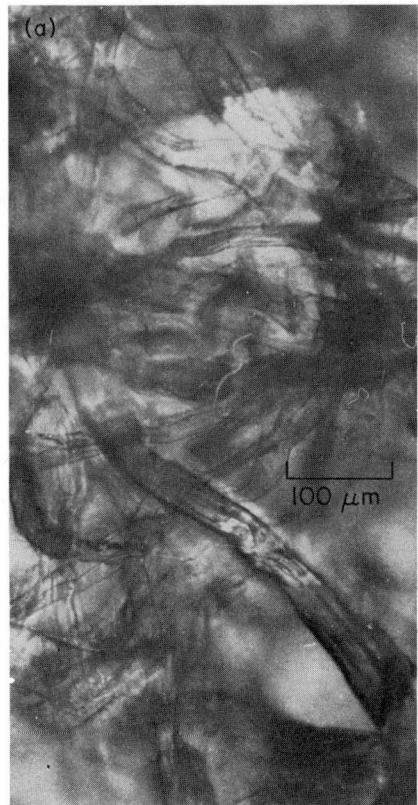
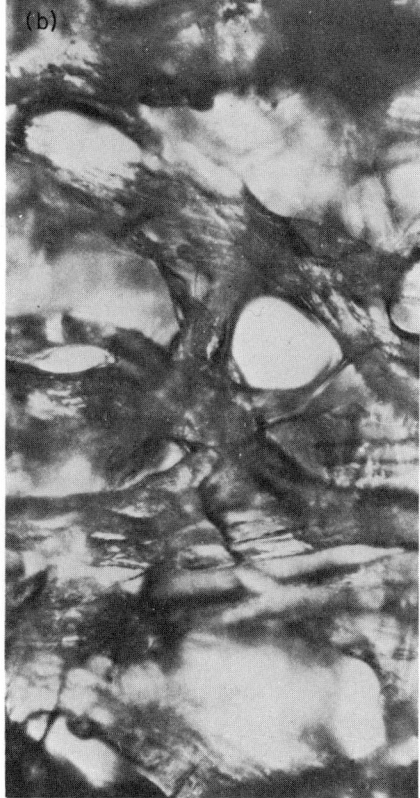

FIG. 8. Photomicrograph of handsheets immediately after formation: (a) water-laid; (b) Radfoam-laid.

with water. The foam-laid sheet (b) contains air bubbles. The foam-laid sheet was allowed to dry at room temperature. The photograph (Fig. 9), taken after 20 min. shows that the bubbles persist and under production conditions may survive into the early stages of drying. It seems likely that the bubbles space fibres apart through the thickness of the sheet so that a

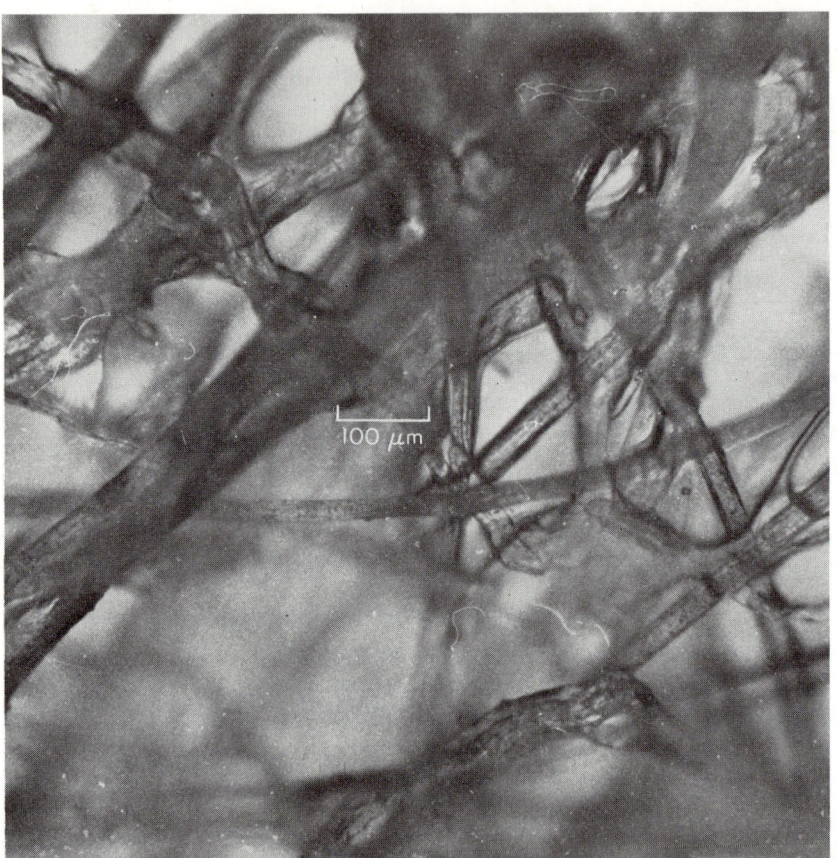

FIG. 9. Foam-laid handsheet 20 min after formation.

multi-layered bubble and fibre sandwich is formed. When the bubbles dry out, a very open bulky structure is left.

Figure 10 shows photomicrographs of cross-sections of a conventional water-laid sheet (a) and a foam-laid sheet (b). The foam-laid sheet is more uniform and the increased thickness, at the same basis weight, is readily apparent.

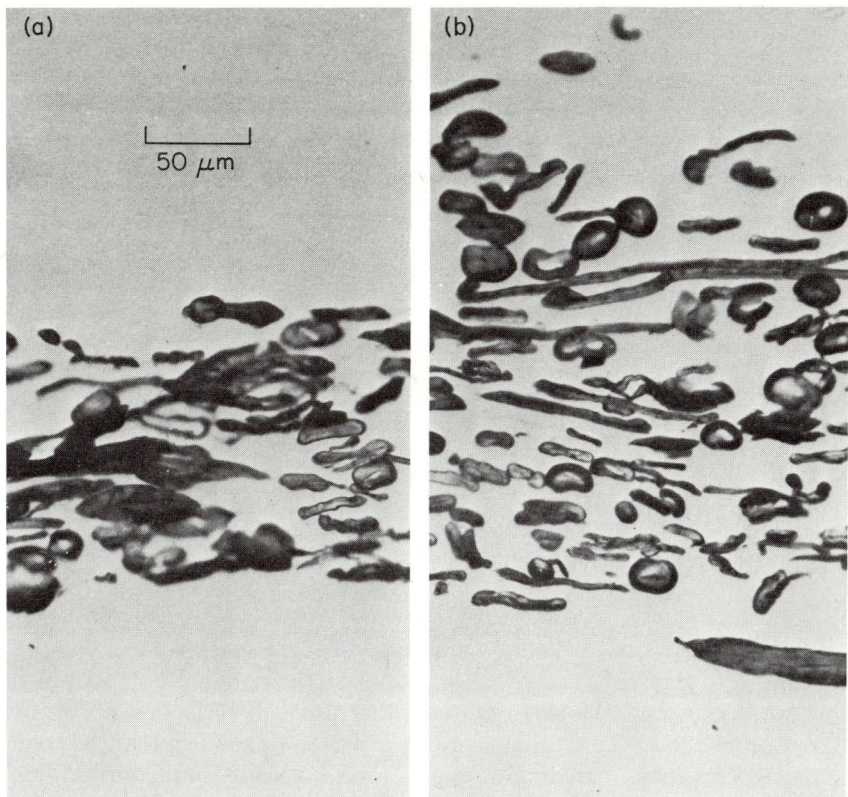

Fig. 10. Micrograph of cross-sections of softwood pulp laid handsheets: (a) water-laid; (b) Radfoam-laid.

Additional modifications to sheet properties arise when the suspending medium is foam and the spacing effect of bubbles is involved as shown in Table 1. The third line (c) shows the results from a foam-laid sheet. The tensile strength is lower than that of the surfactant sheet, while the bulk and porosity have been increased.

5.4 ABSORPTION OF SURFACTANT

The final factor which could play a part in the determination of foam sheet properties is the absorption of surfactant onto fibre surfaces. Various authors[5,7] have examined absorption of surfactants onto papermaking fibres, but their conclusions on the effects of paper properties are varied, in some cases an improvement in strength was indicated, in others strength was lost.

TABLE 1

Effect of various treatments on paper properties
(Softwood sulphite handsheets, $60 \, g \, m^{-2}$)

	Bulk ml g^{-1}	Tensile strength kg per 30 mm	Stretch %	Potts porosity l min^{-1}
	Disintegrated pulp			
a. Water	4·6	0·59	2·0	52
b. Surfactant	5·2	0·24	1·5	57
c. Foam	5·4	0·22	1·3	60
d. Foam, wet pressed (690 kPa, 20°C)	2·4	0·49	1·4	34
e. Foam, hot pressed (690 kPa, 105°C)	1·4	1·39	1·3	6
	Beaten pulp			
f. Surfactant	4·8	0·58	1·9	47
g. Foam	5·5	0·48	1·6	60

While we have not studied a wide range of surfactants, we can definitely say that, with the surfactants we normally use, the bonding potential of the fibre surfaces is not seriously reduced either through surfactant absorption or any other effects. This conclusion is supported by the fact that the trends of sheet properties, including strength, resulting from surfactant addition and foaming can be reversed by pressing to regain the lost areas of fibre contact at which bonding occurs.

Lines (d) and (e) of Table 1 show the properties of pressed sheets. Wet pressing (690 kPa, 20°C) regains almost all the strength lost by surfactant and foaming although bulk is lost as would be expected. By drying foam sheets under pressure (690 kPa, 105°C), the original strength can be surpassed with a corresponding loss in bulk. After hot pressing, the strength of water, surfactant, and foam-laid sheets are identical.

We can conclude that, apart from the improved formation resulting from foam dispersion, the particular combinations of properties associated with the foaming process are the result of physical effects, surface tension and bubble spacing, which are operative during sheet formation, consolidation, and drying. Chemical effects do not appear to be important.

To summarize this effect, it can be said that foam positions fibres further apart than in conventional papermaking so that bulk is generated. However, a similar effect on a smaller scale appears to compensate for the resulting decreased fibre contact area by extending the bonding range of each fibre so that the contacts, or near contacts, which are made are more effective. As a result, bulk is maintained without loss of strength.

References

1. Mason, S. G. (1954). *Tappi*, **37**(11), 494.
2. Radvan, B. and Gatward, A. P. J. (1972). *Tappi*, **55**(5), 748.
3. Corte, H. K. (1969). *Das Papier*, **23**(7), 381.
4. Campbell, W. B. (1933). *Tappi*, **16**, 452.
5. Nicolaysen, V. B. and Borgin, K. (1954). *Nor. Skogind.* **8**, 260.
6. Dixson, H. P., Jr. (1940). *Tappi*, **23**, 192.
7. Touchette, R. V. and Jennes, L. C. (1960). *Tappi*, **43**(5), 484.

Discussion

Creak (*Polystar Technical Service Centre, Antwerp, Belgium*) Could you please tell me if foam collapse occurs when the dispersed fibre is on the wire.

If so, could you imagine that uniform deposition of fibres on the wire by controlled foam collapse would improve the properties of the finished paper.

Punton The foam does not collapse on the wire of the paper machine but is removed by the action of suction boxes beneath the wire in a similar manner to the way water is removed from the paper web in conventional papermaking. Using this process results in the paper having a more even fibre distribution than would be obtained using conventional foaming methods. This improvement becomes more and more noticeable for longer fibre lengths.

The type of controlled foam collapse you talk of would be difficult to achieve at the sort of production speeds currently employed on paper machines.

Akers (1) What techniques were used to measure viscosity?
(2) What rate of agitation is used in the breast-box?

Punton Viscosity was measured using a Brookfield Viscometer. The choice of this instrument was made necessary for two reasons. Firstly, measurements need to be made quickly and easily. Secondly, measurements of foam viscosity had to be made on the paper machine in a mill environment as well as in the laboratory. The viscometer had therefore to be portable and capable of being used in relatively inhospitable situations.

On the two production machines converted to run the Radfoam process no agitation is applied in the breast-box. However, on our high-speed pilot plant we are able to adjust the degree and scale of turbulence to suit the fibre type being used.

Epstein Have you any knowledge or experience in suspending any particulate solids in foam other than fibres? Is the suspending of woodpulp fibres in water–foam stable? Have you any experience in suspension of noncellulose fibres in foam?

Punton Our own experience of suspending particulate solids in foam is restricted to filler materials such as clay and titanium dioxide. However, one of our licencees is using the Radfoam process with the prime objective of achieving uniform dispersion of particulate solids.

The suspension of cellulose fibres in foam is sufficiently stable to enable the foam–fibre mixture to be transported from foamer to wire without a significant degradation in the foam quality taking place.

The noncellulosic fibres we have successfully used in the process include rayon, nylon and other synthetics, asbestos and other mineral fibres. Fibres up to 12 mm are commonly used whilst certain fibres up to 25 mm have been produced into webs.

Roberts, K. What is the drainage rate of "foamed" papers on the paper machine wire compared with "normal" paper?

Punton Foam can only be removed from the web by the action of suction boxes. So in that sense the drainage capacity of the machine needs to be higher than for conventional papermaking.

But the rate of drainage given that capacity is high and we have run our pilot plant at 900 m min^{-1} which is a drive speed limitation. We are confident that higher speeds of operation are possible.

Beresenyi (*Textilipari Kutato Intezet, Budapest, Hungary*) Have you any experience with longer fibres? I mean with fibres for textile webs?

Punton Originally the process was developed so as to be able to manufacture nonwoven textiles on a paper machine. Nonwovens with exceptionally good uniformity of fibre distribution have been produced which included fibres up to 12 mm in length.

Zichy I think that the generation of microbubbles by added alcohol or by dissolving excess air and subsequently reducing the pressure on the liquid proceeds by the same mechanism of spontaneous homogeneous nucleation as a state of supersaturation is reached. Alcohols reduced the solubility of gases in water; perhaps electrolytes would act in a similar manner.

Melville The reasons for the reduction in bubble size on addition of electrolytes or alcohols has not yet been unequivocally established. Prevention of coalescence, as has been shown to occur with salts and alcohols, will obviously be important in reducing mean bubble size although, as stated, there could also be other effects taking place.

If it could be established that the majority of the microbubbles formed throughout the duration of a flotation experiment arose from homogeneous nucleation then obviously microbubble formation would be attributable to supersaturation effects. However it is not clear to me why a nonequibrilium supersaturation situation should exist other than for a short time after the addition of ethanol, whereas the size of the microbubbles remained constant with time over the order of hours.

13
Experimental methods for the study of fire-fighting foams

J. G. CORRIE

Fire Research Station,
Borehamwood WO6 2BL, Hertfordshire, England

Summary

This paper describes studies on the use of foam in the extinction of flammable liquid fires, and experimental techniques which can be used to evaluate its performance. These include four methods for producing foam in the laboratory, and methods for measuring the expansion, drainage rate and shear stress at controlled shear rate of the foam produced. Areas in which further knowledge of foam performance are required are suggested, and 24 references are given.

1 Introduction

Flammable liquid fires are often difficult to extinguish using water alone, and the method most widely used is to apply the water in a foam, which will float upon the surface of the fuel. The foam extinguishes the fire by a combination of effects. The principal effects, in probable order of importance are:

1. interception of the radiant heat from the flames;
2. cooling of the fuel surface;
3. mechanical sealing of the fuel surface preventing the escape of vapour;
4. dilution of the oxygen concentration over the fire by steam from evaporating foam;
5. isolation of the fuel surface from the oxygen supply in the air.

A substantial research effort is being made in many countries to improve the efficiency of fire-fighting foams to reduce the high cost of providing fire protection for petroleum storage and similar risks since the size of foam production equipment needed can be very large, e.g. in order to extinguish a 90-m diameter petroleum tank a rate around 30 $m^3 min^{-1}$ of water as foam is necessary. Such protection against fire must be provided even though it is hoped that it will never be needed.

Foam is produced by using a proprietary foam liquid which is added to the water to give a resultant concentration between 2 and 6 per cent.

The foam liquids are classified according to their principal active constituent. The most important ones are:

1. Protein foam liquid—solution of hydrolysed protein.
2. Fluorochemical foam liquid—solution of several surfactants with perfluorinated groups which is referred to in the USA as aqueous film-forming foam (AFFF).
3. Fluoroprotein foam liquid—protein liquid containing perfluorinated surfactants which are similar, but not identical, to those in the fluorochemical foam liquids.
4. All-purpose protein foam liquid—protein foam with additives to prevent destruction by polar solvents.
5. Synthetic foam liquid—usually a lauryl alcohol derivative.

In addition to the surfactant the proprietary foam liquids contain other ingredients to make them more suitable for their intended use. Examples are antifreeze, viscosity reducers, antibacterial agents, antoxidants, anticorrosives, foam stabilizers, substances to increase resistance to radiant heat or to increase resistance to hot fuel and dyestuff to provide brand recognition. In addition extraneous substances may be present as by-products of manufacture.

The most important method of production of foam in fire-fighting is the self-aspirating branchpipe, which is called a monitor in the larger sizes mounted on vehicles or turrets.

The water, containing 2–6 per cent of foam liquid, is fed into the branchpipe at a pressure rarely below 700 $kN m^{-2}$ and air is induced into the liquid stream by a venturi action; the foam is formed by violent turbulence which occurs in the foam-making section of the branchpipe. The foam has bubble diameters of the order of 1 mm, an air to liquid ratio of around 8:1; it is projected from the branchpipe as a "rope" or in a spray pattern, with a throw which varies from around 15 m in branchpipes discharging 200 $l min^{-1}$ of liquid and which can be held by one fireman, to around 75 m from the large monitors which may have capacity up to 5000 $l min^{-1}$. In the case of branchpipes producing low-expansion foam the rate of foam production is stated in terms of the liquid throughput.

Foams of an entirely different character are also used. These are the high-expansion foams and have a bubble diameter of the order of 10 mm and an air to liquid ratio between 500 and 1000 : 1. They are produced by blowing air through a net onto which the water containing the foam liquid is sprayed. A synthetic foam liquid is always used. High-expansion foam cannot be projected and rolls along as a layer; it can be delivered through trunking.

High-expansion foam is not specifically used for flammable liquid fires but is useful in inaccessible situations such as fires in ducts and basements, tunnels and sewers, and is also used for the protection of stores. Production rates of high-expansion foam are stated as volumes of foam, i.e. cubic metres per minute, and sometimes as metres per minute of depth when the filling of a particular compartment is referred to. Single units producing up to 1000 $m^3 min^{-1}$ are available.

Medium-expansion foam with an air to water ratio between 75 and 200 : 1 is also used, but to a much smaller extent. Other foams are used in special circumstances, such as those containing carbon dioxide from chemically activated extinguishers, or containing nitrogen for tank inerting.

Air to water ratios of foams are defined by the expansion, which is the ratio of the foam volume to the volume of liquid from which it is produced. Thus a foam made from 1 volume of liquid and 9 volumes of air has an expansion of 10.

2 The nature of fire-fighting foam

Before describing the experimental techniques used to study the physical properties of fire-fighting foam, a consideration of their nature will reveal some of the difficulties which arise and add relevance to the comments on the various experimental techniques.

A number of studies of the drainage properties of foams has been made.[1-6] Perhaps because of the current interest in the use of foams in counter-current separation processes, the approach has usually been based upon laws for liquid flow through capillaries or between parallel plates. In spite of the detail of these analyses they do not provide a satisfactory interpretation of the behaviour of foams as encountered in fire studies.

One difficulty is illustrated in Fig. 1 which shows the relationship between expansion and the time for drainage to commence. The results were obtained using a 2 per cent solution of a synthetic foam liquid and a 6 per cent solution of fluorochemical foam liquid in an 800-ml stirred jar. Clearly there are foams which are not draining, and in fact most fire-fighting foams do not start to drain for several minutes—by which time the fire is frequently out anyway. Observations such as those shown in Fig. 1 have led to the division of foams into two groups which can be described as "fully foamed liquids" which do not commence to drain immediately, and "partially foamed liquids" in which draining does occur immediately. This focuses attention on the foam which just begins to drain as soon as it is formed. Then the properties of this foam—its surface area, bubble diameter, drainage rate, and shear stress could perhaps provide a more fundamental basis for comparing different foaming agents just as they are about to drain than comparisons at a fixed expansion and arbitrary time after formation as is now the practice.

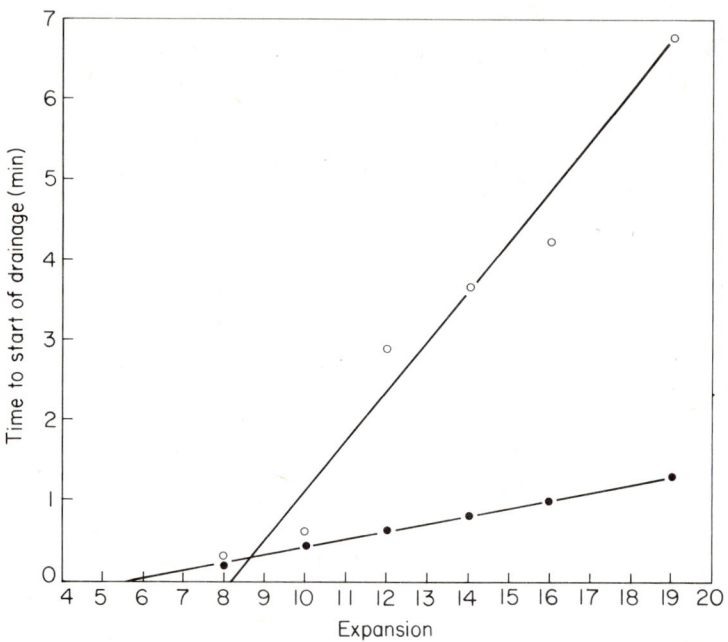

Fig. 1. Relationship between time to start of drainage and expansion for 800-ml stirred jar tests. ●, 6 per cent fluorochemical; ○, 2 per cent synthetic.

A hypothesis which may explain this aspect of foam behaviour is that there is a critical bubble wall thickness, perhaps related to bubble diameter, below which liquid does not drain from the foam. Critical film thicknesses have been identified and studied by a number of workers using single films and optical techniques,[7-10] but no one has related these observations to the drainage behaviour of foams.

Another interesting point is raised by the work of Clark and Blackman[11] in 1948. They studied protein and soap foams made with three different gases—air, oxygen and carbon dioxide—and followed their decay by photography and calculation of the specific surface of the foam. They recorded that the small bubbles shrank and the large bubbles expanded. They derived a formula based upon the pressure difference between the bubbles. The three gases used have markedly different solubilities and markedly different rates of surface decay were observed. It must be realized that a mathematical interpretation of the decay of foam and drainage of liquid, from a static sample, *must* be based upon a mass transfer process and decrease in bubble wall area upon which the superimposed fluid-flow functions which influence the rate at which the liberated liquid will percolate through the sample to the base of the container.

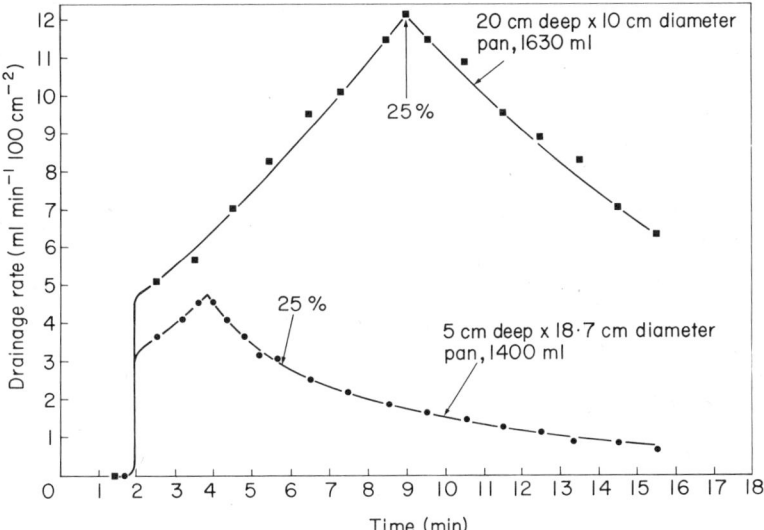

FIG. 2. Drainage rate per unit area of protein foam of expansion 9·1 in two pans with different depths.

Figure 2 shows the drainage rate curves for protein foam in sample pans of two different depths. The same protein foam was used but slight differences may have existed between the two samples because of practical difficulties such as temperature control. There are three distinct phases of drainage. Firstly there is a period of 2 min when no drainage occurs followed by a period during which liquid accumulates below the foam at a progressively increasing rate. During this period fluid flow properties are important and limit the rate at which liquid liberated by film decay percolates downwards through the foam.

After this period, depending upon the depth of sample a pronounced "maximum" occurs in the drainage rate curve and the rate of accumulation of liquid at the base begins to fall progressively (instead of increasing). During this third phase the rate of liberation of liquid by film decay becomes controlling, rather than the fluid flow properties influencing the time for percolation to the base. It must be noted that there is a major difference in the times for 25 per cent of the liquid to drain from the pans with different depths. In the 20-cm deep pan the 25 per cent drainage time is composed of

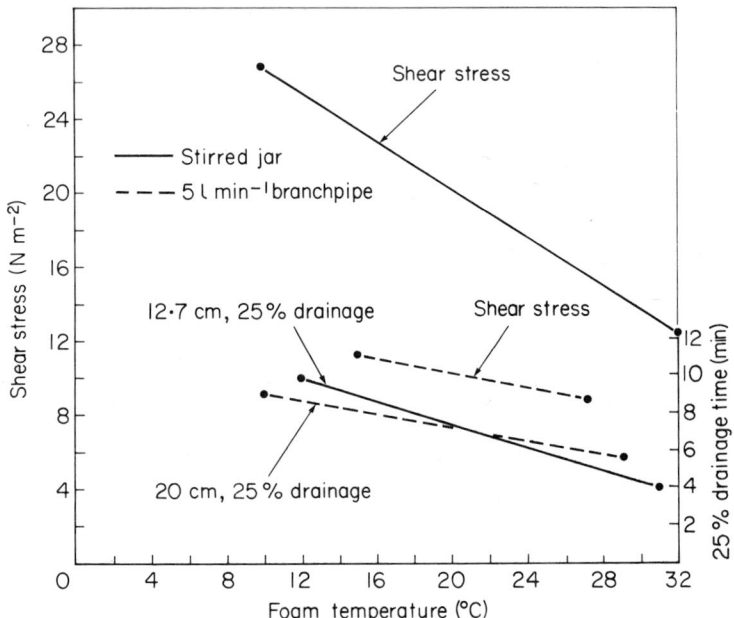

FIG. 3. Shear stress and 25 per cent drainage time with a fluoroprotein foam produced in the stirred jar and the 5 l min^{-1} branchpipe at various temperatures.

only two phases, "no drainage" and "increasing rate", while in the shallow pan it is composed of three phases, "no drainage", "increasing rate", and "decreasing rate". Thus the two pans are not measuring the same properties.

A careful comparison made at the Fire Research Station[14] showed that pan diameter did not affect the drainage rate.

Several studies of the effect of temperature upon foam properties have been made.[13, 14] Figure 3 summarizes some results obtained at FRS. Care is required in interpreting this data. Since it is impractical to adjust the temperature of foams after they are made, the tests were made by producing foam at different temperatures. Thus identical foams may not have been produced at each temperature, but any effects would be small, particularly in the stirred jar tests. Figure 3 provides adequate proof that temperature has an important effect upon the shear stress and drainage rate of foams—as expected.

3 Behaviour of a sample of fire-fighting foam

A simplified description of the behaviour of a foam sample can be made using the above data.

1. If the foam is produced with sufficiently small bubbles, or sufficiently high expansion, no drainage will occur immediately.

2. From the time of formation all foams will decay by a process of gas transfer—the small bubbles will shrink and disappear and the large bubbles will become progressively larger and fewer. The total bulk volume of the sample will show little change.

3. During this stage no liquid is liberated : the specific surface is decreasing and the remaining film thickens. Liquid is being redistributed but not released.

4. Later, usually after several minutes with fire-fighting foam, the film becomes too thick to incorporate further liquid, the critical film thickness has been reached, and the excess liquid starts to drain away.

5. Thus a sample of foam is changing continuously from the moment it is made, but is uniform throughout its depth until drainage commences. When drainage has commenced the foam properties are different throughout the depth of the sample.

6. The rate at which liquid accumulates at the base of the sample depends upon the depth of the sample, its age and temperature, but is independent of the diameter of the sample.

This simplified description provides a valuable guide to assist the improvement of experimental techniques, but omits some aspects which should be mentioned.

Foam has different properties when in motion than when at rest, and this was noted as long ago as 1948 by Blackman.[15] This effect is observed when measuring the shear stress, particularly with some protein foams which exhibit large differences between the yield stress and the continuous shear stress. Most laboratory measurements of the physical properties are made with samples of foam which are at rest; but in fire-fighting the foam begins to function while moving. Since fire control is often achieved in times of less than 30 s the changes which occur in the few seconds after foam comes to rest are of particular interest and this is a subject which has received little study. How these changes might explain one important aspect of the behaviour of foam on fires has been suggested by the author.[16] No mention is made of the bubble-size distribution of foams. This is important because it causes changes in the shear stress of a foam without altering the drainage rate or the expansion. In a fire, a foam with low shear stress which will spread rapidly is necessary, and in addition it must not lose its water content. Simpson and Hodkinson[17] described experiments in which bubble-size distribution was investigated. They measured the compressibility of floating rafts of bubbles, composed of uniform or mixed sizes, a novel experimental technique which promises to assist the investigation of bubble-size distribution effects. Nor has the simplified description included reference to the readiness with which a liquid can be converted into foam. The synthetic and fluorochemical liquids foam readily in branchpipes, while some protein liquids are much more difficult to foam, although it can be demonstrated in the laboratory that they will make excellent foams.

4 Laboratory tests

Since the objective is the extinction of fires the standard approach is to conduct experimental fires, as at FRS in large numbers, varying in size from 0.25 m^2 area to 100 m^2 or larger. Fire tests are expensive and give rise to smoke pollution problems and so the numbers are reduced by correlating laboratory measurements with fire control and extinction times. Three principal measurements are made: expansion, shear stress, 25 per cent drainage time (in 20-cm deep pan).

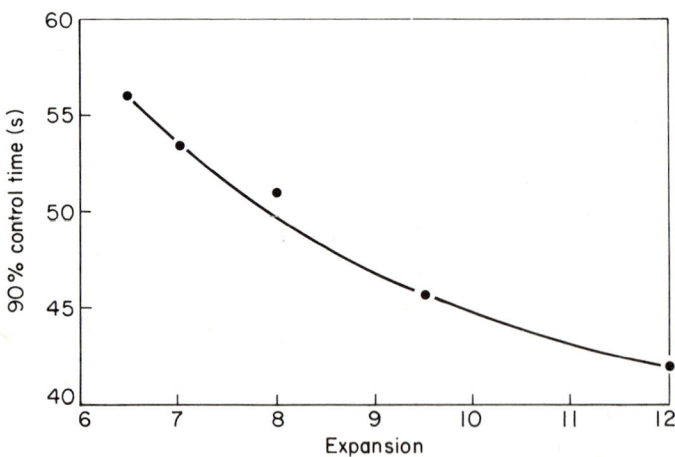

FIG. 4. The effect of expansion on the time to control 0.28 m^2 area petrol fires with synthetic foam of 10 N m^{-2} shear stress. Gentle surface application of foam at 0.041 m^{-2} s^{-1}.

A foam is rarely used without first making these three measurements. In addition the temperature of the foam will be noted; and the foam liquid from which it is made and the concentration used will be known. This provides a practical definition of the foam quality and enables its probable fire-extinction properties to be assessed. For example Figs 4 and 5 show how

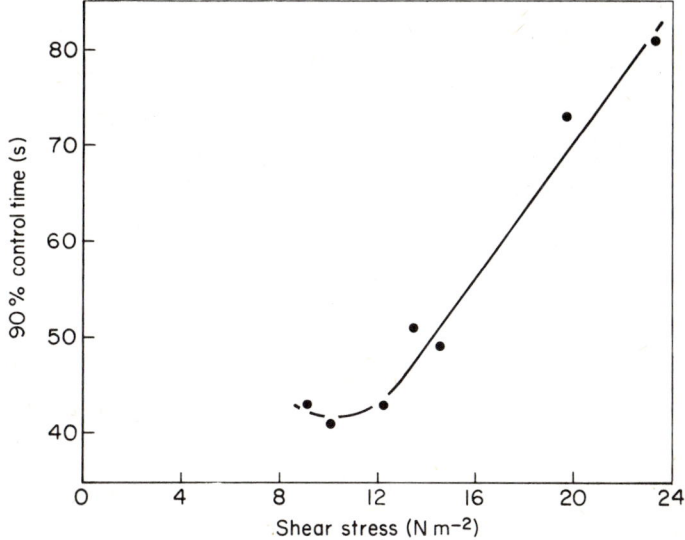

FIG. 5. The effect of shear stress on the time to control 0.28 m^2 area petrol fires with synthetic foam of expansion 12. Gentle surface application of foams at 0.04 l m^{-2} s^{-1}.

expansion and shear stress affected the fire-control properties of synthetic foam in a series of laboratory fires.

Other laboratory measurements are also used in fire research such as resistance to foam destruction by hot fuel, resistance to radiant heat, propensity for foam to mix with the fuel, transmission of fuel vapour, etc, but these are specific to the particular problem of fire control, while the three principal measurements have a wider application.

4.1 EXPANSION

Expansion is determined by filling a measure with foam and weighing it. Measures varying from 1 to 200 l are used. For high-expansion foams a 130-l bin is used[18] and Fig. 6 shows a typical arrangement. Electrical conductivity methods[19] have also been devised but are not in wide use.

Fig. 6. High-expansion weighing cabinet.

4.2 TWENTY-FIVE PER CENT DRAINAGE TIME

For many years the quarter drainage times of fire-foams have been deter-mined using a pan 18·7 cm (7$\frac{3}{8}$ in.) diameter and 5·1 cm (2 in.) deep. Details are given in UK Defence Standard 42–3.[20] An alternative of greater accuracy can be used and is described elsewhere.[14] This employs a brass pan 20 cm deep of either 20 cm or 10 cm diameter. When stating results the pan depth should be indicated to avoid confusion, and the foam temperature should also be recorded. In the 20-cm deep pan, low-expansion branchpipe foams usually have quarter drainage times between 4 and 10 minutes at 20°C.

For high-expansion foams it is usual to measure the 50 per cent drainage time and this is done in conjunction with the expansion test illustrated in Fig. 6. When the 130-1 bin of foam has been weighed a small bung in the conical base is removed to allow draining liquid to fall away. The time is noted for the bin to lose half its net weight.

4.3 SHEAR STRESS

This measurement is only used for low-expansion foams, but will probably be extended to medium- and high-expansion foams. The torsional vane viscometer has been described.[20] The measurement is made 1 minute after collecting the sample. Figure 7 illustrates the principle of the viscometer. The vane

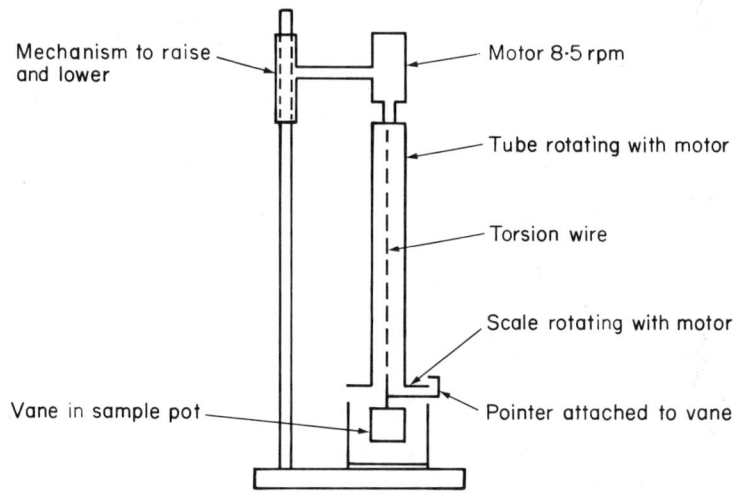

FIG. 7. Diagram of torsional foam viscometer.

is 31·8 mm high, 31·8 mm wide, and 1·22 mm thick and is just fully immersed in the foam. The top of the wire rotates at 0·141 Hz (8·5 r.p.m.). Various diameters of torsion wires are used for the range of fire-fighting foams now available.

An alternative construction is also employed in which the sample pot is rotated instead of the torsion wire. This does not alter the fundamental characteristics; it has some advantages and some disadvantages.

The Defence Standard[20] describes the method to be used for calibrating the wire and calculating the viscometer constant so that the scale deflection can be converted to $N\,m^{-2}$ of shear stress. Only the area of the vertical cylinder sheared by the vane is considered in the calculation. This method originated from the work of Clark[21] who found the end effect to be negligible. This was on the basis of a private communication and is a point which merits re-appraisal.

When this shear stress measurement is examined critically it is apparent that it is not fundamental and it is related to the design of the instrument and the method of calculation. An investigation has begun at FRS on shear stress measurement and the first aims are to obtain curves showing the relationship between shear stress and rate of shear. Preliminary results indicate that this relationship will vary significantly for the different fire-fighting foams: protein foams having substantial yield values at zero rate of shear stress, while synthetic foams have negligible yield values.

The shear stress measurement is of great practical value. It is quick, highly reproducible, and correlates well with the rapidity of fire extinction.

4.4 LABORATORY METHODS OF FOAM PRODUCTION

One requirement of successful foam research is ability to reproduce, in the laboratory, successive samples of foam with the same selected quality and temperature, so that repeated tests are possible for experiments such as the variation of decay rate with sample depth, or the quantity required to extinguish fires of different fuels.

A slightly different requirement is to be able to produce foam in an accurately defined manner and to investigate the effect of changing the type or quantity of surfactant upon the properties of the foam produced. In this case the design of the equipment will affect the results obtained and it should have similar characteristics to large-scale equipment to which the results are to be applied.

There are no chemically pure fire-fighting foam liquids which can be used as reference standards but most foam liquids retain their properties with negligible change for at least several months and this permits a single batch to be used throughout a lengthy experiment, and the interchange of samples between laboratories for collaborative tests, checking of viscometer constants, etc.

4.5 THE STIRRED JAR

A standard stirred jar has been developed at FRS and is described elsewhere.[22] It produces 800 ml of foam at a preselected expansion in a period of 6 min. Figure 8 illustrates the stirred jar assembled for operation and shows the essential feature of positioning the jar in a horizontal position. If the stirrer blades are made 3·2 mm diameter ($\frac{1}{8}$ in.) instead of 2·4 mm ($\frac{3}{32}$ in.) the jar will operate more effectively with liquids which do not foam readily.

This is a very useful apparatus for small-scale tests in the laboratory for expansions up to 20 and is the only apparatus which provides a sample at

Fig. 8. Stirred jar assembled for operation.

zero time, i.e. when the agitator is stopped, it is therefore valuable for some investigations such as drainage properties, the sample being allowed to drain without transferring from the jar.

An example of the useful data it will produce has already been referred to by Fig. 1 which shows the relationship between expansion and time for drainage to start. Figure 9 is another example which shows the effect of pH upon the shear stress of protein foam—data of immense practical importance obtained more accurately and economically than by any other method.

The stirred jar has two limitations. Firstly the temperature of the foam approximates to the room temperature during the 6-min stirring period, and

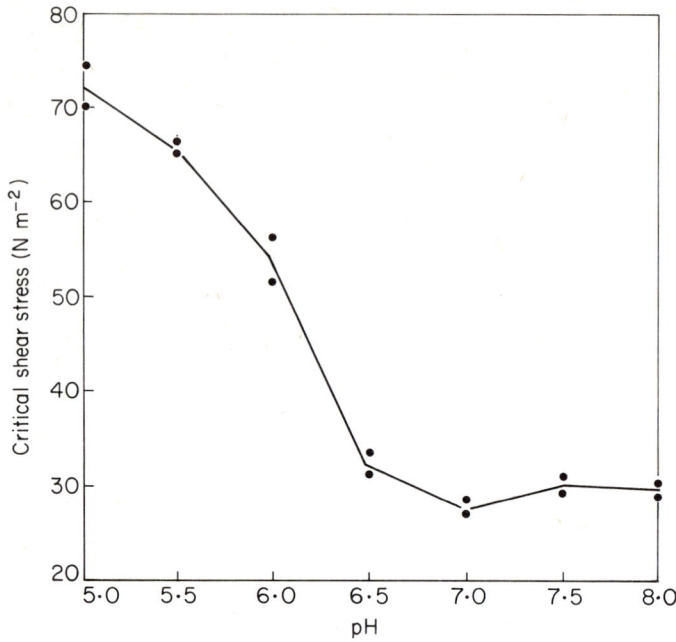

FIG. 9. Relationship between critical shear stress and pH of protein foam produced in 800-ml stirred jar—4 per cent solution at expansion 8.

thus it is troublesome to produce foam samples at preselected temperatures. Secondly, the shear stress of the foam is fixed by the agitator speed and dimensions, and is independent of the concentration of surfactant, above a small minimum requirement; an energy equilibrium is established during the 6-minute stirring period. This is an important limitation for fire research because when foam is produced in branchpipes a much higher concentration of foam liquid is required before the shear stress is limited by the energy level. The short residence time in the branchpipe does not permit equilibrium to be achieved unless an excess of surfactant is present.

4.6 CONTINUOUS FOAM GENERATORS

Continuous foam generators are widely used in fire research and several descriptions have been published.[20] Figure 10 shows the principles employed. The premixed foaming solution and an air supply are each controlled to any required rate through rotameters. The liquid and air then pass along a packed column in which the foam is formed. The liquid flow rate is set first. The air flow is next adjusted to give the required expansion which is verified by sampling the foam. The packing in the foam column is varied until foam with the required shear stress and drainage rate is produced. The column packing consists of discs of wire gauze which are assembled in canisters containing 1 to 10 discs so that any number can be used. Usually between 5 or 25 discs are required, but up to 100 have been used. Canisters containing gauze discs of several different mesh sizes are provided. The discs are secured

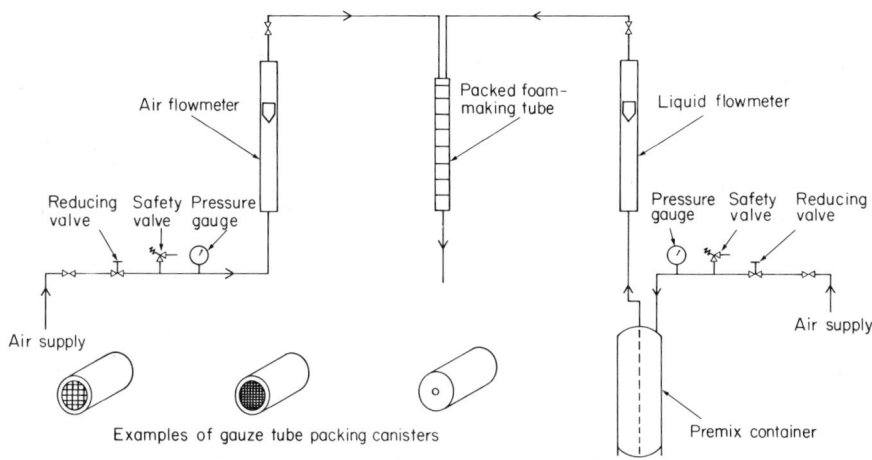

FIG. 10. Continuous foam generator.

by thin brass orifice plates at each end of the canister, and canisters with orifices of various sizes are provided. By adjusting the number and mesh size of the discs, and the sizes of the orifice plates, the shear stress and drainage rate of the foam can be altered over a wide range. It is thus possible by trial and error to adjust the generator to produce foam with the same measured properties as that from a large branchpipe—but this can sometimes be a tedious process.

Generators with air flows up to 30000 l min^{-1} and liquid flows up to 1200 l min^{-1} have been constructed.[23] A convenient size for many laboratory uses is 0–15 l min^{-1} of air and 0–2 l min^{-1} of liquid.

The foam temperature can be varied by adjusting the premix temperature.

The generator works smoothly in the expansion range 5–30, but at higher expansions problems can arise from large bubbles of air occurring in the foam, i.e. incomplete mixing. With care, however, uniform foams up to 200 expansion can be produced with some foam liquids.

The characteristics of the generator are closer to the stirred jar than to the branchpipe, shear stress being controlled by the energy equilibrium in the packed column, rather than the surfactant concentration.

4.7 LABORATORY-BENCH FOAM GENERATOR

Figure 11 illustrates a modified generator in which the packed tube has been replaced with a vertical variable-speed agitator. This was constructed with

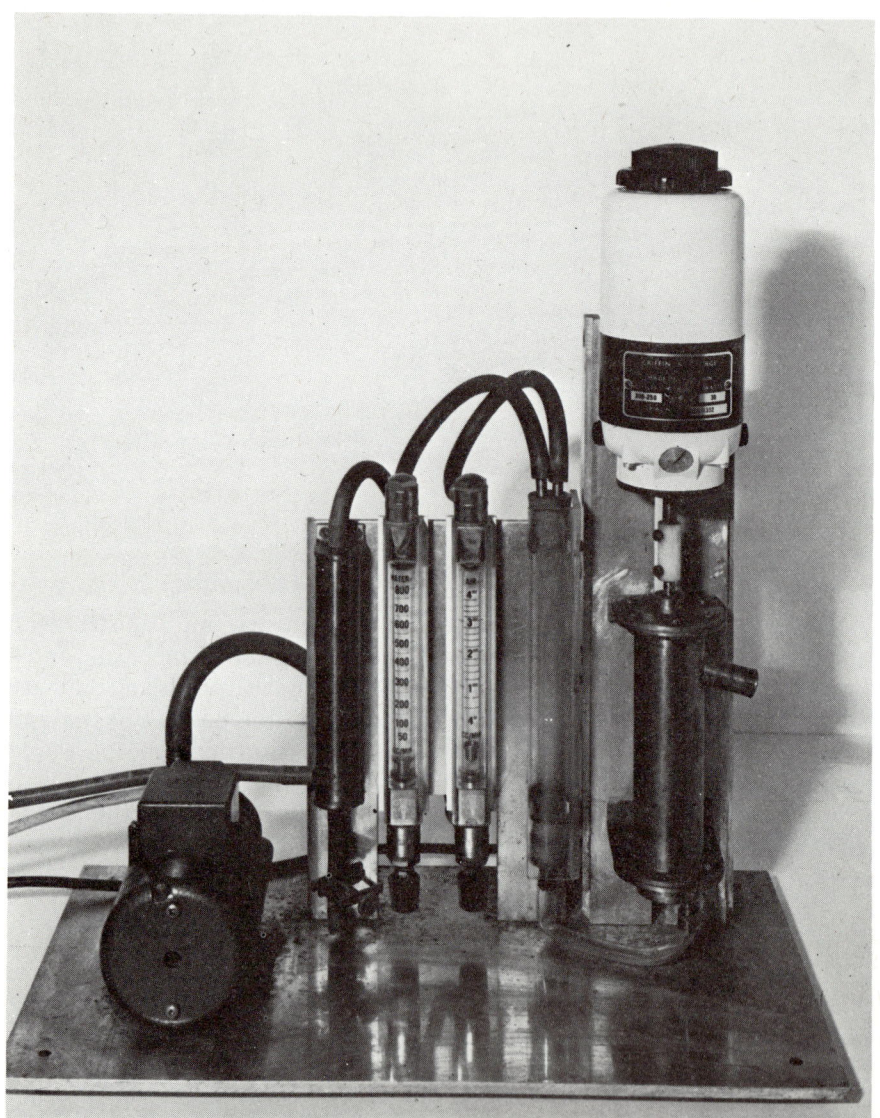

FIG. 11. Laboratory bench generator.

the objective of making a generator with similar characteristics to branch-pipes in which the foam is formed by a rapid impact. Although not completely successful the generator is of interest because it is small enough to be used over a laboratory sink and is assembled almost entirely from commercially available laboratory equipment. The air rotameter has a range of 0–4000 ml min^{-1} and the liquid rotameter 0–800 ml min^{-1}; the 60-W motor has a speed range of 600–6000 r.p.m. The liquid feed pump provides a head of 6 kN m^{-2}. An air supply pressure of 10 kN m^{-2} is adequate. If premix and foam temperature are close to room temperature an apparatus of this type serves well for experiments in which numerous small samples of identical foam are required, and it avoids the 10-min interval required to prepare a new sample using the stirred jar.

4.8 FIVE-LITRE PER MINUTE STANDARD FOAM BRANCHPIPE

The construction of a 5 l min^{-1} foam branchpipe has been described in detail[24] and a standard procedure for using it to test a foam liquid. The

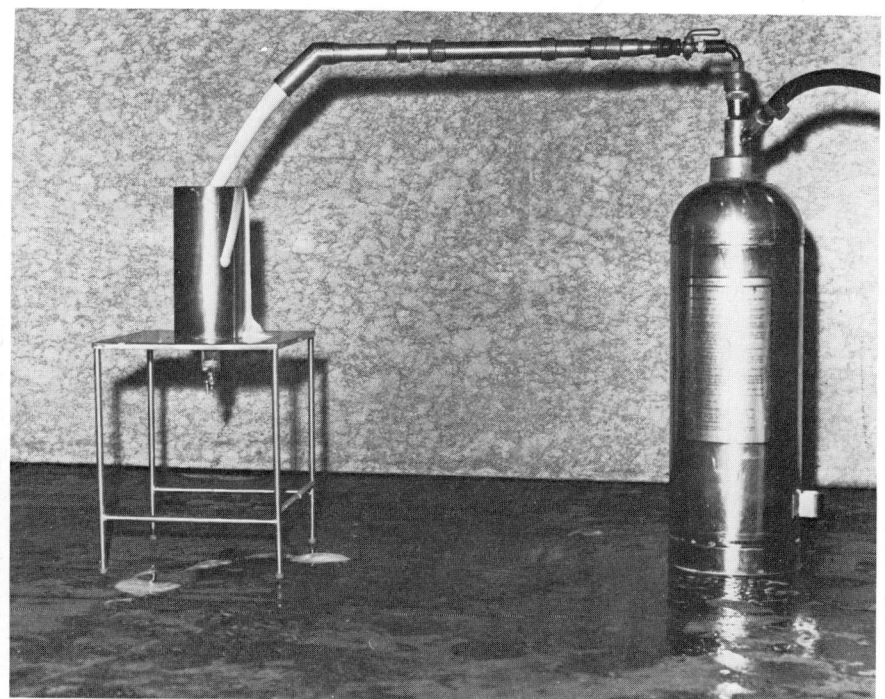

FIG. 12. Branchpipe in use for foam testing.

branchpipe is simple to construct and use, and permits comparison between laboratories on the foaming properties of solution. By adjusting the temperature of the premix solution (10 l is a convenient volume) the temperature of the foam can be controlled, and so it is possible under normal laboratory conditions to produce foam at temperatures ranging between 12–30°C $\pm$ 1°C. The branchpipe is normally operated at 690 kN m^{-2} on the premix supply. Figure 12 shows the branchpipe in use for foam testing. A sample is being

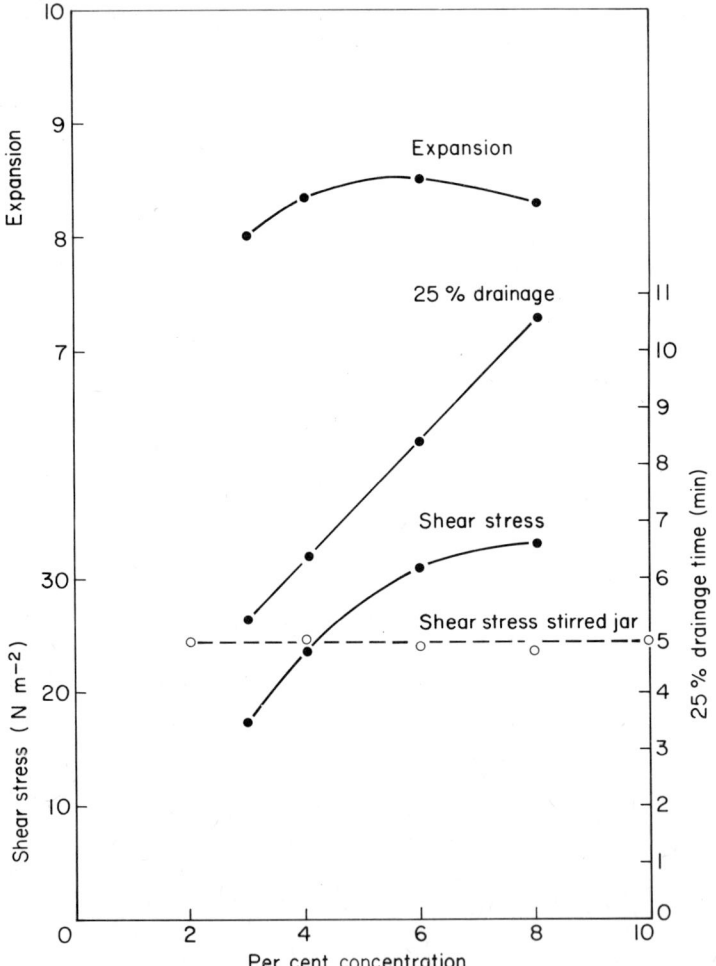

FIG. 13. Effect of concentration on foam properties from 5 l min^{-1} branchpipe and shear stress of stirred jar foam—4 per cent protein foam liquid.

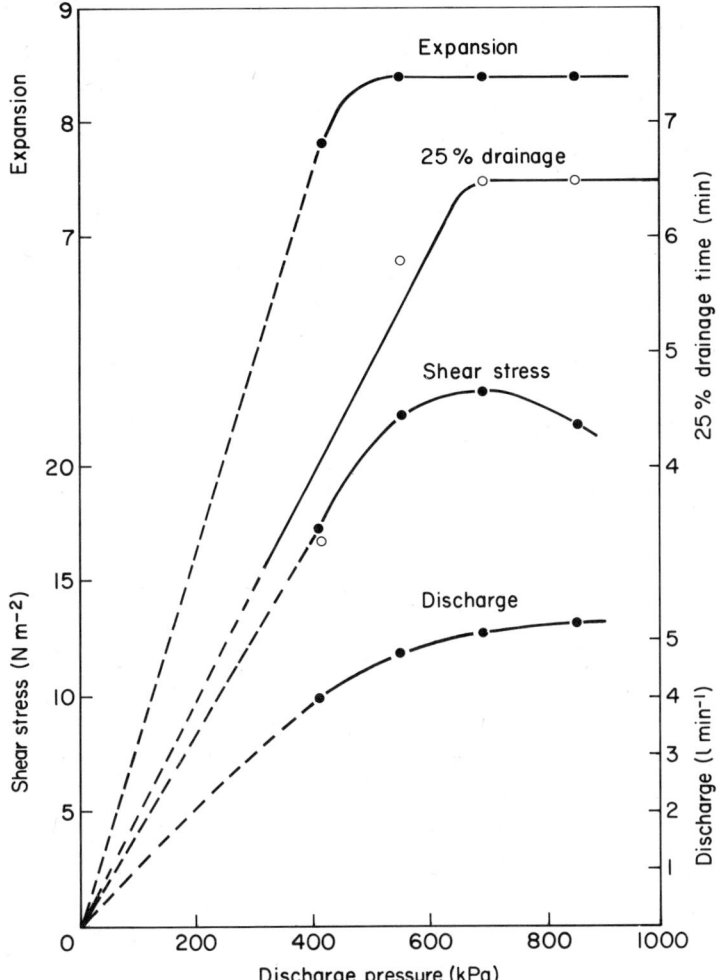

FIG. 14. Effect of discharge pressure on foam properties from $5\,l\,min^{-1}$ branchpipe—4 per cent protein foam liquid.

collected in a 20-cm drainage pan. Figure 13 shows how the foam properties varied with concentration using a protein foam liquid, and Fig. 14 with the discharge pressure. From Fig. 14 it can be seen that $700\,kN\,m^2$ is a preferred discharge pressure for this branchpipe, because the properties become independent of pressure above this point. Figure 13 shows that the shear stress is influenced by the concentration until a 6 per cent solution of protein is used, above which level it does not increase further. For comparison a similar shear stress curve for the stirred jar is shown in Fig. 13; the shear stress is independent of concentration down to the lowest level tested—2 per cent. Concentration does not greatly change the expansion, while the drainage time increases progressively with the concentration because of the increasing viscosity of the liquid and the shear stress of the foam. The expansion and other properties cannot be varied at will, but the branchpipe provides an index of the readiness with which a liquid will foam. Because it has been shown[25] that this 5 l min^{-1} branchpipe has similar characteristics to larger branchpipes used by the fire services it is particularly useful in fire research. It is well suited to laboratory programmes in which many successive samples of identical foam are required at room temperature, and permits a rapid assessment of the quality of different batches of foam liquid.

5 Concluding observations

1. Effective foam research depends upon the ability to reproduce uniform samples of foam with selected properties, and to assess the properties of foam samples produced from different liquids in a standard manner. The procedures and equipment described serve these purposes well in the case of fire-fighting foams and should assist in other fields of foam research.

2. The simplified interpretation of the decay and drainage of static samples of foam is a useful guide to the correct handling of foam samples.

3. The classification of foams into "fully foamed liquids" and "partially foamed liquids", depending on whether or not they are losing liquid by drainage, merits general adoption.

4. There are many properties of foams which require further study and their use in fire-fighting draws attention to the following subjects.

a. A theoretical treatment of foams which includes transition from non-draining to draining foam and the origin of all drainage in film decay, as well as fluid flow considerations.

b. The effect of bubble-size distribution on fluidity and decay, and techniques for controlling it.

c. A more fundamental understanding of the shear stress properties of foams and their measurement.

d. Attention to the "readiness" with which a liquid foams as distinct from the quantity or quality of foam it is capable of producing.

e. Elucidation of the indications that there are two types of foam: those typified by expansion 10 readily made by impact devices such as branch-pipes, and those typified by expansion 500 + which can only be made by net devices.

References

1. Jacobi, J. M., Woodcock, K. E. and Grove, C. S. Jr. (1956). *Ind. Eng. Chem.* **48**(11), 2046–2051.
2. Lemlich, R. (1968). *Ind. Eng. Chem.* **60**(10), 16.
3. Leonard, R. A. and Lemlich, R. (1965). *AIChE J.* **11** (1), 18–29.
4. Haas, P. A. and Johnson, H. F. (1965). *AIChE J.* **11**, 319.
5. Leonard, R. A. and Lemlich, R. (1965). *AIChE J.* **11**, 18; (1965). *AIChE J.* **11**, 25.
6. Fang-Shung Shih and Lemlich, R. (1971). *Ind. Eng. Chem. Fundam.* **10** (2), 25.
7. Mysels, K. J. *et al.* (1959). "Soap Films—Studies of their Thinning and a Bibliography". Pergamon Press, New York.
8. Lyklema, J. and Mysels, K. J. (1965). *Chem. Soc.* **87** (12), 2539.
9. Prins, A. and Van Dem Tempel, M. (September 1970). Faraday Society Special Discussion.
10. Corril, J. M. *et al.* (1961). *Trans. Faraday Soc.* **57** (1), 821.
11. Clark, N. O. and Blackman, M. (1948). *Trans. Faraday Soc.* **44**, 1.
12. Benson, S. P., Morris, K. and Corrie, J. G. (1973). An improved method for measuring the draining rate of fire-fighting foams. Joint Fire Research Organisation Fire Research Note 972.
13. Jablonski, E. J. (March 14, 1961). The influence of temperature on the viscosity and drainage rates of mechanically produced fire-fighting foams. US Naval Research Laboratory, Washington DC. NRL Report 5598.
14. Benson, S. P. (1973). The effect of temperature on the physical properties of foam produced in the stirred jar and in the 5 l/min branchpipe. Joint Fire Research Organisation Fire Research Note 976.
15. Blackman, M. (1948). *Trans. Faraday Soc.* **44**, 205.
16. Corrie, J. G. (April 1974). Problems with the use of foam. International Fire and Security Exhibition and Conference, London. To be published.
17. Simpson, A. W. and Hodkinson, P. H. (1972). *Nature (London)*, **237** (5354), 320.
18. Specification No JCDD/28. (September 1971). High expansion foam liquids. Home Office Fire Department.
19. Thorne, P. F. and Tucker, D. M. (1971). Some electrical properties of high expansion foam. Joint Fire Research Organisation Fire Research Note 911.
20. Defence Standard 42–3/Issue 1/24 (January 1969). Foam liquid—Fire Extinguishing Directorate of Standardisation, Ministry of Defence, First Avenue House, London WC1.
21. Clark, N. O. (1947). A study of mechanically produced foam for combating petrol fires. Chemistry Research Special Report No. 6. HMSO, London.
22. Corrie, J. G. (1971). A stirred jar for the production of standard foam in the laboratory. Joint Fire Research Organisation Fire Research Note 863.
23. Fittes, D. W. and Nash, P. (1965). The design and development of an experimental gas turbine operated foam generator. Joint Fire Research Organisation Fire Research Note 583.
24. Benson, S. P., Griffiths, D. J. and Corrie, J. G. (1973). A 5-litre per minute standard foam branchpipe. Joint Fire Research Organisation Fire Research Note 971.
25. Benson, S. P., Griffiths, D. J. and Corrie, J. G. (1973). Foam branchpipe design. Joint Fire Research Organisation Fire Research Note 970.

H

14

Microbubbles: generation and interaction with colloid particles

JAMES B. MELVILLE* and EGON MATIJEVIĆ

Unilever Research, Port Sunlight Laboratory,
Wirral, Merseyside L62 4XN, England

and

Institute of Colloid and Surface Science and Department of Chemistry,
Clarkson College of Technology, Potsdam, New York 13676, USA

Summary

The preparation of microbubbles, their interaction with solid particles of different surface characteristics, and the application to flotation of colloidal systems have been discussed. It is now recognized that one of the essential parameters in separation of dispersed matter by foam is the size of the gas bubbles used. The most efficient removal of colloidal particulates has been achieved by using microbubbles of the order of ∼ 50 μm in diameter (microflotation).

Conventionally bubble size is reduced by addition of ethanol. This work describes the generation of microbubbles in various electrolyte solutions, with special reference to the effect of electrolyte type and concentration on bubble diameter. The interactions of microbubbles with hydrophobic (polystyrene latex) and hydrophilic (copper hydrous oxide) particles, respectively, have been studied. The results were related to colloid stability of the systems as a function of pH and electrolyte concentration. Heterocoagulation of microbubbles with copper hydrous oxide occurred whenever the latter was colloidally unstable. Heterocoagulation of microbubbles with polystyrene latex, as indicated by flotation efficiency, did not occur whatever the surface potential on the particles.

1 Introduction

Modern technology has produced a number of so-called adsorptive bubble separation techniques.[1,2] in which dissolved or suspended matter is removed

* Work carried out in Potsdam while JBM was visiting Fellow.

from a liquid by means of attachment to gas bubbles and subsequent trans-
portation to the surface of the liquid. Most of the work reported in the literature has been concerned with the phenomenological aspects of removing a
particular substance, rather than with the basic principles of the flotation
processes. It would seem a more meaningful approach to consider the
individual stages which make up a flotation process, and these may be conveniently subdivided as follows :

1. Generation of bubbles.
2. Collision between bubbles and particles.
3. Adhesion of bubbles to particles.
4. The behaviour of bubble–particle aggregates at the liquid–air interface
at the surface of the flotation vessel.

The most efficient removal of colloidal particulates has been achieved by
using microbubbles of the order of 50–100 μm in diameter (microflotation)[3-7].
Bubbles of this size can be produced in several different ways, e.g. electrolytically, by shearing gas streams, by the passage of gas through fine pores
into solutions containing frothers such as ethanol, etc, but one of the problems
encountered at present is in finding cheaper methods of producing them in
high concentrations for large-scale applications. During earlier work on
removal of metal ions from solution by microflotation we observed[8] that
the bubbles produced by passing nitrogen gas through a fine sintered glass
frit into a synthetic "sea-water" were of microbubble dimensions even
before the addition of a frother. Thus the first part of the present work
describes the generation of microbubbles in various electrolyte solutions,
with special reference to the effect of electrolyte type and concentration on
bubble diameter. Comparison is made with microbubbles produced using the
more conventional ethanol frother. The second part of this work is concerned
with the interaction of microbubbles with hydrophilic (copper hydrous oxide)
and hydrophobic (polystyrene latex) particles. Because of the intent to
relate bubble–particle interaction to the colloid stability of the systems at
different pH and salt concentrations, the microbubbles used in these experiments were produced conventionally with a standard level of ethanol frother.

2 Experimental

2.1 MATERIALS

All chemicals were of analytical reagent grade and were used without further
purification. Water was doubly distilled from an all-Pyrex apparatus.

The polystyrene latex dispersion was prepared according to the method
of Ottewill[9] et al. using a potassium persulphate initiator. The dispersion,

consisting of uniform spherical particles of diameter 1·8 μm, was dialysed for two weeks before use.

2.2 TECHNIQUES AND PROCEDURES

2.2.1 *Microflotation apparatus*

The microflotation apparatus (shown in Fig. 1) has been described in detail elsewhere.[2, 3] The cell consisted of a 600-ml Büchner funnel. Humidified

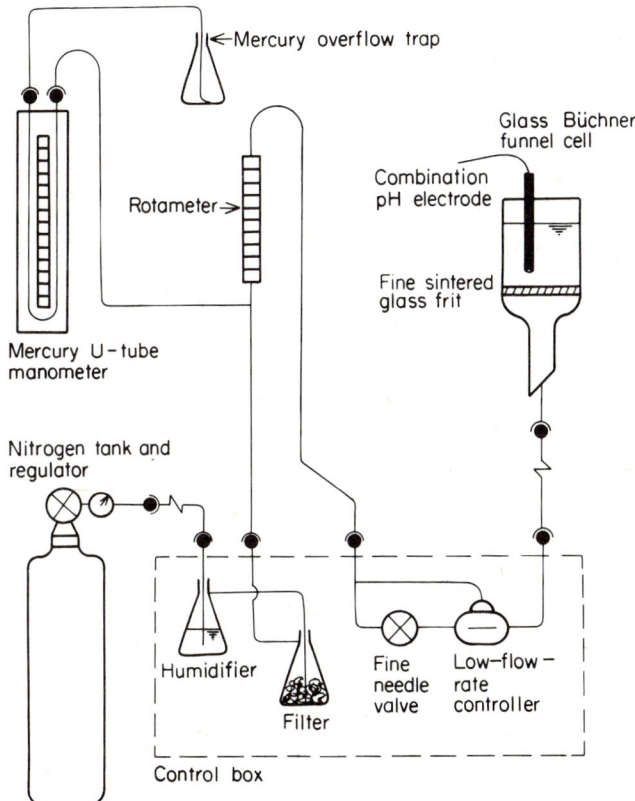

FIG. 1. Schematic diagram of microflotation apparatus.

nitrogen gas was introduced at a carefully controlled rate of flow through a fine sintered glass frit at the bottom of the funnel.

A modified photomicrographic apparatus (Vickers Instruments Ltd) as described in detail by Kaufman *et al.*[10, 11] was used for photographing the microbubbles. This consisted of a Polaroid roll film back, eyepiece assembly, shutter mechanism and adapter, fitted to a Unitron Model MPH microscope body. The necessary illumination was supplied by an electronic flash unit mounted over a lucite fibre optic.

All photographs were taken at the same point in the cell as marked by a line scribed on the inside wall of the cell.

2.2.2 *Bubble size measurement*

Nitrogen was passed through 400 ml of a solution at a rate of flow of 150 ml min^{-1}. After a few minutes a series of photographs were taken. The diameters of the bubbles appearing in the photographs were measured using a Zeiss TGZ3 Particle Analyser. All bubbles which were clearly defined were counted. The analyser was calibrated using a Baush and Lomb stage micrometer slide with 0·1-mm and 0·01-mm divisons which was suspended in distilled water in the microflotation cell and photographed.

2.2.3 *Electrophoresis*

Electrophoretic mobilities were measured with a Rank Brothers microelectrophoresis apparatus using a van Gils cell and platinized platinum electrodes. Ten particles were timed over a given distance in both directions. From the twenty measurements an average value for the absolute mobility was calculated.

2.2.4 *Interaction of microbubbles with copper hydrous oxide*

The pH of a 400-ml aliquot of a cupric nitrate solution was adjusted to the required value with NaOH. The system was then agitated for 10 min by the passage of nitrogen gas at 30 ml min^{-1}. Using the photomicrographic/fibre optic equipment the state of aggregation of the precipitated copper hydrous oxide at the end of this time was noted. Then 1 ml of ethanol was added to the cell and the timer was started. The system was inspected for heterocoagulation between microbubbles and particles, again using the photomicrographic/fibre optic apparatus. A visual semi-quantitative estimate of the amount of solid material floated after 5 minutes' passage of gas was made.

2.2.5 *Interaction of microbubbles with polystyrene latex*

The flotation efficiency of polystyrene latex dispersions was used to indicate the nature of the interactions between the polystyrene latex particles and microbubbles.

400-ml aliquots of dispersion containing 7×10^6 particles per ml were used in the microflotation cell and 1 ml of ethanol was added as a frother. Where necessary pH was adjusted using NaOH or HNO$_3$ and ionic strength was controlled with KNO$_3$. Nitrogen gas was passed through the dispersion at a rate of 30 ml min^{-1} for 10 minutes. The degree of flotation was obtained by measuring the turbidity of the dispersion at a wavelength of 500 nm, before and after an experiment using a Cary 14 spectrophotometer with 10-cm cells. The polystyrene latex dispersion was earlier shown to obey the Lambert–Beer law. Because of the low particle number concentration the system was not complicated by coagulation during the length of time of an experiment.

3 Results

3.1 GENERATION OF MICROBUBBLES

The size distribution of bubbles formed by passing nitrogen gas through a fine sintered glass frit at a rate of 150 ml min^{-1} into 0·36 mol dm^{-3} Fe(NO$_3$)$_3$

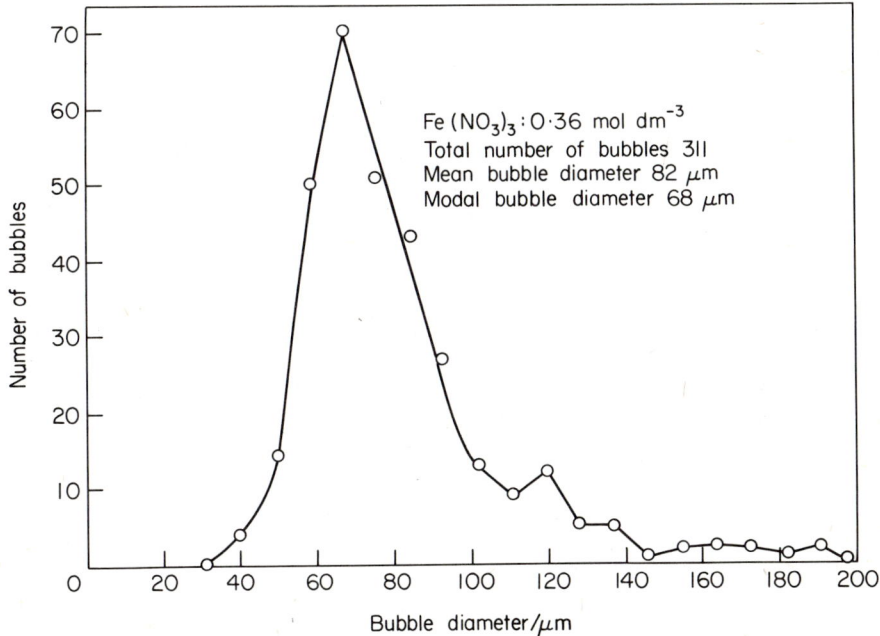

FIG. 2. Bubble size distribution obtained by passing nitrogen gas into 0·36 mol dm^{-3} Fe(NO$_3$)$_3$ at a flow rate of 150 ml min^{-1}.

is shown in Fig. 2. It is clear that the bubbles generated were in the "microbubble" size range with a mean diameter of about 80 μm and a modal diameter of about 70 μm. Similar microbubbles were produced in a variety of other salts.

The effect of different cations on the mean diameters of microbubbles is shown as a function of electrolyte concentration in Fig. 3. Bubble diameter decreased with increasing concentration towards a limiting value of about 80–90 μm. The greatest effect was observed in the presence of $Fe(NO_3)_3$.

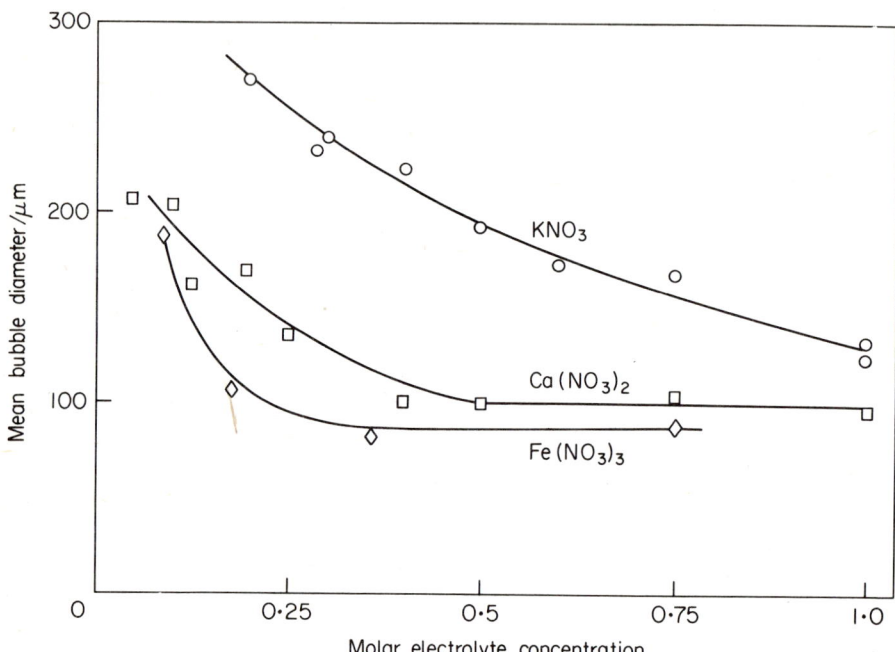

FIG. 3. Mean bubble diameter as a function of electrolyte concentration. ○, KNO_3; □, $Ca(NO_3)_2$; ◇, $Fe(NO_3)_3$. Gas flow rate 150 ml min^{-1}.

which under the conditions of these experiments produced hydrolysed ferric ions. Divalent Ca^{2+} ions were somewhat less effective but they produced smaller bubbles than monovalent K^+ ions. Figure 4 shows that anions also have a similar effect upon bubble size. When the same cation (Na^+) was used, sulphate ions were much more efficient in reducing the bubble size than the two monovalent anions. Interestingly, $NaNO_3$ had a considerably larger effect than KNO_3, the curve for which is reproduced in Fig. 4 for comparison purposes.

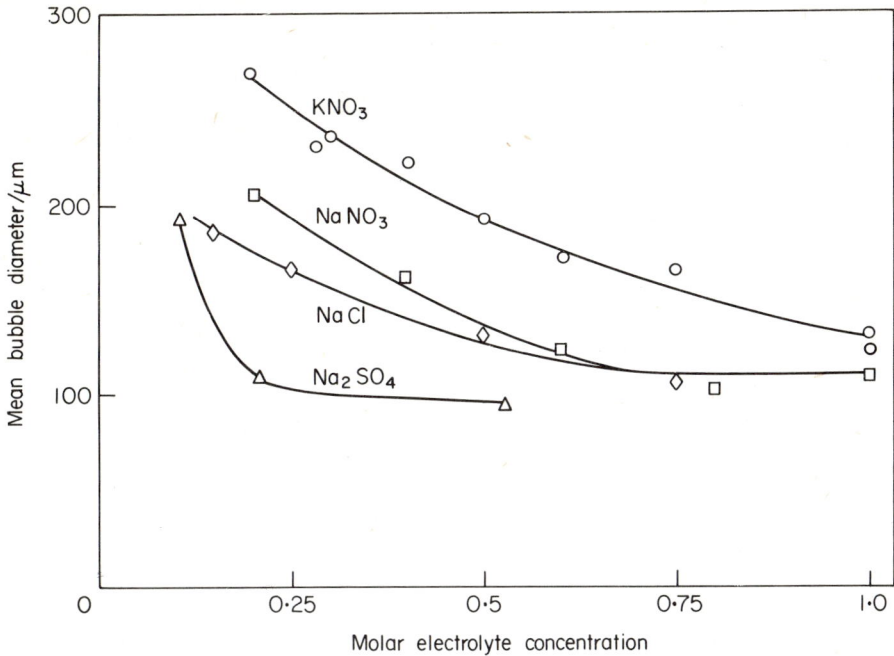

FIG. 4. Mean bubble diameter as a function of electrolyte concentration. ○, KNO_3; □, $NaNO_3$; ◇, $NaCl$; △, Na_2SO_4. Gas flow rate 150 ml min^{-1}.

3.2 INTERACTION OF MICROBUBBLES WITH COPPER HYDROUS OXIDE

Figure 5 gives the precipitation domain of copper hydrous oxide from cupric nitrate solutions as a function of pH. Various boundaries indicate regions of different precipitate characteristics. These are related to interaction with microbubbles as follows. The boundary separating regions I and II represents the solubility boundary and to the left of this there was no precipitation; hence accordingly no interaction with microbubbles would be expected. The boundary is in good agreement with earlier findings.[12] In region II the precipitated solid was in the form of a stable colloidal dispersion and no heterocoagulation between microbubbles and particles was observed. Region III defines the domain in which the precipitated copper hydrous oxide was colloidally unstable and over this entire region heterocoagulation with microbubbles was observed. The double shaded area within region III illustrates greater than 75 per cent removal of solid from dispersion after 5 minutes' passage of microbubbles. This latter region is in good agreement with optimum pH for removal of precipitated copper hydrous oxide from dispersion by conventional microflotation[6].

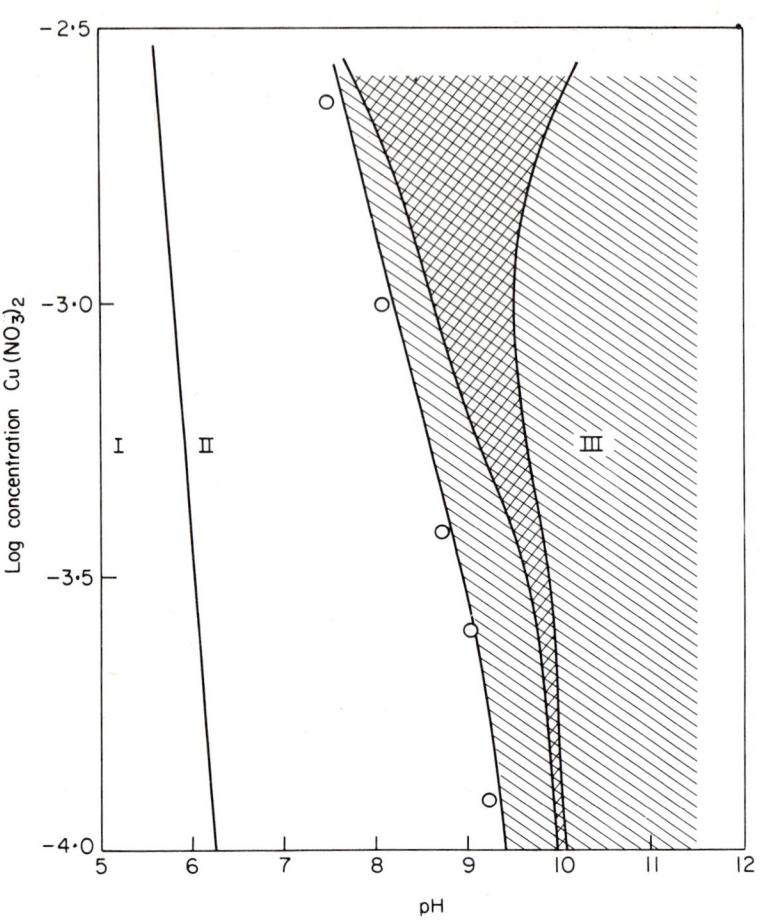

FiG. 5. Precipitation domains of cupric nitrate as a function of pH. Ageing time 10 minutes. Region I: homogeneous solution. Region II: stable colloidal dispersion, no heterocoagulation with microbubbles. Region III: coagulated dispersion, heterocoagulation with microbubbles. Double shaded area within region III denotes greater than 75 per cent removal of solid from dispersion after passage of microbubbles for 5 minutes.

3.3 INTERACTION OF MICROBUBBLES WITH POLYSTYRENE LATEX PARTICLES

Figures 6, 7 and 8 show electrophoretic mobilities and the efficiency of removal by flotation of polystyrene latex as a function of pH, potassium nitrate concentration and tetradecyl trimethyl ammonium bromide (C_{14} TAB) concentration respectively. The surface potential of the negatively charged polystyrene latex particles was clearly reduced by lowering the pH and by addition of KNO_3, and tended towards zero. (Electrophoretic measurements could not be performed above about 0·3 mol dm^{-3} KNO$_3$ or below about pH 1·3 because of gassing at the electrodes.) However there was no observable flotation. When C$_{14}$ TAB, a cationic surfactant, was added, the surface potential was not only reduced but, at concentrations in excess of about

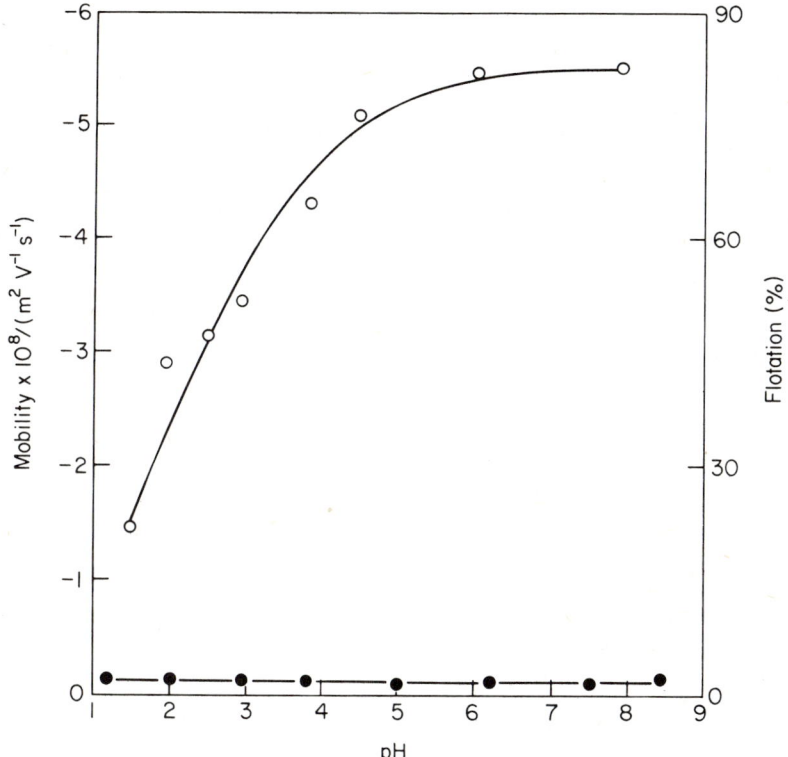

FIG. 6. Electrophoretic mobility (O) and flotation efficiency (●) of polystyrene latex dispersions as a function of pH. KNO$_3$ concentration 0·01 mol dm^{-3}.

4×10^{-6} mol dm^{-3}, was rendered positive. Again, however, no particles were separated by flotation. Thus for polystyrene latex particles surface potential had no influence on flotation.

4 Discussion

Although the effects of electrolyte solutions upon bubble size have been investigated before, usually with a view to increasing mass transfer by increasing available surface area, the use of electrolytes to generate micro-bubbles does not seem to have been reported hitherto.

Marucci and Nicodemo[13] studied the effect of various electrolytes at concentrations up to 0.7 mol dm^{-3}, on nitrogen bubbles. Although the bubbles they described were about one order of magnitude greater in size

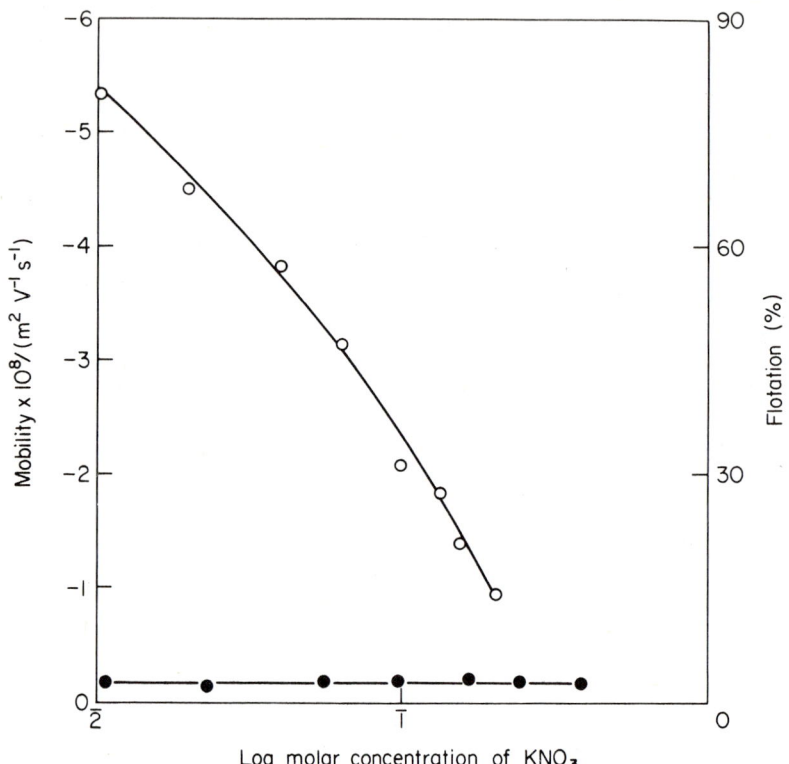

FIG. 7. Electrophoretic mobility (O) and flotation efficiency (●) of polystyrene latex dispersions as a function of KNO_3 concentration, pH 9.

than microbubbles they found a dependence upon concentration and upon valency similar to that found here. They concluded that in all cases the bubble size was the same at formation, but that coalescence near the distributor determined the mean size of bubbles in the column. In view of the charge dependence, Marucci and Nicodemo[13] opined that the effect was due mainly to electric repulsive forces which hindered coalescence between bubbles brought into contact by the movement of the liquid. Increasing prevention of coalescence by increasing electrolyte concentration reduced the mean bubble size. However, repulsive forces arising from an electric double layer are, in general, reduced rather than enhanced by increasing electrolyte concentration. Moreover, neither the similarity in the effects of anions and cations of the same valency upon bubble size, nor the differences exhibited

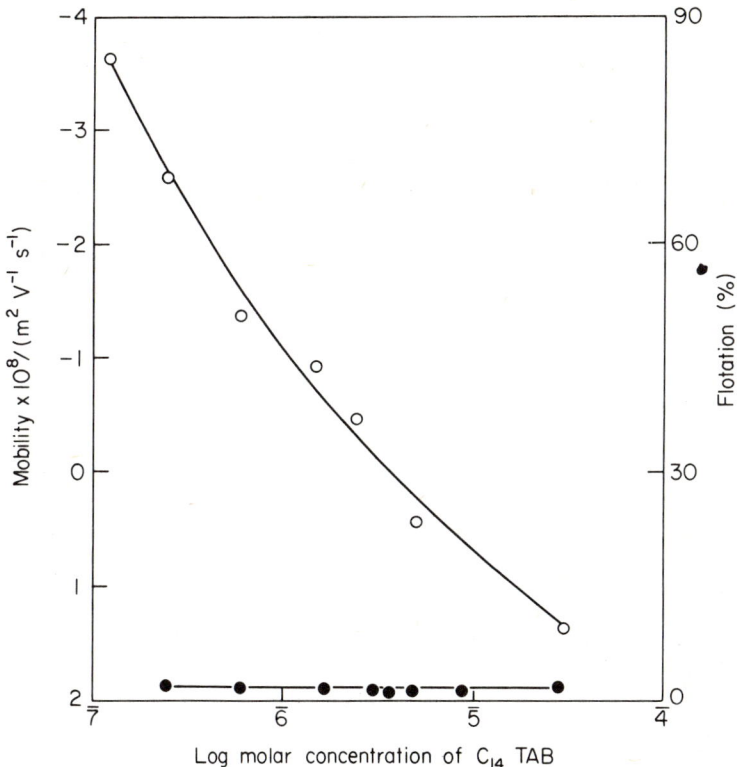

FIG. 8. Electrophoretic mobility (O) and flotation efficiency (●) of polystyrene latex dispersions as a function of tetradecyl trimethyl ammonium bromide (C_{14} TAB) concentration. KNO_3 concentration 0.01 mol dm^{-3}, pH 9.

by $NaNO_3$ and KNO_3 in the present work can be explained on the basis of double-layer repulsion.

That coalescence was increasingly retarded by the presence of increasing amounts of electrolyte was confirmed by Lessard and Zieminski[14] from experiments with two capillaries immersed side-by-side in electrolyte solutions. However, as Zieminski and Whitmore[15] found that KCl is much more effective than KI in decreasing bubble size, i.e. preventing coalescence, despite the fact that KI produces a higher surface charge than KCl,[16, 17] factors other than electric charge are important in controlling bubble size. Zieminski *et al.*[14, 15] ascribe the major part of the effect to the drainage properties of the thin film between two interacting bubbles and the influence of the presence of inorganic salts on local viscosity. Salts containing small or highly charged ions are "structure makers", and polarize the nearby water molecules, increasing the rigidity of the surface film, whilst other ions —"structure breakers"—actually lead to decreased local viscosities in the film. This explanation could account for the observation in the present work that $NaNO_3$ has a greater effect on bubble size than KNO_3, as Na^+ is a structure maker and K^+ a structure breaker.

Lessard and Zieminski[15] found a critical concentration for each electrolyte, depending upon valency and size of the ions, above which bubble coalescence was essentially eliminated. They were able to correlate these values with values of entropy of solution[18] and also with self-diffusion parameters for water.[19]

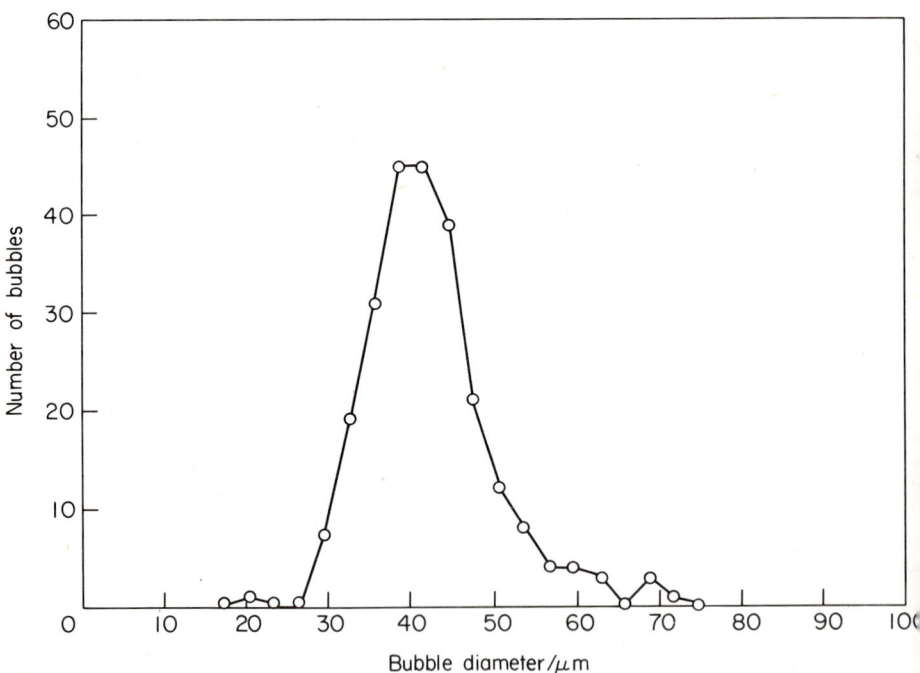

FIG. 9. Bubble size distribution obtained by passing nitrogen gas through water containing 2·5 ml l⁻¹ ethanol and 25 mg l⁻¹ lauric acid. Gas flow rate 30 ml min⁻¹; total number of bubbles 243; mean bubble diameter 42·1 μm; median bubble diameter 41·5 μm; modal bubble diameter 41·5 μm; standard deviation 7·7 μm. (After Kaufman, ref. 10).

Thus the nature of the liquid layer between interacting bubbles would appear to be the dominant parameter in bubble coalescence. Certainly similar viscous effects have been considered for Brownian coagulation of particulates, and Spielman[20] has calculated that Hamaker constants estimated from measured coagulation rates can be greatly in error if viscous forces are ignored.

Whilst the work of Marucci and Nicodemo[13] and Zieminski et al.[14,15] illustrates the importance of eliminating coalescence in order to retain primary bubble size, it provides no information regarding those factors which determine primary bubble size. It was noticeable in the present work that bubbling occurred from many more parts of the frit in the presence of electrolyte solutions than with pure water and this may suggest an effect on initial bubble size. Although primary bubble size is generally considered to be determined by the orifice diameter and the wettability of the orifice, other factors may well be important in a system as complex as a submerged frit. The bubbles described in the above work[13–15] were considerably larger than the microbubbles under consideration here but it is likely that similar principles will apply.

It is interesting to compare the microbubbles generated in electrolyte solutions with those used previously in microflotation.

Typically microflotation experiments are performed using a lauric acid collector and an ethanol frother at concentrations of 1.25×10^{-4} mol dm^{-3} and 2.5 mol dm^{-3} respectively. The size of bubbles produced under such conditions is shown in Fig. 9 (a size distribution taken from Kaufman[10]). Such bubbles may be produced also in the absence of the collector. Clearly the bubbles produced in the frother–collector mixture would appear to be smaller (~ 40 μm) than the bubbles produced in this work. However, whilst the mean and modal diameters of the bubbles described by Kaufman[10] are almost identical, indicating a high degree of monodispersity, this is not true of bubbles produced in electrolyte solutions. Indeed in this work mean diameter often exceeded modal diameter by up to 20 or 30 μm, indicating considerable polydispersity of bubble size. This is clear from Fig. 10 which shows bubbles produced in (a) a collector–frother mixture and (b) in 0.36 mol dm^{-3} Fe(NO$_3$)$_3$.

Thus coalescence was not entirely eliminated by electrolyte in these experiments and was undoubtedly exacerbated by the high gas flow rate (150 ml min^{-1}) necessary to produce sufficient bubbles at low electrolyte levels and maintained constant throughout the experiments. Preliminary measurements showed that mean bubble diameter decreased with decreasing flow rate due to a reduction in bubble number concentration and hence a reduction in the amount of coalescence. Indeed, primary bubbles were of comparable size to those generated in collector–frother solutions.

The reason for the spectacular reduction in bubble size on addition of alcohol (with or without a collector) during microflotation has not so far been established unequivocally. Zieminski *et al.*[21] reported an increase in efficiency with increasing alcohol chain length and also found similar effects with mono- and dicarboxylic acids. Again, as they suggest, prevention of coalescence is probably important, but, once more, there are likely to be other effects as the addition of ethanol greatly increases the number of sources of bubbling from the frit.

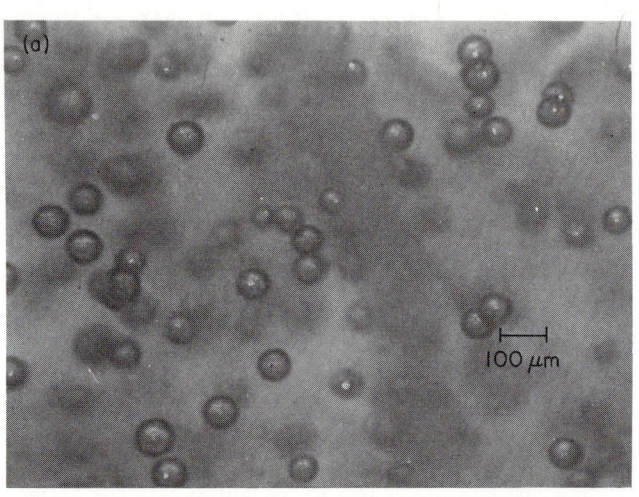

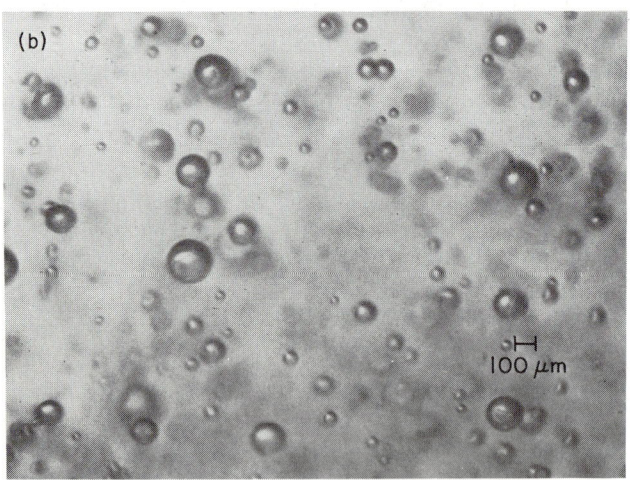

FIG. 10. Photomicrographs of microbubbles obtained (a) in $2 \cdot 5$ ml dm^{-3} ethanol and 25 mg dm^{-3} lauric acid; (b) in $0 \cdot 36$ mol dm^{-3} Fe(NO$_3$)$_3$.

The experiments on bubble interaction with copper hydrous oxide and polystyrene latex provided several interesting features which have a direct bearing on flotation. The influence of coagulation upon removal efficiency is one aspect of flotation which is not clearly understood. It is generally supposed that destabilizing a colloidal dispersion improves flotation not only from hydrodynamic considerations but also from a surface chemical point of view. Certainly with copper hydrous oxide heterocoagulation between bubbles and particles coincided with instability of particles, but with polystyrene latex this was certainly not the case.

It was observed that sometimes where adhesion between bubbles and copper hydrous oxide was good and gave large bubble–particle aggregates the latter were brought to the surface of the flotation vessel only to be redispersed by the turbulence caused by the gas flow. Actual flotation did not occur until the passage of nitrogen had ceased; then many of the composite aggregates rose to the surface. In practice such redispersion is prevented by the introduction of a surfactant to provide a stable foam; however this may not be desirable in cases where the floated material is to be recovered. Thus it is interesting to note that, in general, as can be seen from Fig. 5, flotation efficiency increased with increasing amount of solid. At the higher concentrations, where there was sufficient solid to completely cover the surface producing a quiescent layer, removal was total. Similarly, if two glass plates were suspended to a depth of about 0·5 cm in the liquid in the flotation vessel, again forming a quiescent layer, complete removal could be achieved even with low solid content.

Adhesion between microbubbles and polystyrene latex particles, as indicated by flotation efficiency, did not take place under any conditions tested.

The possibility of flotation having been masked by redispersion of floated particles by turbulence may be discounted, as stable foams which would have supported any floated material were produced in the presence of C_{14} TAB. Furthermore, addition of lauric acid to latex dispersions containing KNO_3 produced stable foams but no flotation.

That the lack of flotation of latex particles could not be attributed to a low collision efficiency was illustrated by allowing a concentrated latex sample to stand overnight in 0·5 mol dm^{-3} KNO_3 before attempting flotation. The large aggregates of latex particles ensured optimum collision[22] but again flotation did not ensue.

Under conditions where the ζ-potentials of the polystyrene particles were highly negative, lack of flotation could be attributable to electrostatic repulsion. However, when the ζ-potential on the particles was reduced or eliminated by lowering pH or adding KNO_3 or C_{14} TAB, or reversed by adding higher concentrations of C_{14} TAB, flotation efficiency did not improve. Although particle number concentration was so low that effectively coagulation did not take place during the time scale of a flotation experiment, it was confirmed in separate experiments that such changes in ζ-potential did indeed lead to coagulation. Clearly, under such conditions nonfloatability cannot be explained in terms of electrostatic repulsion. Although it is usual to discuss flotation in terms of the stability of the dispersion, i.e. particle–particle interaction, a more meaningful approach can be made if particle–bubble interactions are studied directly, and a possible explanation lies in a consideration of the van der Waals attraction between bubbles and particles. In order for flotation to take place there must be an attraction between bubbles and particles which is sufficiently strong to overcome the disjoining tendency of turbulence forces.

The Hamaker interaction constant between two different particles embedded in a medium is given approximately by

$$A = (A_{11}^{\frac{1}{2}} - A_{22}^{\frac{1}{2}})(A_{33}^{\frac{1}{2}} - A_{22}^{\frac{1}{2}}) \tag{1}$$

where A_{11}, A_{22} and A_{33} are the Hamaker constants of particle, water and bubble respectively.

For a bubble A_{33} is zero, hence:

$$A = A_{22} - (A_{11}A_{22})^{\frac{1}{2}}. \tag{2}$$

Generally accepted values for water and polystyrene are 3.6×10^{-20} J and 6.4×10^{-20} J respectively:

Thus

$$A = 3.6 \times 10^{-20} - (3.6 \times 10^{-20} \times 6.4 \times 10^{-20})^{\frac{1}{2}}$$
$$A = -1.2 \times 10^{-20} \text{ J}.$$

Although it is now well known that the Hamaker treatment does not take account of the contributions from all frequencies this neglect is minimal in the presence of moderate or high concentrations of electrolyte, and the above oversimplification is probably not too serious. Thus, the limitations of this simple approach notwithstanding, it is likely that the overall Hamaker constant will be negative. This would mean that the London–van der Waals forces between bubble and polystyrene latex particle would be repulsive instead of attractive, hence there should be no flotation of polystyrene, whatever the stability of the dispersion. This was indeed established in these experiments which confirm that prediction of flotation from a consideration of particle–particle interaction alone is not valid. Certainly coagulating a colloidal dispersion would probably improve collision frequency[22] and this in itself can aid flotation, but to actually predict the nature of bubble particle interaction (and hence flotation) bubble–particle interaction must be considered directly.

Reliable information on the Hamaker constant of copper hydrous oxide is not at present available so there can be no direct comparison of this system with polystyrene latex. It is interesting to note that according to equation (1) strong van der Waals repulsion between bubbles and particles will occur only with those materials whose Hamaker constant is considerably greater than that of water. However experiments with Teflon dispersions,[23] where this is not the case,[24] have also resulted in a lack of flotation, whereas silver iodide (Hamaker constant appreciably greater than that of water[24]) has been readily floated.[25]

It is our intention in the future to study many more materials in order to develop a greater understanding of the situation.

The above findings on the floatability of "hydrophilic" particles and the nonfloatability of "hydrophobic" particles dispel the often-expressed idea that in order to improve flotation a colloid must be made more "hydrophobic" and also confirm a need for more fundamental studies on flotation and its related topics.

References

1. Karger, B. L., Grieves, R. B., Lemlich, R., Rubin, A. J. and Sebba, F. (1967). *Separ. Sci.* **2**, 401.
2. Rubin, A. J. (1968). *J. Amer. Water Works Ass.* **60**, 832.
3. Rubin, A. J., Cassell, E. A., Henderson, O., Johnson, J. D. and Lamb, J. C. (1966). *Biotechnol. Biol.* **8**, 135.
4. Rubin, A. J. and Cassell, E. A. (1965). *Proc. 14th Southern Water Resources and Pollution Control Conf.* **14**, 222.
5. Mangravite, F. J. Jr., Buzzell, T. D., Cassell, E. A., Matijević, E. and Saxton, G. B. (1975). *J. Amer. Water Works Ass.* **67**, 88.
6. Mangravite, F. J. Jr., Cassell, E. A. and Matijević, E. (1972). *J. Colloid Interface Sci.* **39**, 357.
7. Cassell, E. A., Matijević, E., Mangravite, F. J. Jr., Buzzell, T. D. and Blabac, F. (1971). *AIChEJ,* **17**, 1486.
8. Melville, J. B. and Matijević, E. (Submitted). *J. Colloid Interface Sci.*
9. Goodwin, J. W., Hearn, J., Ho, C. C. and Ottewill, R. H. (1973). *Brit. Polym. J.* **5**, 347.
10. Kaufman, K. M. (1971). M.S. Thesis, Clarkson College of Technology, Potsdam, NY.
11. Cassell, E. A., Kaufman, K. M. and Matijević, E. *Water Res.* In press.
12. McFadyen, P. and Matijević, E. (1973). *J. Inorg. Nucl. Chem.* **35**, 1883.
13. Marrucci, G. and Nicodemo, L. (1967). *Chem. Eng. Sci.* **22**, 1257.
14. Lessard, R. R. and Zieminski, S. A. (1971). *Ind. Eng. Chem. Fundam.* **10**, 260.
15. Zieminski, S. A. and Whittemore, R. C. (1971). *Chem. Eng. Sci.* **26**, 509.
16. Frumkin, A. (1974). *Z. Phys. Chem.* **109**, 34.
17. Jarvis, N. C. and Schuman, M. A. (1968). *J. Phys. Chem.* **72**, 74.
18. Latimer, W. (1955). *J. Chem. Phys.* **23**, 90.
19. McCall, D. and Douglass, D. (1965). *J. Phys. Chem.* **69**, 2001.
20. Spielman, L. A. (1970). *J. Colloid Interface Sci.* **33**, 562.

21. Zieminski, S. A., Caron, M. M. and Blackmore, R. B. (1967). *T. & E. C. Fundamentals*, **6**, 233.
22. Reay, D. and Ratcliffe, G. A. (1973). *Can. J. Chem. Eng.* **51**, 178.
23. Kratohvil, S. and Matijević, E. Unpublished results.
24. Visser, J. (1972). *Advan. Colloid. Interface Sci.* **3**, 331.
25. Melville, J. B. Unpublished results.

Discussion

Sheiham (*Water Research Centre, Medmenham, England*) Is there any evidence from kinetic data or any other source that microbubbles do not easily escape from the solution during flotation. Has a relation been sought between the amount of material recovered and surface area of gas–water interface.

Melville In our experiments we observed that whilst some microbubbles rose directly the trajectory of others was affected by turbulence to the extent that they were recirculated once before escaping. It is not known at this stage what proportion of bubbles were thus affected.

Secondly, it is known that for particles of colloidal dimensions flotation efficiency is greater when the bubble size is smaller, i.e. when the surface area is higher.

Roberts, K. On Scandinavian economics, the cost of generation of microbubbles is not significant in the total cost of a cleaning plant. A dissolved air technique can and has been used successfully for over 30 years, so I am not sure where a large economic advantage occurs if a new generation system for bubbles is developed.

Secondly, in systems where a few ppm of surface-active or other materials can change microflotative removals drastically, what do you think of adding 2500 ppm ethanol in the system as a quasi "nonactive" material.

Melville The fact that dissolved air flotation (DAF) is in use at present does not exclude the possibility that both technological and economic improvements can be made in microbubble production. The capital costs involved in installing pressure vessels, compressors etc. for DAF are appreciable. Moreover, it follows that the considerably lower pressures involved in the passage of gas through a frit represents a proportionately lower expenditure of energy, thus reducing running costs.

Secondly, the effect of ethanol on colloid stability has been extensively studied[1,2,3], and at concentrations as low as 0·25 per cent has been found to be negligible. Similarly the effect on surface tension is negligible.[4] In our experiments the aim, as stated, was to correlate bubble–particle adhesion with colloid stability. In the polystyrene latex work correlation was direct as ethanol was present in both flotation and electrophoresis experiments. In the copper hydrous oxide work precipitation domains prepared in

[1] Tezak, B., Matijevic, E. Shulz, K., Mirnik, M., Merak, J., Vauk, V. B., Slunjski, M., Babic, S., Kratohvil, J. and Palmer, T. (1953). *J. Phys. Chem.* **57**, 301.
[2] Kratohvil, J. P., Orhanovic, M. and Matijevic E. (1960). *J. Phys. Chem.* **64**, 1216.
[3] Matijevic, E., Ronayne, M. E. and Kratohvil, J. P. (1966). *J. Phys. Chem.* **70**, 3833.
[4] Kaufman, K. M. (1971). M. S. Thesis, Clarkson College of Technology, Potsdam, N.Y.

the presence and absence of 0·25 per cent ethanol were found to be identical; the region in which heterocoagulation between bubbles and particles occurred was also unaffected by the presence of alcohol.

Akers You draw attention to that fact that collision is a necessary prerequisite for flotation. Particle–bubble collision is controlled essentially by hydrodynamic factors and these must be considered in any analysis. Is heteronucleation at the particle surface a significant factor in dissolved air flotation as it obviates the need for capture? In the presentation you referred to the bubbles being ejected from the frit. If this were to occur, and surely capillary detachment is more likely, the stop distance, i.e. the distance for the bubble to come to rest due to viscous drag, would be of the order of 10^{-3} — 10^{-4} μm. Hence the bubble would spend most of its time rising at its Stokes's velocity of about 1·4 mm s^{-1}.

Melville Flotation studies on a wide range of colloidal materials have been carried out[1] in Port Sunlight using the dissolved air technique. There was no evidence of significant bubble nucleation at the particle surface and it would seem that the collection mechanism was that of particle capture by rising bubbles. Calculations of the Stokes's velocity are probably not relevant to the practical situation in view of the turbulent motion of the liquid. Indeed, in our experiments, the microbubbles rose to the surface more rapidly than would have been predicted by the Stokes's equation.

[1] F. Jones. Private communication.

15

The conformation of proteins at the air–water interface and their role in stabilizing foams

D. E. GRAHAM and M. C. PHILLIPS

Biosciences Division, Unilever Research Laboratory,
Colworth/Welwyn, The Frythe, Welwyn AL6 9AG, Hertfordshire, England

Summary

The foaming behaviour of solutions of β-casein (a flexible more-or-less random coil molecule) and lysozyme and bovine serum albumin (globular proteins of different stabilities) are correlated with the adsorption characteristics of these proteins at the air–water interface. Foamability is related to the rate of decrease of the surface tension of the air–water interface by protein molecules whereas foam stability is related to the structure of the adsorbed protein films. Thus, flexible protein molecules which can rapidly decrease the surface tension of the air–water interface give good foamability whereas highly ordered globular molecules which are difficult to surface-denature give poor foamability. Rapid build-up of a film pressure by protein tends to lead to formation of a coarse foam (i.e. large air bubbles) whereas a slow increase favours small air bubbles (i.e. a creamy foam). Prior denaturation (e.g. by heat or extremes of pH) of globular proteins under conditions where no precipitation occurs leads to enhanced foamability. Once formed, the foams stabilized by proteins which are globular in the native state are more stable than foams prepared with the flexible β-casein molecule. The rheological properties of the protein films adsorbed at the surfaces of the air cells are of overriding importance in determining the stability of a foam. Stability is enhanced when the films are resistant to shear and incompressible (i.e. have a high dilatational modulus). Provided there is no precipitation of protein, the foam stability is greatest at the isoelectric point of the protein.

1 Introduction

It has been known for many years that protein solutions foam very effectively[1] and protein-stabilized foams have many industrial applications. However, because complex mixtures of proteins are normally foamed, the foaming characteristics of different types of pure proteins are as yet unknown. Also, the mechanisms by which adsorbed protein layers at the air–water interface stabilize foams are not understood; the main stumbling block here has been that basic data, such as the adsorption isotherms of the proteins, have not been available. In an earlier study[2] we measured the adsorption isotherms for β-casein and lysozyme and here we build on this information to obtain a qualitative correlation between the foaming behaviour of the proteins and their molecular structure.

Three proteins with different tertiary structures have been used, viz. the flexible more or less random coil β-casein molecule, the rigid globular lysozyme molecule (temperature of denaturation[3] $T_{denat} = 72°C$) and the more labile globular bovine serum albumin (BSA) molecule ($T_{denat} = 65°C$, ref. 3). The foamabilities of solutions of these pure proteins and the stabilities of the resultant foams have been monitored. The effects of prior denaturation of the globular proteins are also discussed. Knowledge of the rheological properties of the adsorbed protein layers and the effects of pH are used to indicate which properties of the surface-denatured protein films are important for foamability and foam stability. It seems that rapid adsorption of protein is necessary for good foamability, and that surface layers of protein which are resistant to shear and are incompressible promote foam stability.

2 Experimental

2.1 MATERIALS

The proteins. bovine serum albumin (BSA, crystallized and lyophilized puriss grade, Koch Light Laboratories Ltd, Buckinghamshire, UK), lysozyme (3 × crystallized, dialysed and lyophilized from egg white. Sigma Chemical Co., London, UK) and β-casein A (supplied by Dr M. T. A. Evans) and α-lactalbumin (supplied by Dr R. A. Badley of this laboratory) were stored at 5°C and used without further purification. The proteins used in tracer studies were 1–[14]C-acetyl derivatives. The acetylation and purification of BSA was carried out in an identical fashion to that described in detail elsewhere[4,5] for β-casein A and lysozyme. The acetylation was shown to have no effect on the surface activity of β-casein, although with lysozyme and BSA a slight increase was observed.[4]

The pH 7 buffer solutions have been described previously;[4] for experiments at other pH's, NaOH or HCl were added to the subphase of 0.1 mol dm^{-3} NaCl so that a constant ionic strength was maintained.

2.2. APPARATUS

2.2.1 Adsorption studies

The use of a gas flow counter for measuring the surface radioactivity resulting from adsorbed ^{14}C-labelled protein has been described else-where.[4]

2.2.2 Shear viscoelastic studies

The air-bearing viscoelastometer used for measuring the shear properties of adsorbed protein films has been described fully before.[6] In these studies the creep compliance curve was measured after the protein had adsorbed to a steady-state surface pressure. A constant shearing stress of either 0·007 or 0·016 mN m^{-1} was applied. The total displacement of the glass disc in the interface was less than 5×10^{-2} radians; displacements of about 2×10^{-4} radians could be detected.

2.2.3 Dilatational modulus

The surface dilatational modulus was measured with a longitudinal wave method as developed by Lucassen et al.[7, 8] Sinusoidal waves were generated by a movable barrier at the surface of water contained in a Langmuir trough made of Teflon (capacity $\sim 100 \text{ cm}^3$ and surface area/volume $= 1 \text{ cm}^{-1}$). This periodic compression and expansion of the surface at variable frequency and amplitude deforms the mean trough area, A, by an amount dA. The change in the surface pressure (dπ) caused by a relative change in surface area (dA/A) is related by

$$\varepsilon = -A \left(\frac{d\pi}{dA}\right)_T \tag{1}$$

where ε is the surface dilatational modulus at temperature T.

The protein was injected below a clean interface as described elsewhere,[4] and the change in surface pressure (π) followed using a roughened glass Wilhelmy plate (perimeter 4·8 cm) suspended from one arm of an electro-microbalance (EMB 1, R & I Instruments, London). After adsorption to a steady state π, the force changes observed by the electromicrobalance during the movement of the barrier were recorded on a chart recorder (Telsec 700, Oxford Instruments, UK).

2.2.4 Foam stability

5-ml aqueous solutions of each protein were agitated in 25-ml graduated cylinders in a horizontal position using a mechanical shaker (Gallenkamp, London, UK) at a preset rate of 30 Hz. The volume of foam created was measured as a function of the shaking time. In order to compare the stabili-ties of foams, the solutions were first shaken for 60 min, and then the gradu-ated cylinders were stood upright and the foam volumes recorded over periods of several hours. We used a shaking technique rather than a gentle bubbling technique to prepare the protein foams because we wished any effects of protein coagulation to become apparent. Since there is no standard procedure for making foams,[9] the foam parameters discussed in this paper are self-consistent but they are only qualitatively, rather than quantitatively, comparable with the results of other foaming techniques.

3 Results

3.1 ADSORPTION CHARACTERISTICS OF PROTEINS

The determinations of surface concentration (Γ) and surface pressure (π) at the air–water interface were carried out with $1-^{14}C$-acetyl derivatives of the native proteins (for simplicity, the proteins are referred to below as lysozyme,

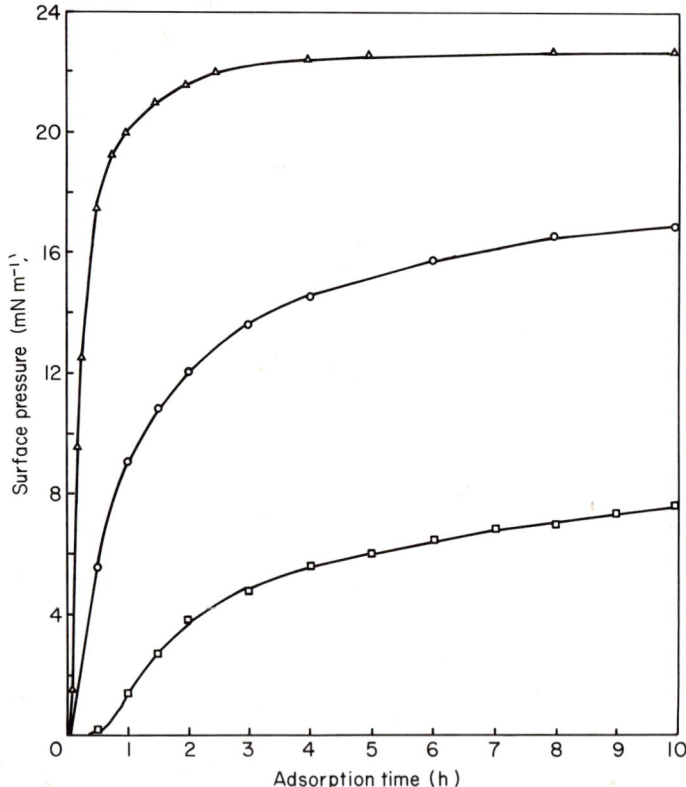

FIG. 1. Rate of increase of surface pressure at an air–water interface during the adsorption of $1-^{14}C$-acetyl derivatives of β-casein A (△), BSA (○), and lysozyme (□). The initial substrate protein concentration was 5×10^{-4} wt % and the aqueous phase was phosphate buffer. pH = 7·0 and $I = 0·1$.

BSA and β-casein). The rates at which the three proteins decrease the surface tension of the air–water interface are shown in Fig. 1. Clearly β-casein adsorbs fastest to give steady state π values whereas the globular proteins change the surface tension more slowly; this applies for a range of substrate protein concentrations (C_p).

Prior heating of solutions of BSA and lysozyme causes an increase in the rate at which the molecules adsorb to the air–water interface, i.e. π reaches steady-state values faster than for solutions of the corresponding native, unheated protein solutions. Also, provided C_p is less than that which provides for saturated monolayer coverage, the steady-state π and Γ values are greater than those obtained with native BSA and lysozyme solutions for a given C_p (see Table 1).

TABLE 1

The effects of prior heat denaturation on the adsorption of lysozyme and BSA from solutions containing 10^{-4} wt % protein

	Lysozyme		BSA	
	Native	Heat denatured	Native	Heat denatured
Surface pressure π/mN m^{-1}	8	21	14	17·5
Surface concentration Γ/mg m^{-2}	1·9	2·5	1·7	2·8

Variations in the substrate pH do not significantly change the adsorption behaviour of either BSA or lysozyme. The steady-state π and Γ values for BSA, lysozyme and β-casein with $C_p = 1 \times 10^{-4}$ wt % do not alter over the pH range 2–10 (data for β-casein could not be obtained at pH < 5 because the protein was insoluble).

3.2 FOAM STUDIES

Native proteins The foamabilities of solutions of β-casein, BSA and lysozyme are demonstrated by the foam expansion as a function of the time of shaking (Fig. 2). The shaking times required to obtain half the maximum expansion are about 4 ± 1, 12 ± 2 and >30 min for 0·1 wt % β-casein, BSA and lysozyme solutions respectively. The lysozyme time is only an indication, because the volume of foam continued to increase with shaking for periods up to several hours. The times quoted assume that a maximum expansion occurs after two hours' shaking. We can deduce that β-casein solutions can be foamed more rapidly than BSA solutions, which in turn foam faster than lysozyme solutions. However, the maximum foam volumes attained with protein solutions of the same C_p are ranked as follows; BSA > β-casein > lysozyme.

The lifetimes of the foams prepared with 0·1 wt% protein solutions are shown in Fig. 3. The times $(t_{\frac{1}{2}})$, for the foams to decay to half their initial volumes are 16 ± 5, 32 ± 8 and > 200 min for β-casein, BSA and lysozyme

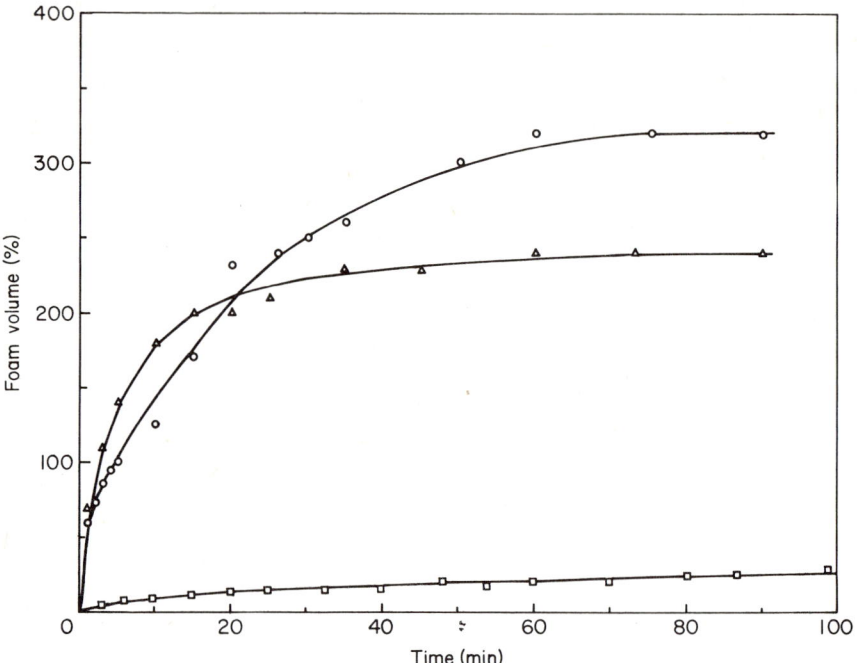

FIG. 2. Rate of increase of foam volume during shaking 0·1 wt% protein solutions in 0·1 mol dm^{-3} NaCl (pH ~ 5·5) of β-casein (△), BSA (○), lysozyme (□). The foam volume is given as the per cent increase in the volume of initial aqueous solution (e.g. 100 per cent foam volume = 5 ml foam).

foams, respectively. The decay of foams stabilized by both β-casein and BSA (in the presence of 0·1 mol dm^{-3} NaCl) follows first-order kinetics. However, different kinetic rates can be obtained by changing the substrate conditions, e.g. in distilled water the same BSA-stabilized foam decays at a pseudo second-order rate.

Effects of heat and pH denaturation Foams were made from protein solutions which had been previously heated; unless otherwise stated, the protein solutions were not turbid, i.e. large-scale protein aggregation and precipitation had not occurred. The results for foams stabilized by BSA and lysozyme are shown in Fig. 4. The heated lysozyme solution generates a greater foam volume than the solution of native protein but $t_{\frac{1}{2}}$ is smaller than that of the native protein solution (i.e. $\sim 150 \pm 10$ min as compared to > 200 min). Similar effects were observed with heated 0·01 wt% BSA solutions. In contrast, a heated 0·1 wt% BSA solution did not have significantly different foam characteristics from the native BSA, and both the foam volume and $t_{\frac{1}{2}}$

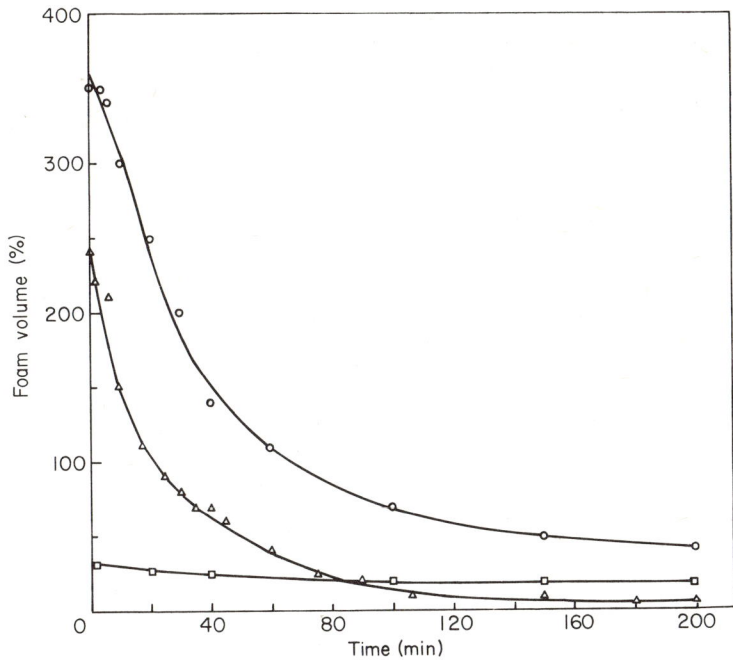

FIG. 3. Rate of collapse of foams prepared by shaking for 60 min 0·1 wt% solutions in 0·1 mol dm^{-3} NaCl (H $\sim$ 5·5) of β-casein (△), BSA (○), and lysozyme (□).

are reduced when a 1·0 wt% BSA solution is heated before foaming (Fig. 5). In the latter case the solution became turbid due to the aggregation of heat-denatured protein, indicating that the presence of aggregated protein is detrimental to the foam volume and foam stability.

The foam characteristics of 0·1 wt% BSA (isoelectric point pI $\sim$ 5), lysozyme (pI $\sim$ 10·5) and α-lactalbumin (pI $\sim$ 4·5) solutions as a function of the substrate pH are given in Fig. 6 (a), (b) and (c), respectively. It is interesting to note that the initial foam volumes obtained with 0·1 wt% protein solutions do not generally appear to depend upon the substrate pH over the range from about 2–11. However, 0·1 wt% hysozyme solutions

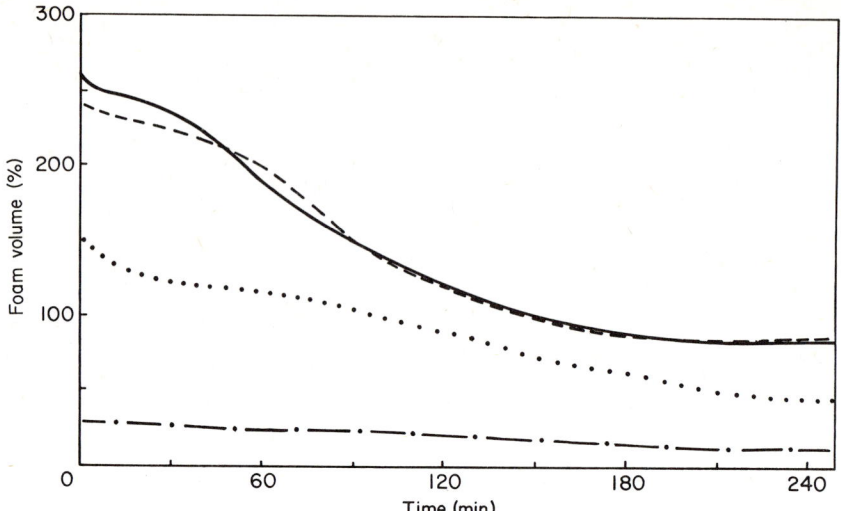

FIG. 4. Rate of collapse of foams prepared by shaking 0·1 wt% solutions in 0·1 mol dm^{-3} NaCl of native BSA (—), heat-denatured BSA (– – –), native lysozyme (– · –), and heat denatured lysozyme (.......). The proteins were heat denatured by holding the solutions at 80°C for 20 min; the solutions were cooled to 20°C prior to foaming.

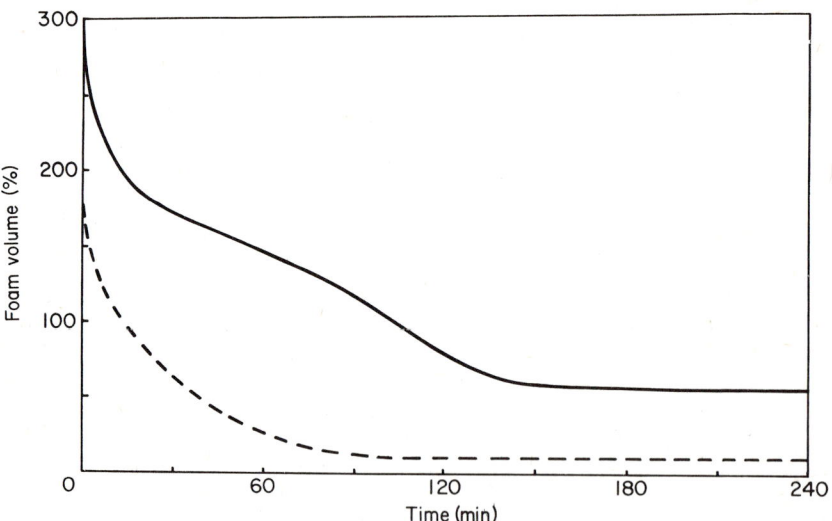

FIG. 5. Rate of collapse of foams prepared by shaking 1·0 wt% solutions in 0·1 mol dm^{-3} NaCl of native BSA (—), and heat denatured BSA (– – –). The conditions for heat denaturation were as given in Fig. 4.

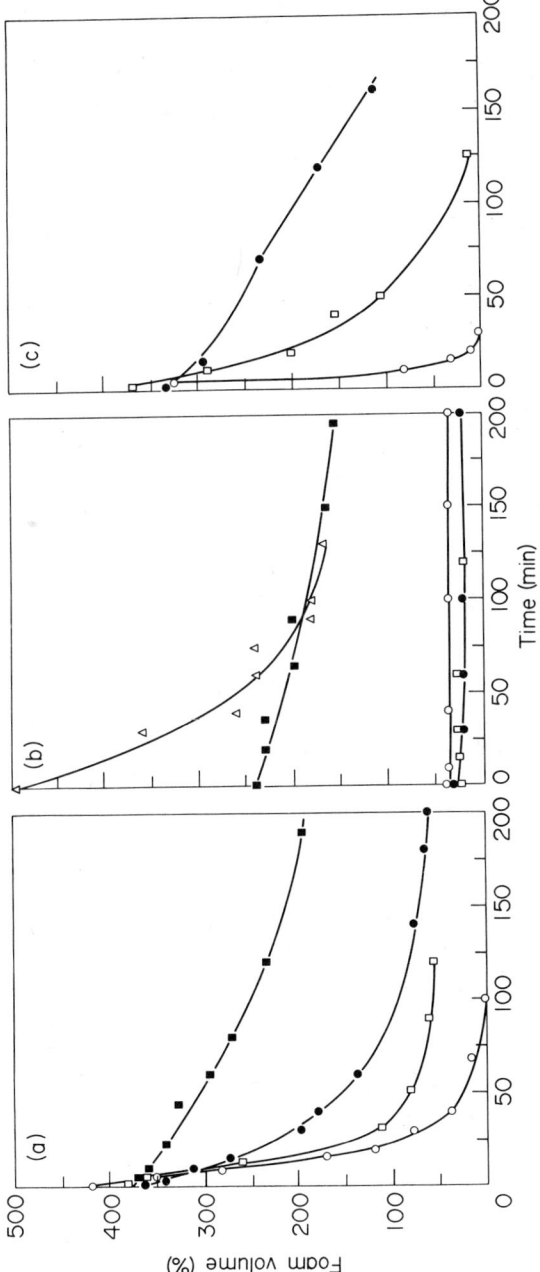

Fig. 6. Rate of collapse of foams prepared by shaking 0·1 wt% solutions of protein in NaCl (I = 0·1) of various pH's. (a) BSA at pH = 2·5 (O); 5·5 (●); 8·0 (□) and 12·0 (■). (b) Lysozyme at pH = 1·9 (O); 7·0 (●); 11·0 (□); 11·6 (■) and 12·4 (△). (c) α-Lactalbumin at pH = 1·3 (O); 7·0 (●) and 11·0 (□). The solution pH was manipulated by addition of HCl at NaOH to maintain a constant ionic strength of 0·1.

are an exception at pH values >11; the foam characteristics change (i.e. the initial foam volume increases whereas the stability decreases) and the aqueous solution becomes turbid after foaming. This is likely to be due to some conformational change in the lysozyme structure which also causes the protein to be surface coagulated. In contrast to the general foamabilities, the foam stabilities are sensitive to pH. In particular, there are marked effects at the extremes of pH, e.g. at very high pH's the stability of BSA-stabilized foams increases ($t_{\frac{1}{2}} \sim 200 \pm 10$ min) and at low pH's the stability is very small ($t_{\frac{1}{2}} \sim 12 \pm 2$ min). Highly alkaline solutions of lysozyme foam very well and give comparatively stable foams (Fig. 6(b)). It is interesting to note that α-lactalbumin-stabilized foams are generally less stable than BSA- and lysozyme-stabilized foams and are more comparable to β-casein-stabilized foams.

3.3 RHEOLOGICAL PROPERTIES OF ADSORBED PROTEIN FILMS

The dilatational modulus (ε) and shear viscoelastic moduli of adsorbed films of β-casein have been discussed fully elsewhere.[10, 11] The dilatational modulus is elastic at low $\Gamma (<1$ mg m^{-2}) but exhibits a small viscous element at higher Γ. Similarly, the shear properties are comparatively low, the viscosity being <1 mN s m^{-1} and the elastic modulus <0.1 mN m^{-1}. Adsorbed films of both BSA and lysozyme contain a viscous component and so ε becomes complex. However, values of ε have been measured at several oscillatory frequencies and have been extrapolated to zero frequency for comparison. ε as a function of C_p exhibits maxima for both proteins at C_p's $\sim 8 \times 10^{-5}$ wt% for BSA and at $\sim 8 \times 10^{-5}$ and $\sim 8 \times 10^{-4}$ wt% for lysozyme. Generally the ε values are greater for the adsorbed films of lysozyme than BSA, which are in turn greater than those for the β-casein films.

The shear viscoelastic properties of the protein films are analysed from creep compliance curves using the method of Inokuchi[12], and this indicates that the films behave as Burger bodies,[10, 13] i.e. they can be represented as being comprised of dashpots (viscous parts) and springs (elastic parts). The principal viscous (η_1) and elastic (G_1) components both exhibit a maximum at the air–water interface when $C_p \sim 5 \times 10^{-3}$ and 6×10^{-4} wt% for BSA and lysozyme,[10] respectively. The maximum η_1 and G_1 for lysozyme are $>10^4$ mN s m^{-1} and 15 mN m^{-1} (cf. ref. 10) whereas the equivalent figures for BSA are 2×10^2 mN s m^{-1} and ~ 4 mN m^{-1}.

In summary the highly ordered lysozyme molecule adsorbs to the air–water interface to form a cohesive film which, compared to those formed from BSA or the disordered highly flexible β-casein molecule, is very resistant to mechanical deformation. The rheological properties are summarized in Table 2.

TABLE 2

Comparison of the surface characteristics of the proteins used in this study

Protein	Surface pressure $\pi/(\text{mN m}^{-1})$	Surface concentration $\Gamma/(\text{mg m}^{-2})$	Dilatational modulus‡ $\varepsilon/(\text{mN m}^{-1})$	Shear viscosity $\eta_1/(\text{mN s m}^{-1})$	Shear elastic modulus $G_1/(\text{mN m}^{-1})$
β-casein	5	0·6	26		
	10	1·0	17	<1†	$<0·1$†
	20	2·3	11		
BSA	5	1·3	~30	~8	$<0·1$
	10	1·6	~35	10	0·2
	17*	2·4	34	55	1·2
Lysozyme	5	1·7	~45	~10	$<0·1$
	10	2·9	~41	$>10^4$	15
	20	7·0	~30	20	1·0

* Comparison is made at $17\,\text{mN m}^{-1}$ because the collapse pressure for BSA is $\sim17{\cdot}5\,\text{mN m}^{-1}$.
† These values are the limit of sensitivity of the apparatus.
‡ These values are measured 9 cm from the moving barrier in a trough 20 cm long.

4 Discussion

4.1 FOAMABILITY

The correlation between the data in Figs 1 and 2 (the difference of several orders of magnitude in the surface to volume ratio in the foam and Langmuir trough situation allows this comparison to be made) indicates that rapid changes in π favour the formation of a foam of large volume. This arises because unless the fresh air–water interface created around the surface of air bubbles during shaking is not quickly stabilized by an adsorbed protein film, rapid phase separation of the air and water ensues. The presence of adsorbed protein not only reduces the free energy of the interface but also increases its dilatational modulus and its resistance to shear (Table 2). As a result the thin water layers between adjacent air bubbles are better able to withstand the mechanical deformation they are subjected to while shaking continues. The overall volume of foam increases as long as the rate of creation of new air cells is greater than their rate of collapse. When these two rates are equal the foam volume becomes constant with time of shaking; under our conditions this occurred after approximately 60 min with β-casein and BSA (see Fig. 2). Protracted shaking can lead to surface coagulation[14] and precipitation of protein; when this occurs the foam volume starts to decrease with shaking time.

I

From earlier work in this laboratory[2, 4, 5, 10, 11] we know that rapid changes in π with time are characteristic of the adsorption of flexible proteins which can easily change their conformation on arrival at the air–water interface. Thus the more or less random coil β-casein molecule adsorbs quickly to give a film which at low Γ is comprised largely of trains of amino-acid residues in the plane of the interface (π for such a film can be calculated from theories based on polymer statistics[11]). and at high Γ contains molecules in a looped configuration with a large number of residues out of the plane of the interface (see schematic diagram in Fig. 7). As a result of this

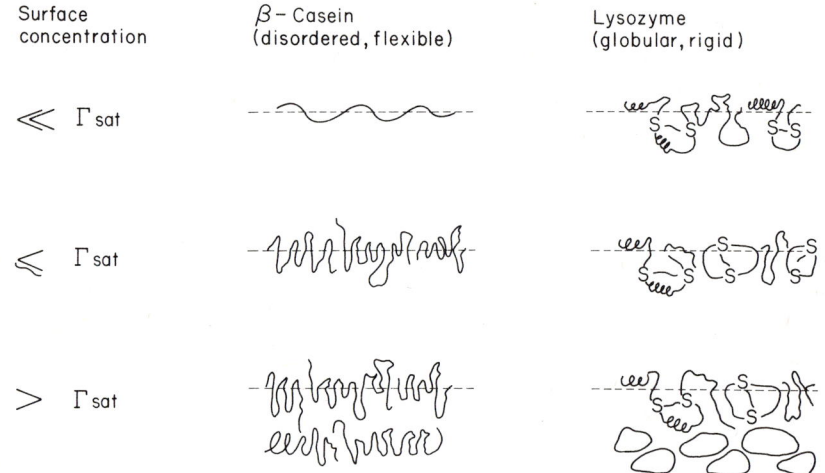

FIG. 7. A schematic representation of the structures of adsorbed films of β-casein and lysozyme at different surface concentrations at the air–water interface.

ability to quickly cover the air–water interface, solutions of β-casein foam very readily on shaking (Fig. 2). Although the rates of arrival of lysozyme and BSA in the interfacial region are similar to that of β-casein (i.e. $d\Gamma/dt$ approximately the same), presumably the proportion of residues in the interface for a globular protein is less than that for β-casein, and so $d\pi/dt$ for lysozyme and BSA is reduced (Fig. 1). Because of their rigidity, the partial unfolding (Fig. 7) and rearrangement of the globular proteins in the interface is relatively slow.[2] Consequently, fresh air cells formed during shaking are not stabilized so effectively and the rate of increase of foam volume with time is relatively low (Fig. 2). An additional effect of slow protein adsorption is that small air bubbles are preferentially stabilized and, for example, the low foam volume formed by lysozyme is finely textured and comprised of small air bubbles (i.e. it is "creamy"). In contrast, the high-volume foam formed by the fast-adsorbing β-casein is relatively coarse and contains large air cells. The stabilities of these two types of foam are markedly different and this is considered in the next section.

4.2 FOAM STABILITY

A previous study[10] has shown that oil-in-water emulsions stabilized by β-casein, BSA and lysozyme, do not differ too greatly in their degree of coalescence when subjected to accelerated ageing in a centrifuge. This behaviour clearly contrasts to that observed when the stabilities of foams, prepared with the same proteins, are compared (Fig. 3). The thin aqueous lamellae (diameter of area of contact of two neighbouring droplets is of the order of 10^{-5} cm) separating the oil droplets in an emulsion are stabilized by protein adsorbed at the oil–water interfaces. These thin aqueous films are transient[15] and probably rupture by the growth of surface corrugations[10,15] (the mode of growth of these corrugations is the "stretching mode"[16]). Thickness fluctuations due to surface waves of wavelengths greater than a certain critical value can grow and cause rupture. This mechanism can operate as long as the attractive forces across the thin film are much larger than the forces opposing thinning (electrostatic and steric[17] repulsion arising from the protein layers). In this mode the growth of the corrugations is independent of ε (when $\varepsilon \gtrsim 10^{-4}\,\mathrm{mN\,m^{-1}}$) (ref. 16), i.e. the interfacial films are considered to be effectively rigid. Since β-casein, lysozyme and BSA give surfaces with $\varepsilon \gg 10^{-4}\,\mathrm{mN\,m^{-1}}$ as soon as any significant adsorption has occurred (see Table 2), the absence of any correlation between the rheological properties of adsorbed protein films and the stability of emulsions[10] is to be expected.

In marked contrast to this observation with protein-stabilized emulsions, there is a good correlation between the rheological properties of films of protein adsorbed at the air–water interface and the stabilities of foams made with the same proteins. Thus, β-casein films have the lowest resistance to shear, the lowest dilational modulus (Table 2) and give the least stable foams (Fig. 3), whereas lysozyme films are the most resistant to mechanical deformation and give very stable foams. Consistent with this dominant role for the rheological properties of the protein films in stabilizing the foam, the foam stability tends to be maximized at the pI of the protein (Fig. 6), where the rheological properties of the adsorbed films of protein are also maximized.[10,18,19] This variation with pH is again quite different to that observed with protein-stabilized emulsions where coalescence was enhanced at the pI (ref. 10). The above data suggest that the mechanism of stabilization of the relatively large aqueous lamellae (diameter of the area of contact $\sim 10^{-2}$ cm) between the air cells in a protein-stabilized foam is quite different to the equivalent thin films between oil droplets in an emulsion. In contrast to the aqueous lamellae in an emulsion, those in a foam are unlikely to be transient. This arises because the attractive forces across the lamellae in foams are no longer much greater than the repulsive forces. Since the Hamaker constant for water–air is greater than that for water–oil[27] there must be an increase in the repulsive forces at the air–water interface to give rise to this effect. The increase presumably arises from differences in the conformations of the adsorbed protein molecules at the two types of interface. Hence, the collapse of the protein-stabilized foams probably involves the rupture of metastable[15] thin water films between air cells.

There are two possible ways in which the rheological properties of the adsorbed protein layers can contribute to the overall stability of a foam: (1) the rate of drainage of water from the thin aqueous lamellae is dependent on the rheological properties of the surface films; (2) rupture of a lamella involves overcoming the cohesive forces within the two surface layers. Here we shall not discuss all the processes (e.g. bubble disproportionation) involved in foam collapse,[9, 20] but will discuss mechanisms which directly involve the interfacial protein layers as mentioned in the above two factors. In order to relate the information gathered on the rheological properties of adsorbed protein films at the air–water interface in a Langmuir trough to the foam situation, it is necessary to know the degree of coverage of the surfaces of air bubbles in a foam. Unfortunately, because of complications such as surface coagulation this information is not available and so for the purposes of this comparison we assume that $\pi \sim 10\,\mathrm{mN\,m^{-1}}$ at the surface of the protein-covered air cells. This film pressure is roughly half the maximum π exerted by adsorbed protein films.[2, 4, 5, 10, 11] Under these conditions it is clear from the data in Table 2 that films of β-casein are relatively compressible (i.e. ε is low) and very easily sheared. These properties arise because the β-casein molecules are highly flexible and do not cohere strongly. At the same π, films of BSA and lysozyme are relatively incompressible and difficult to shear (Table 2); the particularly stable globular lysozyme molecule gives rise to films containing surface-denatured molecules[2] which have a lot of residual structure (Fig. 7) and which are highly cohesive (when $\pi = 10\,\mathrm{mN\,m^{-1}}$, the shear viscosity is $>10^4\,\mathrm{mN\,s\,m^{-1}}$ and the shear elastic modulus is $15\,\mathrm{mN\,m^{-1}}$).

The rate of drainage of free liquid films stabilized by surfactant is reduced when the adsorbed films are viscous and incompressible.[10, 15] Earlier studies[10] on thin aqueous films stabilized by β-casein and BSA show that the films stabilized by mobile layers of flexible β-casein molecules drain relatively rapidly to become black. In contrast, the higher ε and η of adsorbed BSA causes the thin films stabilized by this protein to have the slow drainage characteristics of rigid films. In this case drainage occurs by viscous flow of the aqueous phase from between the two motionless surfaces of the film.[20, 21] Since this difference in rate of film thinning will also hold for the lamellae in foams, the lamellae will be thicker in the lysozyme and BSA foams than in the β-casein foams at any given time after shaking is stopped. As a result the lysozyme and BSA foams are more stable because the probability of spontaneous hole formation is much less for thicker films; the activation energy (E) opposing growth of a hole is a function of the film thickness (H) and the interfacial tension (γ).[22]

$$E \alpha (H^2 \gamma) \tag{2}$$

In addition to the reduction in thinning caused by highly ordered surface layers of protein, the greater lateral cohesion in such layers also tends to stabilize the aqueous lamellae. The latter effect arises because film rupture via formation of a hole by cavitation is opposed by an energy barrier which is a function of the lateral cohesion of the molecules in the adsorbed layers at the surface of the thin film.[23] It should be noted that spontaneous formation of a hole can only occur in a region where the film is very thin (i.e. black and < 100 nm thick).

The data in Fig. 5 indicate that the presence of coagulated and precipitated protein reduces the stability of a foam; this effect probably arises because the presence of precipitated protein reduces ε of adsorbed protein layers (R. D. Bee, unpublished observation). Because of this, when adjustment of the pH to the isoelectric point (pI) leads to insolubilization of the protein, the foam stability is decreased. Otherwise, the general effect of operating at the pI is to enhance the foam stability (see Fig. 6 and refs 24–26).

Since the stabilities of the protein foams are greatest at the pI of the proteins (when no precipitation occurs), at which point the resistance to shear and dilatation of protein films at the air–water interface are maximum and the repulsive interactions between two such layers are probably minimized,[10] it is clear that the lateral interactions within the adsorbed layers play a more important role in stabilizing the foams than the disjoining pressure. In summary, it seems that the strength of the gel-like adsorbed protein layers[2] is crucial in determining the stability of a foam.

5 Conclusions

1. Flexible protein molecules which can adsorb rapidly at the air–water interface give good foamability whereas highly ordered globular molecules which are difficult to surface-denature give poor foamability. Rapid adsorption tends to lead to formation of a coarse foam (i.e. large air

bubbles) whereas slow protein adsorption favours small air bubbles (i.e. a creamy foam).

2. Prior denaturation (e.g. by heat or extremes of pH) of globular proteins under conditions where no precipitation occurs leads to enhanced foamability.

3. Once formed, the foams stabilized by proteins which are globular in the native state are more stable than foams prepared with the flexible β-casein molecule.

4. The rheological properties of the protein films adsorbed at the surfaces of the air cells are of overriding importance in determining the stability of a foam. Stability is enhanced when the films are resistant to shear and incompressible (i.e. have a high dilatational modulus).

5. Provided there is no precipitation of protein. the foam stability is greatest at the isoelectric point of the protein.

Acknowledgements

We are indebted to Messrs I. K. Barker and J. W. B. Paynter for technical assistance, to Dr M. T. A. Evans for supplying the radioactively labelled proteins and to Dr J. de Feijter (Unilever Research Laboratory, Vlaardingen) for helpful discussions.

References

1. Cumper, C. W. N. and Alexander, A. E. (1951). *Rev. Pure Appl. Chem.* 1, 121.
2. Phillips. M. C., Evans. M. T. A., Graham, D. E. and Oldani, D. (1975). *Coll. Polymer Sci.* 253
3. Bull, H. B. and Breese. K. (1973). *Arch. Biochem. Biophys.* 158. 681.
4. Adams, D. J., Evans, M. T. A., Mitchell, J. R., Phillips, M. C. and Rees, P. M. (1971). *J. Polym. Sci.* C34, 167.
5. Phillips, M. C., Evans, M. T. A. and Hauser, H. (1973). "Proc. VIth Internat. Congress on Surface Activity", Vol. II, Part 1, p. 381. Carl Hanser Verlag, Munich.
6. Boyd, J. V. and Sherman. P. (1970). *J. Colloid Interface Sci.* 34, 76.
7. Lucassen, J. and Van Den Tempel, M. (1972). *J. Colloid Interface Sci.* 41, 491.
8. Lucassen-Reynders, E. H. and Lucassen, J. (1969). *Advan. Colloid Interface Sci.* 2, 347.
9. Bikerman, J. J. (1973). *In* "Foams". Springer-Verlag, New York.
10. Graham, D. E. and Phillips, M. C. (1976). *In* "Theory and Practice of Emulsion Technology". S.C.I. Symp. Brunel University 1974 (A. L. Smith, Ed.), p. 75. Academic Press, London and New York. In press.
11. Benjamins, J., de Feijter, J. A., Evans. M. T. A., Graham, D. E. and Phillips, M. C. (1975). *Discuss. Faraday Soc.* 59. In press.

12. Inokuchi, K. (1955). *Bull. Chem. Soc. Jap.* **28**, 453.
13. Joly, M. (1972). *In* "Surface and Colloid Science" (E. Matijević, Ed.), Vol. 5, p. 34. John Wiley, New York.
14. Henson, A. F., Mitchell, J. R. and Mussellwhite, P. R. (1970). *J. Colloid Interface Sci.* **32**, 162.
15. Vrij, A. (1966). *Discuss. Faraday Soc.* **42**, 23.
16. Vrij, A., Hesselink, F. Th., Lucassen, J. and Van Den Tempel, M. (1970). *Proc. Kon. Ned. Akad. Wetensch.*, Ser. B. **73**, 109.
17. Phillips, M. C. (1975). *In* "Water A Comprehensive Treatise" (F. Franks, Ed.), Vol. 5, p. 133. Plenum Press, New York.
18. Biswas, B. and Haydon, D. A. (1962). *Kolloid Z.* **185**, 31.
19. El'Shimi, A. F. and Izmailova, V. N. (1971). *Colloid J. USSR.* **33**, 237.
20. Kitchener, J. A. (1964). *In* "Recent Progress in Surface Science" (J. F. Danielli, K. G. A. Pankhurst and A. C. Riddiford, Eds), Vol. 1, p. 51. Academic Press, New York and London.
21. Prins, A. and Van Voorst Vader, F. (1973). "Proc. VIth Internat. Congress on Surface Acticity", Vol. II, Part 2. p. 441. Carl Hanser Verlag, Munich.
22. de Vries, A. J. (1958). *Rec. Trav. Chim.* **77**, 383.
23. Sheludko, A. (1967). *Advan. Colloid Interface Sci.* **1**, 391.
24. Cumper, C. W. N. (1953). *Trans. Faraday Soc.* **49**, 1360.
25. Thuman, W. C., Brown, A. C. and McBain, J. W. (1949). *J. Amer. Chem. Soc.* **71**, 3129.
26. Perri, J. M. and Hazel, F. (1947). *J. Phys. Chem.* **51**, 661.
27. Ninham, B. W. and Parsegian, V. A. (1970). *J. Chem. Phys.* **52**, 4578.

Discussion

Tadros (*ICI Plant Protection Division, Bracknell, Berkshire, England*) In your paper you showed that the foam stability depends on the pH of the protein solution. Have you measured adsorption and rate of adsorption as a function of pH? Does this correlate with foam stability?

If the foam stability is governed by thinning of the liquid film between the foam bubbles, is it possible to treat the kinetics of collapse of foam quantitatively as is done in describing emulsion stability. If the collapse of the foam is governed by thinning of the liquid film, the rate of collapse should follow a first-order equation. Was this the case?

Graham and Phillips As we have discussed in the paper, a change in the adsorption rate of the proteins is more likely to affect the foamability rather than the foam stability. Preliminary observations of the rates of adsorptions of BSA (initial $C_p = 1 \times 10^{-4}$ wt% in 0.1 mol dm^{-3} NaCl) at the air–water interface suggest that there is no significant dependence of $d\Gamma/dt$ or $d\pi/dt$ on the substrate pH over the range from about 3 to 10 (at less than saturated monolayer coverage Γ and π are more or less independent of pH over the range 2 to 10). This is also reflected in the observations shown in Fig. 6(a) where the initial foam volumes of BSA-stabilized foams are unchanged over the pH range studied.

As Tadros points out, if the rate of foam decay is determined by the rate of bubble coalescence (i.e. the probability of rupture of the lamellae between adjacent bubbles) then the rate of foam collapse should follow first-order kinetics (cf. Van den Tempel's treatment of the rate of coalescence of oil-in-water emulsions[1]). As mentioned in the paper, the decays of the foams stabilized by both β-casein and BSA (in the presence of $0 \cdot 1$ mol dm^{-3} NaCl) follow first-order kinetics over about the first hour; this effect is not observed with all foams because a change in the composition of the aqueous phase can change the rate of foam decay. Consequently, it is not generally true that the rate-determining step for foam collapse is rupture of the thin films between bubbles. This is likely to be the result of other factors, such as bubble disproportionation, playing a significant role in the collapse of protein-stabilized foams; this will be discussed in more detail elsewhere.

Hansen The positive curvature in the spreading pressure as time plot for lysozyme in Fig. 1 suggests that the unfolding of lysozyme at the interface must have a kinetic order greater than one. Is this plausible?

Graham and Phillips When monitoring the adsorption of various proteins at the air–water interface, we have often observed induction periods in the plots of surface pressure π against time when the substrate protein concentration is low. Under these circumstances, when π starts to increase there is a positive curvature to the π–t curve as Hansen has noted for the lysozyme data in Fig. 1. Since π is a function of the surface concentration Γ of protein, in order to understand the form of the π–t curve we need to know (1) the variation of Γ with t (see reference 1) and (2) the surface equation of state for the monolayer so that we have the dependence of π on Γ.

First-order kinetics would result if the monolayer behaved like an ideal gas because then

$$\pi = RT\Gamma^n \qquad (1)$$

where $n = 1$. As Hansen suggests, the form of the lysozyme curve in Fig. 1 implies that $n > 1$ for this system. There is some evidence to suggest that this is reasonable.

Thus, it has been found[2,3] for various protein films at the air–water interface that at low π a linear relationship exists between the dilatational modulus ε and π with a proportionality constant b of about $3 - 8$. We can therefore write

$$\varepsilon = \frac{d\gamma}{d\ln A} \equiv \frac{d\pi}{d\ln\Gamma} = b\pi \qquad (2)$$

and Lucassen (unpublished work) has shown that integration of (2) gives

$$\pi = k\Gamma^b. \qquad (3)$$

Since $b = 3$–8 it is clear from (3) that kinetics of higher order than 1 are possible and π–t curves of the form shown for lysozyme are to be expected.

Buscall In the paper you refer to de Vries's nucleation theory of the rupture of very thin lamellae. Equation 2 was derived for the case where the film tension was much greater than the surface elasticity. In your work these two quantities are similar and γ in equation 2 should be replaced by $(\gamma + 2\varepsilon)$. For simple systems the de Vries equation

[1] Van den Tempel, M. (1957). Proc. 2nd Internat. Congress on Surface Activity, Vol. 1, p. 439. Butterworths, London.

[2] Benjamins, J., de Feijter, J. A., Evans, M. T. A., Graham, D. E. and Phillips, M. C. (1975). *Discuss. Faraday Soc.* **59**. In press.

[3] Blank, M., Lucassen, J. and Van den Tempel, M. (1970). *J. Colloid Interface Sci.* **33**, 94.

appears to provide a qualitative description of film rupture. For example in recent work on the coalescence of oil-in-water emulsions stabilized by a low molecular weight nonionic surfactant I found that the logarithm of the first-order rate constant for coalescence was proportional to the equilibrium interfacial dilational modulus. (In this case $\varepsilon \gg \gamma$.) This correlation held for rate constants in a range greater than two decades.

Graham and Phillips Buscall is correct in pointing out that for our protein films the complete form of equation 2 in the paper, which relates to the nucleation theory described by de Vries, should read $E = 0.73\, H^2\, (\gamma + \varepsilon)$ where ε is the dilatational modulus of the adsorbed protein layer around the air bubbles.

The data in Fig. 6(a) for the decay of BSA-stabilized foams at several different aqueous pH values have been analysed for first-order kinetics. During the initial period of decay (~ 40 min) all the curves exhibit first-order kinetics with rate constants varying over an order of magnitude ($\sim 5 \times 10^{-4}$ at pH 2.5 to $\sim 3 \times 10^{-5}\,\mathrm{s}^{-1}$ at pH 12). These rate constants do not appear to have any particular relationship with the dilatational moduli (as measured under our experimental conditions; see Table 2) for adsorbed BSA films at the corresponding pH values.

Russo I was interested in the effect of pH on the foamability of lysozyme and in particular the effect of pH > 11.5. I take it that the objective of this work is to work on edible foams. Surely a pH of 11.5 would make a product completely inedible. Have you therefore investigated the effect of subsequent neutralization to say pH 7.0?

Graham and Phillips Our objective in working at various pH levels was to explore the effects of protein denaturation on foam properties. Results presented in the paper indicate clearly that pH-denaturation can lead to enhanced foamability and foam stability.

Although we have by no means systematically studied the effect of post-forming pH manipulations, we can quote the following isolated observations. Two 0.1 wt% lysozyme solutions at pH 11.5 were shaken for 60 min to create a foam. Then, sufficient HCl was added carefully to one to adjust the pH to ~ 5 and the solutions plus foam were reshaken for a further 10 min. The resulting foams had similar volumes and lifetimes; the properties of the foam created at pH 11.5, but subsequently brought to pH 5, did not revert to the characteristics expected of a foam created at pH 5 (Fig. 6). This suggests that the absorbed protein layers which stabilize the thin aqueous films between air cells are not readily altered by manipulation of the substrate pH *after* adsorption has occurred.

16

Some aspects of the stability of beer foam

R. T. ROBERTS

The Brewing Industry Research Foundation,
Nutfield, Redhill, Surrey, RH1 4HY, England

Summary

Two aspects of the stability of beer foam are considered in this paper. Beer foam is essentially a polypeptide foam. One type of polypeptide, that bound to polysaccharide, has been isolated and studied. This protein–polysaccharide complex or glycoprotein is found to have a greater stabilizing effect on beer foam than does polypeptide free of carbohydrate. The properties of the polypeptide foam are, however, greatly affected by the presence of small molecules, not themselves surface active, which interact with the polypeptides. We report on a study of the effects of certain hop constituents and metal ions on the stability of beer foam.

1 Introduction

In physical terms beer is a solution of alcohol, carbohydrates and many quantitatively minor but qualitatively important complex constituents, all in a solution containing carbon dioxide. The release of this gas caused by disturbance during pouring from a bottle or dispensing from a tap brings surface-active material out of solution to create foam. The quality of foam on a glass of beer has come to be of great importance to the consumer. Too little foam, or the wrong kind, that is large uneven bubbles, detract from the appeal of a glass of beer. Too much and too stable a foam is an obvious embarrassment and can cause problems in serving the beer. Hence there is a need to understand the action of materials acting in producing and stabilizing beer foam.

Briefly the brewing process consists of:

1. *Mashing,* in which starch is converted into fermentable sugars and soluble dextrins, and in which nitrogenous constituents from cereal

protein having a wide range of molecular sizes are brought into solution at levels of ~ 500 mg nitrogen per litre ($\times 6 \cdot 25 = 3 \cdot 125$ g protein per litre).

2. *Boiling*, in which the enzymic action so important in mashing is terminated. Some proteins and tannins are eliminated as precipitates and the bitter principles of hops are transformed into the soluble bitter substances (isohumulones) of beer.

3. *Fermentation*, where the greater part of the carbohydrate is converted into alcohol and carbon dioxide.

4. *Finishing operations*, where the main feature for this paper is that the content of carbon dioxide may be adjusted to levels appropriate for the type of beer. Thus in draught beers the carbon dioxide level is of the order of $0 \cdot 9 - 1 \cdot 1$ volumes and for bottled beers the level is $2 \cdot 3 - 2 \cdot 7$ volumes. As the saturation solubility of carbon dioxide at 20°C is $\sim 0 \cdot 88$ volumes, beer is usually supersaturated with the gas and would be expected to change readily from a single phase system to a two-phase gas–liquid system on release of pressure.

Provided that beer contains sufficient carbon dioxide and is dispensed into a clean, dry glass there is rarely any problem in creating a sufficient amount of foam. Occasionally there can be a problem because, on releasing the pressure in a bottle of beer, the majority of the gas comes out of solution almost instantaneously creating unwelcome large volumes of foam and causing serious loss of beer. This defect, known as *gushing*, is referred to later. In normal circumstances, however, the main concern is that the beer should have a good *head-retention*, i.e. ability to retain a head of foam as the drink is consumed, and a good *adhesion*, i.e. residues of foam should adhere to the glass above the level of liquid. These properties can vary because of seasonal variations in raw materials, because of differences in methods of processing, and for quite unknown reasons. Hence the beer constituents which have surface activity and the interactions between them are being studied in detail.

On pouring a normal beer, about 10 per cent of the liquid is turned into foam by about 10 per cent of the carbon dioxide. The remainder of the gas continues to evolve in streams of bubbles originating at hydrophobic centres on the glass surface, thus providing a small supply of fresh foam. Hence the foam-head consists of bubbles in three stages. Firstly, those bubbles near the bulk of liquid are spherical and well separated by draining liquid; secondly, those in the middle region are hexagonal and the liquid is draining from the lamellae; and thirdly those near the air surface are fully drained and rigid.

The beer constituents which are of interest in the present context are those which influence the strength of bubble-walls and those which affect either the surface viscosity or the bulk viscosity of the liquid. The latter is significant in the rate of drainage of liquid from the foam. There is considerable evidence that the main foam-active substances in beer are polypeptides and there is circumstantial evidence that glycoproteins are especially important. Also there is evidence that the bitter substances, the isohumulones, give beer foam-adhesive properties and render it more stable, especially in the presence of heavy metals. This report confirms the efficacy of the glycoproteins and deals with their mode of action and considers the interaction of proteins, isohumulones and certain metals.

2 Methods

2.1 FOAM STABILITY

This was measured by a simple drainage method. The apparatus used was first described by Rudin[1] and it is widely accepted and used in quality control laboratories throughout the brewing industry. It assumes that the collapse and drainage of beer foam follows a logarithmic path so that the return of half of the foam to the liquid phase can be taken as the half-life. The apparatus consists of a graduated glass cylinder having a sintered glass disc at its lower end, and is jacketed to maintain constant temperature. The cylinder is 28 mm in diameter and has graduations at the 50, 75, 100 and 325 mm levels as measured from the sintered disc. A sample of beer is poured in up to the 100 mm mark and gas, either carbon dioxide or nitrogen, is allowed through the sinter until foam reaches the 325 mm mark in about one minute. For beer this usually means that all the liquid has been converted into foam which then commences to collapse. The time taken for the liquid meniscus to travel from the 50 mm mark to the 75 mm mark in these circumstances is approximately equal to the half-life of the foam, and it is this time that is taken as an indication of foam stability and termed the head retention value (HRV). For beer foamed with nitrogen this time is about 300 seconds. In experiments where additions were made to beer, the gas rate was adjusted to form the required volume of foam in one minute, for the control beer, and then kept at this value for measurements on beers with additions. Thus any difference in the time taken to foam was noted as was any visible difference in foam texture. Nitrogen gives a much more stable foam than does carbon dioxide so that the time for collapse is about twice as great when nitrogen is used.

2.2 SURFACE VISCOSITY

Surface viscosity was measured by means of an oscillating disc viscometer (radius of disc 25 mm, weight of bob 0·4205 kg, giving an effective moment of inertia I of $1\cdot344 \times 10^{-4}$ kg m^2). The solution under examination was contained in a large paraffin wax coated petri dish (radius 72·5 mm).

Measurement of the coefficient of surface viscosity (η_s measured in g s^{-1}) generally involves measurement of the damping of the oscillation of the viscometer shearing element on the clean surface and the corresponding damping after addition of surfactant. This method is obviously impossible with beer as the solution has a preformed surface layer. A method devised by Bulas and Kumins[2] to overcome this difficulty was thus employed. Using this method the coefficient of surface viscosity (η_s) is calculated from

$$\eta_s = \frac{2\cdot303\,(\lambda_s - \frac{1}{2}\lambda_h)I}{2\pi T}\left(\frac{1}{r_1^2} - \frac{1}{r_2^2}\right)$$

where λ_s is the logarithmic decrement of the amplitude of swing when the disc is on the surface of the beer, λ_h the logarithmic decrement of the amplitude of swing when the disc is fully immersed in beer, I the moment of inertia of oscillating system, r_1 the radius of disc, r_2 the radius of bounding vessel, and T the period of oscillation.

Before measurement of surface viscosity the beer samples were usually degassed by allowing them to stand quietly in plastic bags overnight. This procedure ensures that the poured beer has the minimum amount of foam which would of course interfere with the measurement. Some exposure to air is unavoidable and the surface of degassed beer ages less rapidly than that of beer poured straight from the bottle. No change was found in λ_h, the damping of the disc when fully immersed, when this was measured at various times from the pouring of the beer. When studying the rate of change in surface viscosity, time was always measured from the moment of pouring.

3 Results

3.1 PROTEIN–POLYSACCHARIDE COMPLEXES

The key step in the isolation from beer of the protein–polysaccharide complex investigated was the use of Concanavalin A (Con A), a lectin derived from the Jack Bean that has the ability to combine with certain types of carbohydrate, to separate a fraction from the high molecular weight material previously isolated from beer. Two techniques have been used to obtain the high molecular weight material, the first being saturation of the beer with ammonium sulphate and the second precipitation with zinc chloride and ethanol, both precipitates then being subjected to extensive dialysis. These high molecular weight materials were then subjected to affinity chromatography on Concanavalin A bound to Sepharose (Pharmacia Fine Chemicals) and that fraction bound to the Concanavalin A after retrieval by elution with

0·1 mol dm^{-3} α-methyl glucoside and dialysis was examined for its effect on foam stability. The material that binds to Con A was isolated from beers brewed with various cereal adjuncts and in all cases was found to improve foam stability. The increases in head retention value are shown in Table 1. The head retention values are for unhopped beer which had been diluted

TABLE 1

Head retention values of diluted unhopped beer measured in the Rudin apparatus but foamed with nitrogen. Material isolated by Con A adsorption added at 3 mg per 100 ml

Origin of glycoprotein	Saturated ammonium sulphate precipitate s	Zinc chloride + ethanol precipitate s
All malt beer	218	234
25 per cent Wheat flour	228	228
25 per cent Maize flakes	224	264
25 per cent Rice flakes	232	232
25 per cent Barley flakes	212	248

Control diluted beer 200 s

with 4 parts of 3·5 per cent (w/v) ethanol and foamed with nitrogen. To simplify comparisons, results are expressed relative to control values of 200 in all experiments. Figures show the increase when 3 mg of the Con A-adsorbed material is added to 0·1 l of diluted unhopped beer. Dilution of beer to 20 per cent has very little (~ 15 per cent) effect on the head retention as measured by drainage but it dilutes the surface-active material present and so, on addition of a given weight of material, the relative effect in the bubble lamella is amplified.

Several of the Con A-adsorbed materials were examined by electrophoresis in 0·1 mol dm^{-3} borate buffer. In this system the polysaccharide becomes charged, and thus if the protein present is weakly bound to the polysaccharide, electrophoresis under these conditions should separate the complex. In each experiment two identical strips were run, one being stained for protein and the other for carbohydrate. It was found that the material migrated as a single band, carbohydrate and protein appearing in the same position. This is taken as evidence that the material that binds to Concanavalin A is a protein–polysaccharide complex in which the binding is quite strong, though it does not in itself prove that the binding is covalent.

Isoelectric focusing in polyacrylamide gels was then employed in an attempt to characterize the polypeptide portion of the complex. The adsorbed material showed at least 20 bands of polypeptide to be present. Most of the bands are grouped together with isoelectric points about pH 4, the pH of beer. The polyacrylamide gel was also stained for carbohydrates and it was found that each band of polypeptide material took the carbohydrate stain. No staining was seen in the region of sample application indicating that there was no free immobile polysaccharide present.

TABLE 2

Reducing sugars formed after 3 h digestion of Con-A adsorbed material by carbohydrases

Enzyme	Reducing sugars liberated $\mu g \, mg^{-1}$ of sample
β-Amylase	0
α-Amylase	13
Amyloglucosidase	24

The chemistry of Concanavalin A binding gives information on the nature of the carbohydrate moiety since Con A is very specific in its binding. It is known that the carbohydrates most strongly adsorbed are those having α-glucopyranosyl or α-mannopyranosyl nonreducing terminal groups. The strength of binding decreases rapidly with increasing length of straight chain polymers, but it will bind strongly those branched polysaccharides which have several nonreducing terminal α-glucose or α-mannose units. Hydrolysis of the protein–polysaccharide complex isolated from beer shows that the only sugar present in any quantity is glucose. This enables a choice

TABLE 3

Effect on head retention of diluted unhopped beer of glycoprotein fraction (3 mg per 100 ml) before and after amyloglucosidase digestion

Addition	Head retention s
Nil	200
Glycoprotein	230
Amyloglucosidase digest	214

TABLE 4

Effect of head retention of diluted unhopped beer
of glycoprotein (3 mg per 100 ml) before and after
proteolytic digestion

Addition	Head retention s
Nil	200
Glycoprotein	234
Proteolytic digest	218

of carbohydrate-splitting enzymes to be made that will digest the poly-saccharide portion of the complex and thus enable its significance in the foam-stabilizing properties of the complex to be ascertained.

The three enzymes chosen were α-amylase, β-amylase and amyloglucosi-dase. Table 2 shows the reducing sugars formed (μg per mg sample weight) as measured by the Nelson–Somogyi method after a 3-hour digestion at the conditions optimal for each enzyme. β-Amylase has practically no effect; amyloglucosidase has the greatest effect.

Table 3 shows the smaller increase in head retention obtained when an equal weight of the complex is treated with amyloglucosidase before addition to an unhopped beer. Considering the amount of reducing sugar formed in the digestion, and thus the amount of polysaccharide that has been removed (5–10 per cent), this reduction is quite significant.

TABLE 5

Effect of amylopectin on head retention of
diluted unhopped beers

Amylopectin mg	Head retention s
0	162
1	158
2	161
3	156

In another series of experiments the polypeptide moiety of the complex was removed by digestion with the proteolytic enzyme Subtilisin. Before addition to beer a simple shaking test showed that the enzyme had apparently destroyed all the foaming potential of the protein polysaccharide complex.

Table 4 shows the effect on head retention of an unhopped diluted beer on addition of the digested complex. This can be compared with the improvement caused by an equal weight of undigested material.

The carbohydrate moiety appears from the results of these experiments to be a highly branched α-glucan and thus similar in structure to amylopectin. To ascertain that this highly branched dextrin on its own had no effect on beer foam, it was isolated from barley and added to beer. Table 5 shows that this polysaccharide has no effect on the stability of beer foam.

3.2 HOP ISO-α-ACIDS AND METAL IONS

Figure 1 shows the effect on the stability of beer foam of the addition of various metal ions to unhopped beer. Metal ions were added to a final

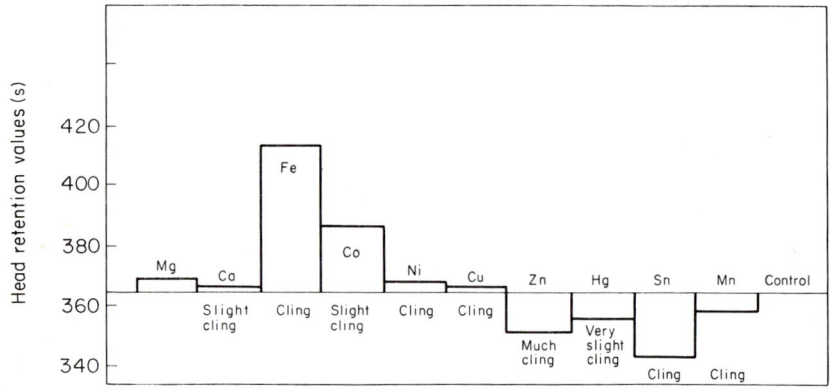

FIG. 1. 10^{-4} mol dm^{-3} metal ions added to unhopped beer.

concentration of 10^{-4} mol dm^{-3} in the undiluted beer. As can be seen in most cases the metal ions enable the collapsing foam to adhere or cling to the walls of the glass tube. In the cases where the metal ions actually reduce the head retention value a visible haze is seen. This haze is caused by precipitation of polypeptide material by the metal ion.

Figure 2 shows the effect on head retention of beer when the same metal ions are added at a concentration of 10^{-4} mol dm^{-3} but this time in the presence of 10^{-4} mol dm^{-3} isohumulone. In this case it is seen that each ion enhances the foam stability, and the improvement in head retention value is far greater than in the absence of isohumulone.

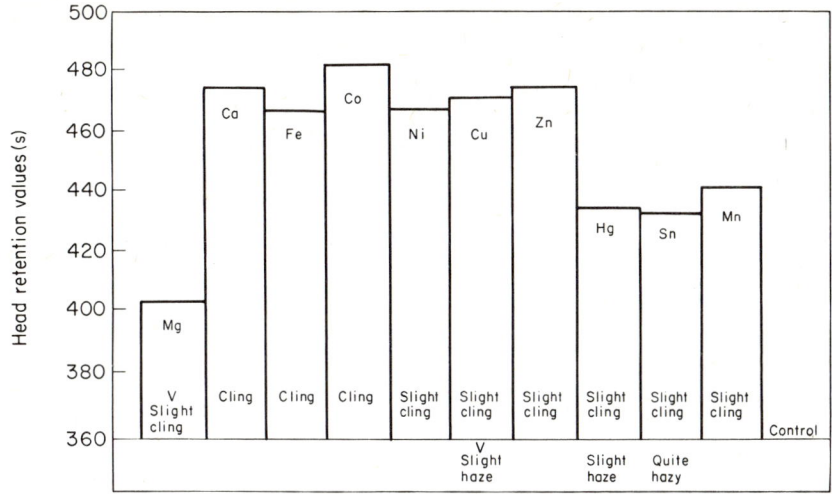

FIG. 2. 10^{-4} mol dm^{-3} metal ions added with 10^{-4} mol dm^{-3} isohumulone to unhopped beer.

The increase in head retention on adding increasing amounts of iso-humulone to beer is shown in Fig. 3. As can be seen this increase levels off after about 40 ppm (10^{-4} mol dm^{-3}). The interaction of metal ions with hop substances has long been noted and it was demonstrated that addition of nickel to beer can serve as a measure of the isohumulone content.[3] A plot of the increase in foam half-life on addition of 10 ppm nickel against isohumulone

TABLE 6

Effects of Ni, Fe and isohumulone on the surface viscosity of unhopped beers

Beer	Additive*	$10^5 \times \eta_s/(\text{N s}^{-1})$	Comments
1	nil	0·05	all nongushing
	Ni	0·08	
	isohumulone	0·04	
	Ni + isohumulone	†	gushing
	Fe	0·07	nongushing
	Fe + isohumulone	2·30	gushing
2	nil	0·03	nongushing
	Ni	0·03	
	Ni + isohumulone	9·0	gushing

* Ni and Fe were added to levels of 5 ppm and isohumulone to 30 ppm.
† Unmeasurably high.

concentration is found to be linear over the range of isohumulone concentrations normally found in beer (up to 10^{-4} mol dm^{-3}).

It has been noted that in the presence of isohumulone the addition of certain metal ions, notably nickel, can cause vigorous overfoaming in the beer. It is known that nickel and isohumulone when present together increase the surface viscosity of beer often to such a value that it becomes impossible

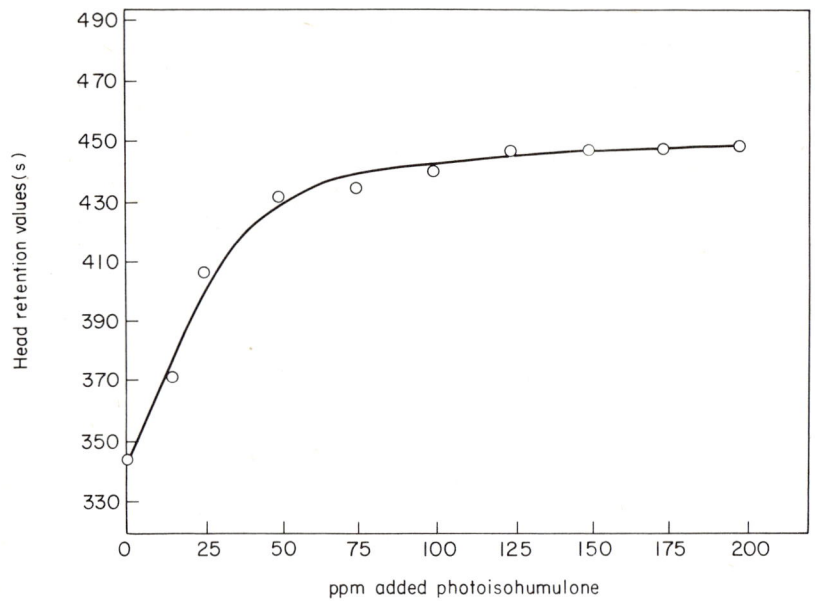

FIG. 3. Graph of head retention values versus ppm of photoisohumulone added.

to measure by the method described. It has been suggested therefore that over-foaming or gushing may be associated with high surface viscosity, but no direct correlation has been found.[4] Whilst nickel ions provide a useful experimental tool in studying foam it must be emphasized that nickel is not a normal component of beer at any significant concentration. Isohumulone itself has no effect on the surface viscosity of a solution of 3·5 per cent ethanol. Table 6 shows the effect of addition of nickel, iron and isohumulone on the surface viscosity of unhopped beers. It is seen that when the metal ion is present with isohumulone a very high surface viscosity is observed. These measurements reported previously by Gardner[4] are for surface viscosity 5 min after pouring.

The effect of isohumulone on the surface viscosity of protein solutions has

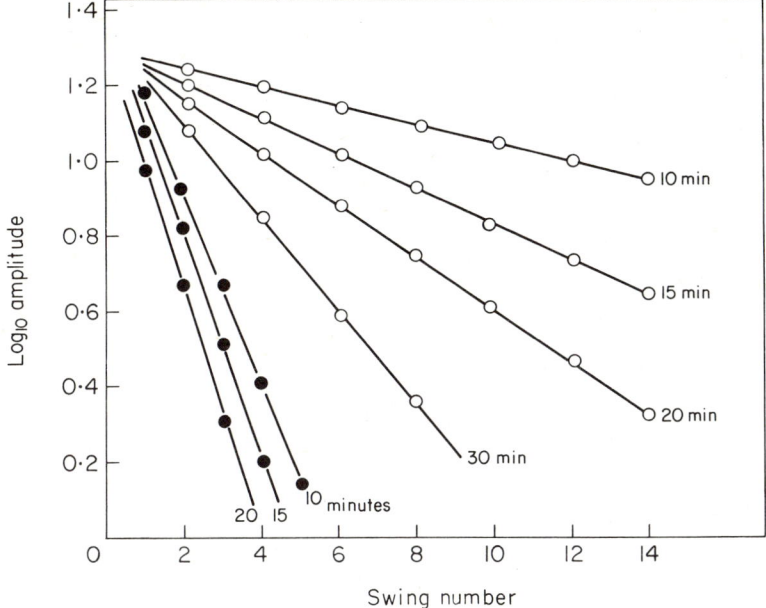

FIG. 4. ○—Damping for ovalbumin; ●—damping for ovalbumin + isohumulone (5 ppm).

also been measured. Figure 4 shows the change in damping with time of a solution of ovalbumin (7·5 mg in 3·5 per cent ethanol, 0·1 l) with and without isohumulone (0·5 mg per 0·1 l). Addition of nickel at the same concentration (0·5 mg per 0·1 l) caused a slight further increase in viscosity. An interesting result is that not only has the surface viscosity increased, but the rate of ageing has decreased.

4 Discussion

4.1 PROTEIN–POLYSACCHARIDE COMPLEXES

The results reported here suggest that protein–polysaccharide complexes, which may be covalently bound as glycopeptides and glycoproteins, have a greater effect on the stability of beer foam than the free polypeptides that are present. Commercially available glycoproteins of known structure have also been added to beer[5] and the foam stability measured. It was found that those showing the greatest improvement in foam stability were those having the highest carbohydrate concentration. Ovomucoid, having between 20 and 25 per cent carbohydrate was found to increase beer foam stability noticeably.

It is apparent, however, from our own work and some previous work[6] on the addition of carbohydrates to beer, that carbohydrates *per se* do not improve the stability of beer foam actively. Large polysaccharides, such as the barley gums, do increase the bulk viscosity of beer and thus, since in the initial stages of foam collapse liquid drainage is the predominant factor, these substances can enhance foam stability by retarding liquid drainage. Increasing bulk viscosity is, however, an impractical method of improving foam stability as it can lead to handling problems such as reduced flow and filtration rates. The carbohydrate moiety of the protein polysaccharide complex that has been isolated from beer, and that of the commercial glycoproteins used in our experiments, is much smaller in size than the polysaccharide gums and therefore will have no effect on beer viscosity at the quantities found present. The highest concentration of the protein–polysaccharide complex isolated was 150 mg from a litre of all-malt beer and this contains no more than 40 per cent carbohydrate.

Polypeptides and proteins are known to improve foam stability on addition to beer,[7] yet, as our results show, a protein–polysaccharide complex or a glycoprotein that contains more than about 15 per cent carbohydrate improves beer foam stability more than an equal weight of polypeptide or protein. It is concluded therefore that when carbohydrate is linked to protein, the complex has great significance in foam stability. The mechanism by which such a complex enhances foam stability is thought to be as follows. The polypeptide moiety enters the bubble wall along with other polypeptides and surface-active material, the polysaccharide moiety remaining essentially in the liquid phase. This polysaccharide moiety is too small to have any effect on foam stability in the initial stages, when the bubbles are spherical and liquid drainage from the foam is rapid. However, when the bubbles start becoming hexagonal and the distance between bubbles becomes much less than the bubble radii the polysaccharide attached to the liquid side of the bubble wall becomes important in affecting the flow of liquid in this region. Liquid flow is retarded earlier than it would be if the polysaccharide were not present. It is also suggested that, since the flow of liquid in this region will tend to align the polysaccharide along the surface of the bubble wall, it will prevent surface-active material from leaving the wall and dissolving back into the draining liquid. This model is independent to a large extent of either structure or chain length of polypeptide or polysaccharide moieties. The polypeptide moiety should be sufficiently surface active to enter the lamella, and hydrophobic interactions should be strong enough to keep it in the lamella against forces such as liquid flow tending to redissolve the complex. The polysaccharide moiety should simply be large enough to show an effect on the stability of the foam.

The results reported here tend to confirm this model. The stabilizing influence of the protein–polysaccharide complex isolated from beer is easily shown. Also removal of part of the carbohydrate moiety is shown to decrease the stabilizing effect, yet the carbohydrate itself is shown to have no effect on foam stability. Removal of part of the polypeptide moiety also reduces the head retention of the beer, and addition of protein to beer increases the foam stability.

It is believed that the mechanism discussed for the action of protein–polysaccharide complex is very similar to the mechanism by which other surfactant–polysaccharide complexes can improve foam stability.

4.2 HOP ISO-α-ACIDS AND METAL IONS

It appears from the results here that isohumulone in the presence of metal ions makes the beer surface viscous, at the same time retarding the ageing of the polypeptide film. The stability of the foam is greatly enhanced and the collapsing foam now has the property of clinging to a glass surface on collapse. It is difficult to assess the effect of isohumulone alone on beer foam as there is inevitably a small concentration of metal ions in beer which are derived from raw materials. Yet, the results reported on ovalbumin suggest that the isohumulone does interact directly with the protein. When metal ions are present they may assist by forming coordination complexes. Oxidative interaction of isohumulone and protein may also occur.

Figure 5 shows the effect on beer foam stability of the addition of ethylenediaminetetraacetic acid (EDTA), a metal chelating agent, with and without

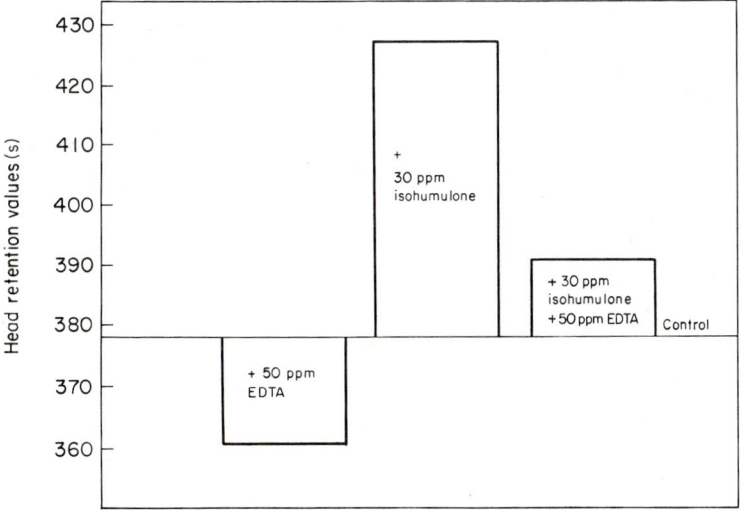

Fig. 5. Addition of EDTA to unhopped beer.

the addition of isohumulone. As can be seen the stability is reduced by the addition of EDTA in the absence of isohumulone, and in the presence of iso-humulone the reduction is even greater.

The results suggest that the probable mechanism for the action of iso-α-acids and metal ions on polypeptide films is a tanning process. The metal ion isohumulone complex cross-links with certain amino acids in the polypep-tides making up the bubble lamella. Thus not only does it prevent the poly-peptide from leaving the lamella and redissolving in the bulk liquid, but it also makes the lamella immune to breakage on thinning. This cross-linking mechanism can also prevent the reorientation of the polypeptide chain in the lamella and thus the rate of change of surface viscosity on ageing is reduced.

As far as overfoaming is concerned, this effect is more difficult to interpret. The metal isohumulone complexes shown here to cause gushing are unlikely to be the culprits in commercial circumstances. Other hop constituents have been reported to affect gushing. Dehydrated humulinic acid encourages it and humulone prevents it.[8] Humulone is more hydrophobic than its isomerized product, and thus its presence in the surface layer may prevent coordination compounds with isohumulone being formed. This observation led to the idea that hydrophobic fatty acids could prevent gushing, and linoleic acid has been found effective.[9] Japanese workers have reported that gushing can be caused by certain proteins derived from moulds that are known to infect barley;[10] here the situation is far more difficult to interpret.

5 Conclusion

It is apparent that beer foam is a complex mixture and its stability is influenced not only by the nature of surface-active material present but also by the interaction with small molecules, not surface-active in their own right. The discussion on surface-active material has been confined to proteins and polypeptides chiefly because they are the major contributors in beer. It has been shown that if the surface-active polypeptide is linked to polysaccharide the complex has greater stabilizing effect on beer foam. It is not suggested that the complex described here is the only such complex in beer nor need it be the most effective. But by showing that protein–poly-saccharide complexes are more effective than protein alone, we open an avenue of research that could lead to control and enhancement of the most effective surface-active material in beer foam.

On the interaction of small molecules discussion has been by necessity limited. We have particularly studied substances derived from hops which can have a considerable effect. It has been suggested that a coordination complex is formed involving metal ions and it is believed that such a mechan-ism could also explain some cases of gushing, as by affecting the charge distribution of surfactant micelles in solution, metal-ion complexes could affect bubble size and thus the kinetics of gas release. This is a subject on which there is yet much work to be done.

Acknowledgement

I thank Mr G. Jackson for his assistance.

References

1. Rudin, A. D. (1957). *J. Inst. Brew.* **63**, 506.
2. Bulas, R. and Kumins, C. A. (1958). *J. Coll. Sci.* **13**, 429.
3. Rudin, A. D. (1958). *J. Inst. Brew.* **54**, 238.
4. Gardner, R. J. (1972). *J. Inst. Brew.* **78**, 391.
5. Roberts, R. T. (1975). *In* "Proc. 15th European Brewery Convention, Nice". To be published.
6. MacWilliam, I. C. Private communication.
7. Anderson, F. B. (1965). *J. Inst. Brew.* **72**, 384; Wenn, R. V. (1972). *J. Inst. Brew.* **78**, 404.
8. Laws, D. R. J. and McGuinness, J. D. (1972). *J. Inst. Brew.* **78**, 302.
9. Gardner, R. J., Laws, D. R. J. and McGuinness, J. D. (1973). *J. Inst. Brew.* **79**, 209.
10. Amaha, M., Kitabatake, K., Nakagawa, A., Yoshida, J. and Harada, T. (1973). *In* "Proc. 14th European Brewery Convention", p. 381.

Discussion

Elliott (*Schweppes Ltd, Hendon, London, England*) When metal ions were used in conjunction with isohumulone to enhance foam levels and stability in what foam were they used? Also were the foam levels affected by changes in pH?

Roberts R. T. The metal ions used in the experiments reported were added to the beer itself as a suitably ionizable salt, first dissolved in 5 ml water, so that on addition, the final metal ion concentration in the beer was 10^{-4} mol dm^{-3}. The isohumulone was added as pure photo-ionized humulone.

The foam level on beer is not affected by more than 10 per cent by change of pH in the range 3·5 to 8·5; above and below these values, the foam stability falls fairly quickly.

17

Food foams—static and dynamic

JUDITH V. RUSSO

2 Hexham Gardens, Isleworth, Middlesex TW7 5JR, England

Summary

The terms "static" and "dynamic" foams as they occur in the food industry are clearly defined. Examples of static foams are marshmallow, filling creams, ice cream and of dynamic foams cakes, bread and soufflés. The techniques for producing the various types of foams are described and discussed. The paper is illustrated with photomicrographs of aerated foams.

1 Introduction

Does a truly static foam exist? By definition a foam consists of a set of bubbles or polyhedra of variable size held more or less firmly within a liquid or solid matrix. In foam-containing foods the gaseous phase is normally present as spheroidal bubbles. This being so, a foam can never be truly static and there is bound to be constant bubble-to-bubble diffusion due to the differing internal pressures between bubbles of differing sizes. However, for the purposes of this paper I am in fact defining two distinct types of foam, which occur both in the food industry and kitchen.

2 Static foams

By a static foam I mean one which is produced at normal temperature such as a "buttercream" formed by a curved blade horizontal mixer (which I shall be describing later), or a marshmallow foam formed from egg white, gelatin and sugar by the action of, say, an Oakes machine (which I shall also be describing later) or a refrigerated foam such as ice cream which is refrigerated and aerated simultaneously. In the case of the marshmallow the setting of the gelatin on cooling forms a stable matrix and in the case of the ice cream the

273

ice crystals and solid fat form the stable matrix. In the case of the "butter-cream" the sugar crystals and partially solid fat form the stable matrix. In all these cases the foams are to all intents and purposes *static* as they do not change *materially* between *production* and *consumption*.

3 Dynamic foams

A dynamic foam is, however, quite different; cakes, sponges, bread, meringues and soufflés are examples. During the preparation of these foods the foam passes through several stages. For instance, sponge cake batter, which con-sists essentially of a mixture of flour, sugar, egg, water and baking powder, is

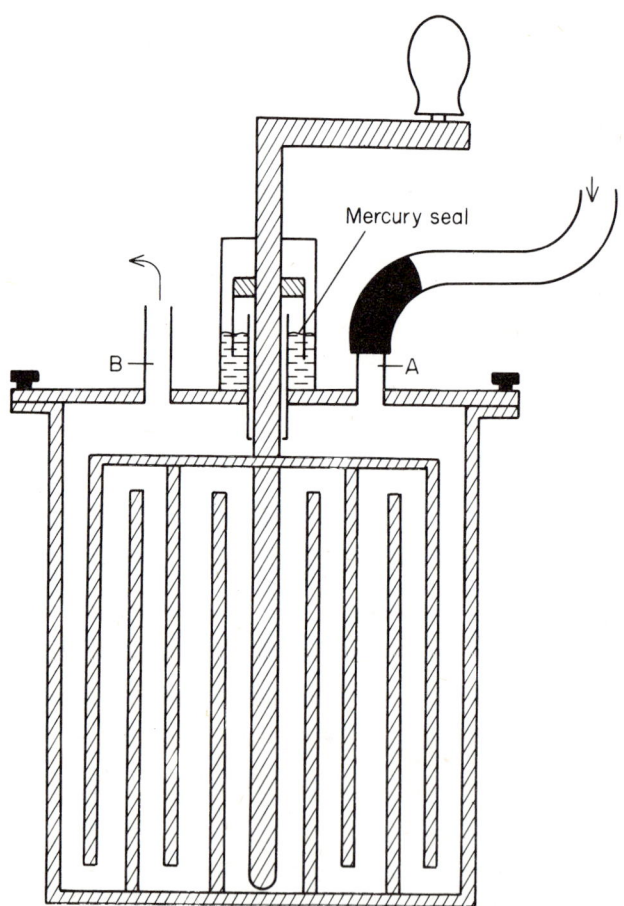

Fig. 1. Mixer for measuring evolution of CO_2 from scone dough.

aerated continuously by an Oakes mixer. The CO_2 from the baking powder, to some extent before cooking but predominantly during the baking period, diffuses into the already formed bubbles, contributing to the dynamic nature of the foam. But baking causes a series of transformations. These processes will be shown in a film during the Symposium illustrating sponge batter being baked on a microscope slide on a hot stage. The important point to note are: (1) bubble expansion, (2) bubble-to-bubble diffusion, (3) the bubbles become polyhedral, (4) eventual loss of bubble movement.

Some experimental work was carried out on the effect of CO_2 on a scone gas burette. Figure 1 shows the apparatus for mixing a dough. Another apparatus is used for baking the dough. During baking a blank is run on a mix without CO_2 so that a correction can be made for expansion effects. (The work I am describing is 20 years old and has never been published before.) Figure 2 shows the combined effect of mixing and baking on the evolution of CO_2. The results of earlier work (1940s) have been reported by Barackman on the rate of evolution of CO_2 at room temperature.

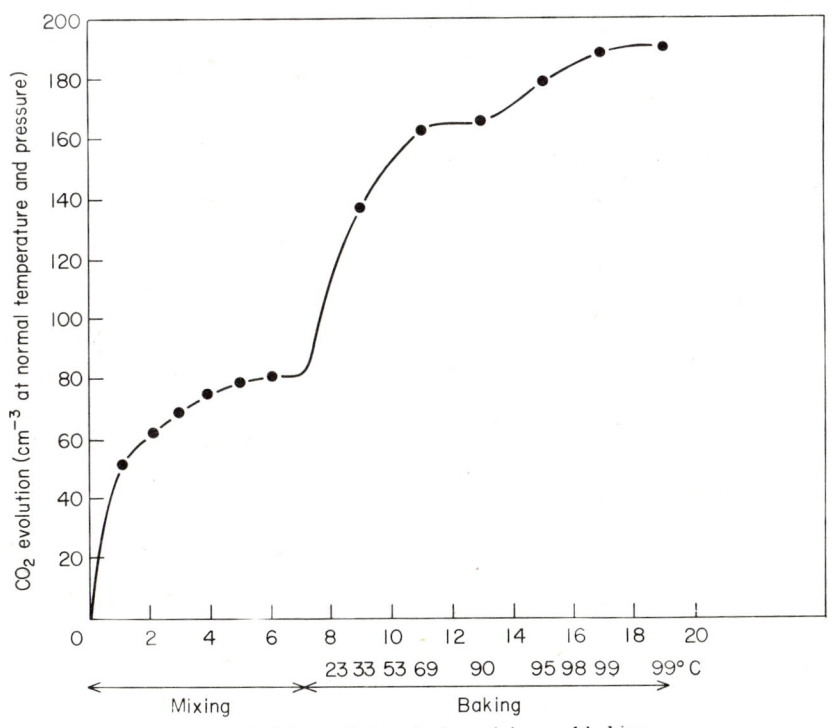

FIG. 2. CO_2 evolution during mixing and baking.

Once the foam has been through these dynamic states a great deal of the expansion can be accounted for by formation of water vapour at elevated temperatures;[4] it is eventually almost static as the surrounding matrix sets due to coagulation of the egg proteins and gelatinization of the starch. Figure 3 shows the final static state of the foam. It is interesting to note the distorted starch granules in the cell walls.

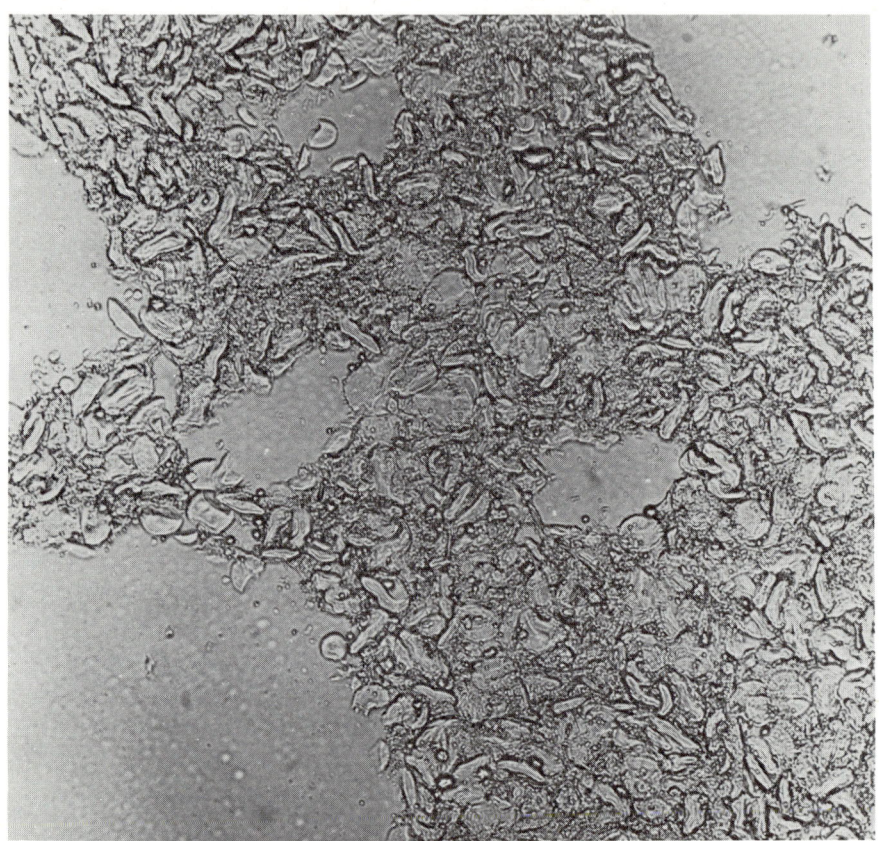

Fig. 3. Microscopic appearance of baked sponge.

Probably one of the most ancient foams is bread. Bread consists essentially of flour, water and yeast. After mixing and incubating the enzyme from the flour, amylase breaks the long starch chain to form the shorter sugar maltose, which is then acted upon by yeast to form alcohol and CO_2. The CO_2 causes the dough to rise during an incubation period at $30°C$; there is then a further increase in size due to expansion and water vapour formation during baking.

An interesting study of gas retention and loss from a fermenting dough can be made using a Chopin Zymotachygraph. This machine measures the total gas evolved during a given time period and on each alternate sequence absorbs the CO_2 in soda lime. By drawing appropriate lines the total and retained gas can be measured. This is shown in Fig. 4. It can be seen from the diagram that for the first few hours of fermentation all the gas is retained

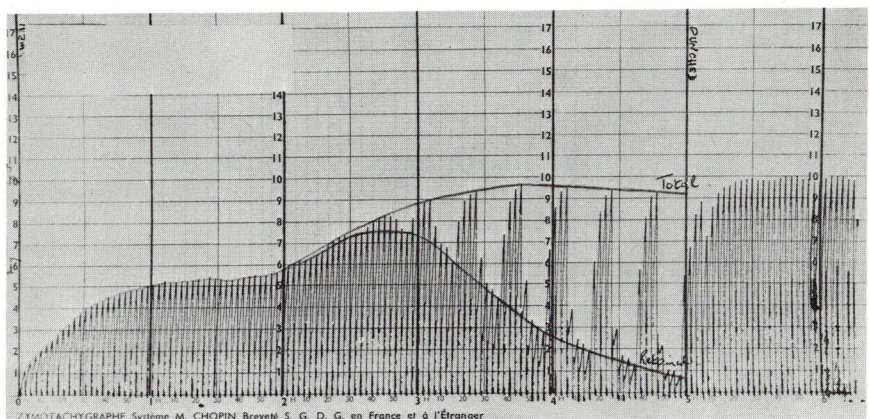

FIG. 4. Gas evolution and retention in bread dough.

and then the dough becomes increasingly porous. The effect of various dough additives such as $KBrO_3$ and of physical treatments such as punching on the retention characteristics of the dough can be assessed using this apparatus. The final set static bread foam is shown in Fig. 5. Here an interesting point to note is that the starch granules in the matrix are far more distorted than they were in the sponge. This is due to the fact that the sugar in the sponge "protects" the starch granules from complete gelatinization.

The soufflé is a particularly interesting example of a dynamic foam. Whatever the type the important constituent in the soufflé is the highly aerated egg white which is folded into the mixture during the last stages of mixing. On baking there is considerable bubble-to-bubble diffusion and a setting or coagulation of the egg white protein. However the matrix in this case is very fragile and the product, unlike cake or bread, very rapidly collapses on cooling. This therefore may be considered as a kind of negative dynamism.

Meringues consist essentially of egg white and sugar and again like the soufflés are aerated either by a hand or electric whisk. The formation of a matrix in the meringue is partly due to coagulation of the egg white protein, and partly to the drying out which occurs during baking.[3]

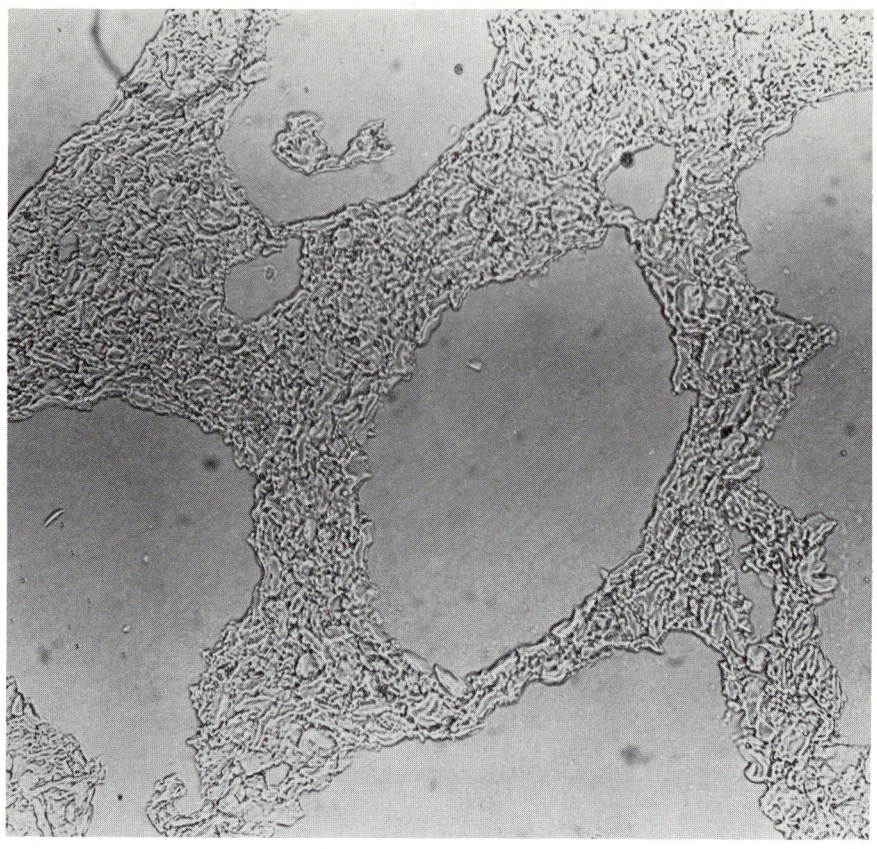

FIG. 5. Microscopic appearance of baked bread.

4 Foaming equipment

Whether the final foam is static or dynamic the equipment used to form the foam is interesting. To start with the very simplest we have the hand whisk, which is capable of producing egg white foams with densities as low as $0.1 \, \mathrm{g \, ml^{-1}}$ (i.e. 9 parts of air : 1 part original). More refined is the Hobart mixer with whisk attachment. A machine of equal refinement is the Morton Gridlap which consists of two contra-rotating horizontal curved blades. This is often used to aerate cake mixes by what is known as the two-stage process, aeration of fat with flour, followed by folding in of other ingredients such as egg and sugar. It is also used for the formation of "buttercream" in fat continuous sweet cake fillings. The blades on this machine rotate at a maxi-

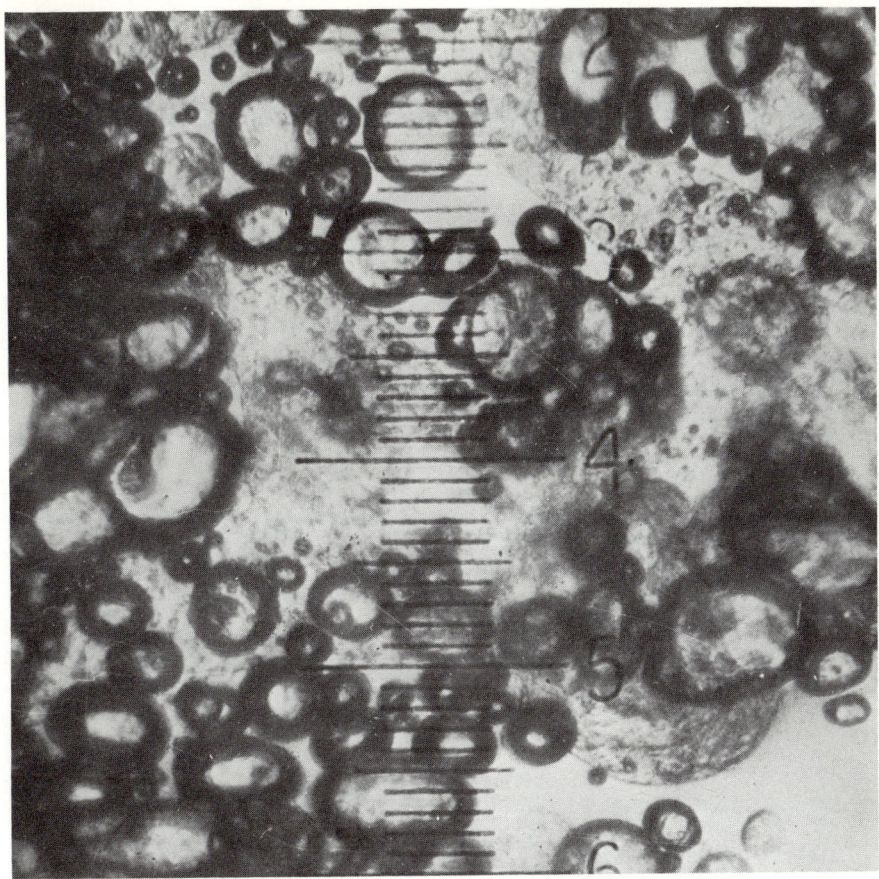

FIG. 6. Microscopical appearance of a "buttercream" filling. One small division = 10 μm.

mum speed of 140 r.p.m. and produce foams with bubble sizes between 10 and 100 μm. A foam of this type is shown in Fig. 6.

More refined still is the Morton pressure whisk which is still used in some bakeries. The advantage of the pressure whisk over the open bowl type whisks is a smaller and more even air bubble size. The most frequent air bubble size in a foam produced on a Morton pressure whisk is 50 μm. Even more refined is the continuous aerator Oakes mixer where a slurry is continuously stirred and aerated between a series of blades against a back pressure of between 30 and 60 p.s.i.g. This machine is shown in Fig. 7. The air bubble size produced by this mixer is about 10 μm. Handlemann, Conn and Lyons[2] have made a very interesting study of the effect of air bubble size on the stability of cake batters. They have shown that the air bubbles obey a Stokes's law relationship, i.e. the velocity of rise of the air bubble in the foam is inversely proportional to the viscosity of the surrounding fluid and proportional to the square of the radius of the bubble. Foams with large air bubbles or in which there has been much bubble-to-bubble diffusion are therefore less stable and there is more likely to be a "layering" (large bubble size) at the top of the cake.

Fig. 7. Oakes continuous aerating machine.

In the equipment described so far the operation of aeration is the prime function and heat exchange plays a very minor role. In fact the Oakes continuous aerator though often supplied with a cooling jacket is a very poor heat exchanger. In aerating ice cream, however, one requires a machine which will both aerate and freeze the ice cream mix. Examples of machines of this type are the CP and the Vogt freezers. In both machines ice cream mix at approximately 5°C is drawn into the barrel of the freezer with excess air and the freezer blades scrape the heat exchanger surface which is usually cooled with liquid ammonia.[1, 5]

The amount of air incorporated is controlled by the differential in capacity between the mix metering pump and the air plus mix pump. All beating takes place under a back pressure as in the Oakes mixer and the amount of work involved is indicated by an ammeter which permits control of the machine.

Machines of this type can similarly be used to aerate mousse. Aeration in this case is normally carried out at a slightly higher temperature than for ice cream as the presence of fairly high concentrations of gelatin would otherwise make processing very difficult—the product could easily set solid before leaving the machine.

I have deliberately been somewhat parochial in this presentation and have kept to products and processes of which I have a special knowledge through my experience in the Lyons Group. I have therefore omitted techniques used in sugar confectionery production such as vacuum aeration.

I would like to thank all my ex-colleagues in the Lyons Group who have contributed "know how" and ideas.

References

1. Arbuckle, W. S. (1966). "Ice Cream", pp. 191–218. AVI Publishing Co., Westport, USA.
2. Handlemann, A. R., Conn, J. F. and Lyons, J. W. (1961), *Cereal Chem.* **38**, 294.
3. Paul, P. C. and Palmer, H. H. (1972). "Food Theory and Application", pp. 543–544. John Wiley, New York.
4. Russo, J. V. (1972). *In* "Aeration of Baking Products", BFMIRA Symposium No. 14, p. 24.
5. Turnbow, G. D., Tracy, P. H. and Raffetto, L. A. (1947). "The Ice Cream Industry", 2nd edn. John Wiley, New York.

Discussion

Pearson (*J. Lyons Ltd, Central Laboratories, Hammersmith, London, England*) In what way does the presence of sugar slow down the rate of staling in cakes and effect the foam formation in baking?

Russo Sugar "protects" the starch from complete gelatinization and therefore degradation. This therefore inhibits sealing.

I have no evidence on the effect of sugar on foaming.

Creak Could you please offer any theory to support the action of potassium bromate and/or punching on the air retention in your foam systems.

Russo Both potassium bromate and punching affect the gluten structure, thus making it more retentive.

Ray (*BP Research Centre, Sunbury-on-Thames, England*) (1) Are there any technical reasons for avoiding the presence of propionic acid when forming food foams, and (2) Could you comment on the mechanism whereby esters such as tripropionin (glycerol tripropionate) decrease water loss from baking or finished products?

Russo (1) Calcium propionate is an allowed preservative in bread and is widely used as such; therefore there seems to be no technical reason for avoiding its use in foams. (2) I have no experience of the effect of tripropionin in bakery products.

18

The use of foams in foods and food production

J. W. MANSVELT
Lenderink & Co. BV, Scheidam, P.O. Box 126, The Netherlands

Summary

Aeration, the incorporation in a food of a great number of small gas cells, leads to foam or sponge structures. It is used to improve texture, consistency and appearance and to reduce the density and thus the cost per unit volume. To create foam structures, it is essential that the systems contain a suitable surface-active agent (whipping agent) and that sufficient energy is supplied, which is usually done by means of beating or whipping.

The most versatile whipping agents are those based on proteins, but for special purposes the products based on partial esters of fatty acids, cellulose derivatives and others find application. Close standardization of performance is essential in all cases.

The equipment used for the beating operation can be either batch operated or continuous. Strict temperature control during beating may be essential. Formulation-wise, aeration is either carried out by means of an all-in or one-step process, but often two-step-aeration must be resorted to, particularly if the formula contains foam-inhibiting substances.

Interesting applications for aeration are found in the bakery field (bread, cakes, meringues), in various types of desserts (instant desserts, chilled cabinet lines and deep frozen products) and in sugar confectionery, where centres for light nougats form a typical example.

1 Introduction

In foods, the incorporation of a gas in finely dispersed form—foam or sponge—has a pronounced influence on a number of aspects, including appearance, texture and consistency, digestibility and, of course, size per unit weight.

The presence of a well-defined volume proportion of gas cells with a specific size distribution may even be an essential condition for the characteristic properties of that particular food. Examples can be found in such widely different foods as bread, ice cream, soufflé and french nougat.

The creation of a gas cell structure presupposes having in the system at some stage of the production process a suitable concentration of a suitable *surface-active agent*, surfactant, and also expending on the system a sufficiently large amount of *energy*.

2 Whipping agents

The surface-active substances, essential in creating and maintaining a foam structure, are usually called *whipping or aerating agents*. Although of widely different origin, all have certain properties in common. Apart from a pronounced surface activity, they must show a definite ability to contribute to the formation of stable interfacial layers. They could well be broadly classified into two groups: one group based on proteins and a second—miscellaneous— group, including such substances as the partial esters of fatty acids, certain glycosides, fatty alcohol sulphates, etc.

Whipping agents either *occur* naturally in one of the ingredients used in preparing the food, or they are *formed* by some mechanism during the production process or they are *added* intentionally.

In the latter case, partially hydrolysed animal or vegetable proteins have been found to be particularly useful.

3 Protein based whipping agents

Although most water-soluble proteins show a tendency to form foams in aqueous solutions, few can be considered as whipping agents in the sense discussed here.

The requirements such a protein or protein derivative must meet can be summarized as follows: (1) soluble in water and sometimes in concentrated sugar solutions, (2) soluble at a wide range of pH; (3) excellent foam formation; (4) active over a wide temperature range; (5) foams formed sufficiently stable; (6) foams formed show minimum stability against the presence of lipids; (7) in all respects acceptable for use in food; and (8) reasonable cost price relative to use level.

A review of the protein products used for practical aeration purposes would in the first place include the traditional *egg white* and the products derived from it: pan-dried or spray-dried egg albumen.[1]

These products are used in a wide range of foods, including meringues, soufflés and nougats. They are typical "natural" products in the sense that in the natural state their technical performance is not standardized at all.

Egg products are often contaminated through microbiological infection, including possibly infection with members of the Salmonellae group.

Modern processing—that includes a pasteurization step—has done much to reduce the danger.[2] Careful pasteurization is essential, as overheating will adversely affect the whipping characteristics.

Solutions of egg albumen in water will form rigid gels when heated to over 70°C. This thermal gelation is strongly affected by the presence of sugar in the solution. With increasing sugar concentration, the temperature at which gelling occurs is progressively increased and at sufficiently high sugar concentrations, the formation of the gel is inhibited completely.[3]

A second group of proteins used extensively for whipping is that of the *gelatins*. Gelatins, hydrolysis products of collagen, are unique in being at the same time whipping and gelling agents. This can be a definite advantage, but can, under a different set of circumstances, easily be a considerable disadvantage.

Raw materials such as eggs, skins and bones inherently show wide variability and it is difficult to standardize within narrow limits the functional properties of ingredients derived from them.

The differences in performance between various lots of egg albumen or gelatin will perhaps not cause trouble in small, batch-type production; they may well be large enough to give serious difficulties in automated continuous production.

Efforts have been and are being made to obtain better standardization. Consequently there is, particularly for industrial application, a wide interest in a group of whipping agents that we, for want of a better expression, will designate as the *newer whipping agents*. These are natural products, factory produced from animal or vegetable proteins by the aid of complicated processes. These processes, usually some form of controlled hydrolysis, transform naturally occurring proteins into whipping agents. By varying the hydrolysis conditions and by other means, the specific characteristics of the products thus obtained can be adapted to any particular set of requirements. Moreover, it is possible to eliminate the variability inherent to all natural products and to produce products that meet given specifications consistently. As starting material for these newer whipping agents, proteins isolated from soybeans,[4] wheat and milk[5] have been proposed.

Foam stability, as mentioned earlier, largely depends on the formation of molecular layers in the gas–liquid interface. In the case of protein-based whipping agents, these layers are stabilized by the formation of two-dimensional networks, consisting of protein chains, interlinked at suitable intervals. This means that the protein, originally present in the solution in some coiled form, unfolds into single chains that are then absorbed into the interface. Next, these chains interact at intervals through chemical or other linkages to form two-dimensional networks.

4 Nonprotein whipping agents

The nonprotein based whipping agents form a rather heterogeneous group of products, including the partial esters of fatty acids, fatty alcohol sulphates, certain cellulose derivatives and natural products, such as lipoproteins, lecithins and some glycosides, notably saponins. Of these various products, only the partial esters of C-16 and C-18 fatty esters with glycerol and, to some extent, sorbitol and sugars are used generally in food production. Most of the others are limited in some way by Food Regulations.

Aqueous dispersions of partial esters of fatty acids—emulsifier-type products—will foam under very specific conditions only: in most cases a combination with a water-soluble protein is essential as is strict temperature control.

Foam stability is in this case thought to be due to interfacial layer formation through side-to-side association of the fatty acid chains.

Systems based on emulsifier-type whipping agents are responsible for the foaming in such natural products as dairy cream.

5 Expending mechanical energy

The second factor essential in creating foam structures is expending sufficient energy on the system. This energy is needed to overcome the interfacial tension during the very considerable increase in the amount of interface, the surface between gas and liquid. The amount of energy actually expended on the system bears no direct relationship to the energy needed for surface expansion. Most of the energy supplied is consumed in overcoming viscous friction.

Various techniques for supplying the energy exist. The more important are: (a) biological or chemical reactions leading to formation of a gas; and (b) direct application of mechanical energy through beating or whipping.

Formation of a gas through a *biological process* is largely used in bakery techniques: the dough or batter is allowed to ferment, permitting yeast cells, added for that purpose, to create a large volume of carbon dioxide. This gas is released to the system in the form of small gas bubbles. During a secondary operation, baking, more gas is released and the gas bubble dispersion is stabilized by the starch gel formed.

Comparable results are obtained by chemical means. The gas is now generated by means of a chemical reaction using a mixture of sodium-bicarbonate and an acid or acid salt.

The direct application of *mechanical energy* may take various forms. In sugar confectionery the pulling process—repeatedly stretching and folding a hot, stiff sugar dough—is used for a large range of products.

There is little doubt, however, that the operation called *whipping* or *beating* is by far the most versatile way of creating a gas cell structure. The liquid or semi-liquid system is brought and kept in violent motion in the presence of the gas. During the process, the amount of gas incorporated steadily increases until a maximum volume is obtained. This maximum volume depends on the whipping agent used, the composition of the system and on the level of energy supplied. During beating, the mean gas bubble size decreases to a limit value and these same factors determine what minimum bubble size is obtained.

6 Equipment for aeration

When considering the equipment available commercially for performing the whipping operation in an industrial manner, the following characterization based on type of equipment and operational conditions is proposed: (1) batch operation at atmospheric pressure; (2) batch operation using compressed air; (3) continuous operation using compressed air and static dispersors; and (4) continuous operation, using compressed air and rotating dispersors.

In batch operation equipment, a certain quantity of the mix to be aerated is placed in the machine and is then beaten until the desired volume of air is incorporated and the desired density has been obtained. The beating can either be performed in an open vessel, using atmospheric pressure, or in a closed vessel, where a higher pressure can be maintained. In the latter case, the foam formed initially will further expand when released to atmospheric pressure at the end of the process.

The typical examples of equipment for batch operation under atmospheric pressure are the small household beater and the planetary mixer. The latter is a most versatile piece of equipment, widely used in all sorts of whipping and blending operations in the bakery, kitchen, small confectionery plant etc. The planetary mixer is badly adapted to industrial processing because of the relatively large amount of labour needed.

Batch operation pressure beating machines are far better adapted to large-scale processing and are easily incorporated into a production line. This is mainly because the aerated batch is transported quickly and cleanly by ejecting it from the bowl by the use of residual air pressure.

Equipment for continuous whipping always consists of some sort of mixing head, into which the batch to be aerated is metered under pressure, together with the required amount of gas. A definite pressure is maintained in the mixing head and again, when leaving the machine, the foam will expand, giving an extra volume increase. The degree of aeration can be controlled conveniently by controlling the volume of gas going into the mixing head.

An important point in choosing continuous equipment is the heat exchange capacity provided. Beating, of course, generates considerable heat and when the batch under consideration needs to be kept below a given temperature, heat exchange surface is an important factor.

Equipment using static dispersors is, due to limitations of the amount of energy available for whipping, restricted to systems that are fairly easy to aerate. Typical examples are the small whipped cream dispensers.

In all those cases, where the output is sufficiently large to merit continuous operation, equipment using motor-driven dispersors is preferred. Continuous aerating equipment of this type, combined with continuous fluid and dry metering devices and continuous mixers, can handle almost any type of aerated food.

7 Formulation

In establishing formulas and techniques for the production of aerated foods, the overall composition of the food under consideration is of prime importance. The performance of whipping agents is greatly influenced by such factors as viscosity, soluble dry solids content and the presence of foam-inhibiting substances, including fats and oils, alcohols, certain flavouring materials, etc. These factors, affecting different whipping agents differently, will determine what whipping agent to use and in which way aeration is best carried out. Formulation-wise, the actual aeration process can be carried out in two basically different ways.

In *one-step or all-in aeration* all or nearly all of the ingredients entering into the recipe are combined and the whole batch is aerated by direct whipping. The system is very simple to use, but restricted in its application to fairly high moisture lines and to those cases where no foam-inhibiting substances are present. If for either of these two reasons the one-step process is not feasible, use must be made of *two-step aeration*. Now, first a part of the ingredients, always including the whipping agent, is aerated to a foam and the rest of the ingredients, in the form of a powder mix, a hot concentrated sugar solution, a batter or a dough perhaps including fat and fat containing ingredients, is then blended in carefully. This process is more complicated to carry out, but its scope is very much wider. Whether in a particular case one-step or two-step aeration is chosen will not only depend on the composition and on the degree of aeration required, but also on the equipment already available and the output desired.

8 Application

It was earlier pointed out that in food processing aeration is proposed for two different purposes: (1) improving texture, consistency and appearance; and (2) reducing the density.

Both considerations are strongly affected by the *degree of aeration* used. The degree of aeration is best expressed as the volume percentage of gas incorporated or—more practically—by means of the density of the aerated product in comparison with that of the nonaerated product.

Considerable improvement in texture, consistency and appearance can often be realized through a quite moderate degree of aeration: even as little as 10 or 20 per cent of gas incorporated may result in considerable improvements of the organoleptic aspects. Too high a degree of aeration will usually have a negative effect.

Introducing gas into an article sold in a predetermined volume means a reduction of the piece weight, the amount of raw materials needed and thus the cost price. Aerating an article sold at a predetermined weight means a bigger piece or pack volume and thus a better sales appeal. Quality aspects and economic aspects can often be combined profitably. It will, however, always be essential to accurately determine and control the optimum degree of aeration.

9 Bakery products

Bread, biscuits, cake and similar lines in many cases derive their unique texture from the presence of a well-defined amount of air. For certain lines, including bread, most of the gas is generated during the baking process; in others, for instance, sponges and meringues, most of the gas needed is incorporated during the preparation of the dough or batter.

In the first case, dough preparation involves a strong kneading or mixing action. During this process a smaller or larger amount of air in the form of very small air bubbles is always incorporated.

In fermentation raised or chemically raised lines, the small initial air cells act during the baking as nuclei for the separation of carbon dioxide. The final structure obtained is strongly influenced by the initial air cells and consequently by the time and intensity of the kneading process and the eventual presence of whipping agent type substances at that stage.

In the case of bakery goods, where the air cell structure is preformed in the dough or batter stage, the situation is different. The major part of the volume is obtained through beating, although, of course, additional volume is created during baking by water vapour expansion or through the use of baking powder.

Here, the role of the whipping agent is often played by the whole egg—

white and yolk—basically used in the formula for reasons of texture, consistency, emulsifying properties and flavour. Using whole egg as whipping agent means using two-step aeration.

In this field, the partial esters of fatty acids have found extensive use as auxiliary whipping agents.[6] Used in conjunction with whole egg, mono-stearates and similar products makes one-step aeration of cake batters possible. This, particularly in the case of industrial production, results in considerable savings in equipment and time. Comparable results can be obtained by using protein-based whipping agents in conjunction with whole egg, however, with the added advantage of better solubility. Monostearates need to be predispersed with hot water and are thus less suited for dry-mix formulation. Here, the protein based whipping agent show a definite advantage.

Emulsifier-type products tend to make the cake crumb age more quickly. This phenomenon, probably due to the interaction of monostearates with starch, is absent in protein-based whipping agents.

10 Desserts

Ordinarily, the term "dessert" is used for that part of the meal which is consumed after the main course and a dessert is, therefore, more eaten for the pleasant sensation than with the idea of stilling the pangs of hunger. Aerated or whipped desserts have been known for a long time. Looking through high-class cookery books, references are found to such delicacies as "mousse au chocolat", "bavaroise" and "soufflé". In most of these classical recipes, the aeration is obtained by two-step aeration: carefully blending in a foam obtained by whipping egg whites or fresh dairy cream. Desserts as discussed here might be classified as: (1) instant products; (2) ready-to-eat products; and (3) frozen desserts.

The *instant desserts* include all those products that are produced either by the housewife or the kitchen chef shortly before consumption from some sort of industrially produced pre-mix. This pre-mix, either in the form of dry powders or of a paste or dough, contains all or most of the essential ingredients, possibly in some treated form. If the preparation is simple and consumes little time, we rightly call such products "instant".

The *ready-to-eat-products* are mainly distributed at chilled cabinet temperature of roughly $+5°C = 40°F$. These products are the real "convenience foods" in the sense that they can be bought and consumed without any preparation whatsoever. They, however, show a limited shelf life (approximately 14 days at $+5°C$).

The *frozen desserts* comprise all products sold in frozen condition at a temperature range below $-20°C$ ($-4°F$). These include the products consumed while still partially frozen—classical ice cream—and the products that have to be thawed completely before consumption. Most desserts are, of course, sweet. The total sweetness impression can be adapted within wide limits by varying the amount and type of sugar and the acidity. Excellent nonsweet products or *savouries* have been prepared which are most adapted to being served as an entrée or appetizer before the main course.

11 Instant products

The scope of desserts obtained through the use of a pre-mix is very wide. Variations in consistency can be handled through judicious choice of type and use level of the stabilizer system and through control over the degree of aeration.

Most of the pre-mixes for whipped instant products are dry mixes, mixes of various ingredients in powder form, and all that is needed for the preparation is adding the indicated amount of liquid—hot or cold water, milk, fruit-juice, an egg or even alcoholic fluid—and aerating the mix by beating. When the desired consistency is reached, the product is filled into suitable cups and placed in the refrigerator for a short time.

Some formulations require two-step aeration: first one pouch from the pack is mixed with water and beaten and then the contents of the second pouch are added. This two-step system is needed in those formulations where ingredients that inhibit or delay aeration are present.

Apart from sweeteners and flavouring material, the pre-mix will always contain a whipping agent and a stabilizer system. As stabilizers, two systems have found general acceptance: the pre-cooked starch–milk protein–phosphate systems and the cold soluble gelatine products. Other hydrocolloids, such as the alginates, may be added for special effects. It is, of course, essential that, whichever stabilizer is used, it hydrates very quickly to give the desired viscosity for easy whipping. Once whipped up, the dessert should show a fairly quick set (15 to 30 min) when placed in the refrigerator. In the case of household mixes, the amount of beating needed to obtain the correct density must not be excessive; a few minutes beating with a simple mechanical or electric beater should suffice. Preparations for the professional kitchen can be expected to be beaten more effectively, using perhaps a planetary beater. In both cases, however, the aerating agent should dissolve almost instantaneously, even in the presence of other ingredients, such as sugars and stabilizers, and should give a quick, smooth aeration with a foam structure sufficiently stable to bridge the period until the stabilizer system comes to a set.

The pre-mixes in powder form are of two different types: one is simply a mechanical mix of dry powders, including sweeteners, stabilizer system, whipping agent, perhaps skim milk powder and flavour. This simple dry mix can, for reasons already explained, only be used in fat-free systems. For those formulations which must be prepared with ordinary full cream milk or where it is desired to have fat in some form included in the dry mix, use must be made of pre-homogenizing. This involves dispersing the basic ingredients, always including the fat and the protein derived whipping agent in warm water or milk and passing this mix through an homogenizer. The resulting emulsion is then spray-dried and mixed with the rest of the ingredients. There is no need to point out that such a mix offers a very definite danger of rancidity if no proper anti-oxidant formulation is provided for.

12 Ready-to-eat desserts

The largest group of convenience desserts and possibly the one now expanding most rapidly is formed by the *chilled cabinet products*. The composition of these desserts is based on milk, either skim or full cream with possibly added dairy cream or vegetable fat (filled dairy products), a stabilizer system, whipping agent, fruit pulps, or other flavouring products.

Products not using milk in some form as the basic ingredient have not, as yet, found an important outlet although interesting nondairy combinations using fruit pulps have been evolved. The same applies to the nonsweet savoury products, using such materials as quark or potato products.

The total dry solid content of the conventional products is usually around 30 per cent; the fat content can, with excellent results, be kept low. From 0 to a maximum of 10 or 11 per cent and a density of 0·50 to 0·80 seems to be preferred generally.

The production of ready-to-eat desserts is simple. In the case of the straight dairy products, the fluid ingredients are heated to approximately 60°C (140°F), the dry ingredients are stirred in and dispersed thoroughly. The complete batch is then homogenized and pasteurized either by holding for 15 minutes at 80°C or by passage through HTST equipment. After cooling to approximately 25°C the batch is aerated and cooled in a scraped surface heat exchanger to the desired density and a temperature of approximately 5°C. The dessert is then filled off into containers and stored at +5°C. Handling products based on fermented milks—yogurt or quark (cottage cheese)—needs some care as the texture of these products is easily damaged by overheating.

The *shelf-life* of chilled cabinet products is all-important. No guaranteed shelf life is possible without a very thorough control over the distribution chain with its weak links, the refrigerated lorry and the display cabinet.

Even under optimum storage conditions the useful life of these products is limited. It can be impaired in two different ways. In the first place the physical stability may be involved. Through syneresis of the stabilizer system or lamellae rupture in the foam structure, the appearance of the product may be effected in such a manner that it is no longer fit to be sold. Using the optimum whipping agent–hydrocolloid combination and proper production techniques, that guarantee complete hydration of the key ingredients, forms the best insurance against failures in this direction.

Secondly, there is microbiological spoilage. Products which have been aerated cannot well be subjected to heating without damaging the foam structure and thus the texture. A sufficiently long shelf-life without pasteurization after filling (or with very limited pasteurization) is only possible if initial infection is minimal. To keep initial infection to a minimum it is essential to use good quality ingredients, efficient heat treatment of the liquid batch and strict hygienic processing.

Using the lowest pH consistent with the flavour used, the highest percentage of low molecular weight sugars consistent with texture and flavour, nitrogen for aerating and antimicrobial chemicals, such as sorbic acid and benzoic acid, when permitted, can help in increasing the shelf-life. Products made from fermented or acidifed milk (pH of 5 or lower) show a longer shelf-life.

Special attention within the range of chilled cabinet products should perhaps be paid to the products derived from *quark or cottage cheese*. These products are not so much eaten as desserts but find, often combined with fruit, considerable popularity as snacks, being low in fat and high in protein. Particularly the young secretary or typist, desiring to maintain her fashionable underweight, has been persuaded to accept the low-fat, high-protein philosophy. In this case, the advantage of aeration is very evident; judicious use of aeration transforms healthy but unappetizing quark into something which is both healthy and pleasant.

In between the ready-to-eat desserts and the frozen desserts is an interesting development worth mentioning: what in France, the country of origin, is designated as "prét à glacer". This dessert, which we have for want of a better name called "all-temperature-dessert", is distributed through the chilled cabinet line and can be consumed at that temperature as a mousse. It is, however, possible to place the cups of "prêt à glacer" in the home freezer compartment for three or four hours and then consume the product as an ordinary ice cream. Formulae for such products on dairy or on fruit pulp base have been worked out.

13 Frozen desserts

In this group have been brought together all those products which are distributed and sold at a temperature of below $-20°C$ ($0°F$). As explained, these include the products eaten while still partially frozen, such as *standard ice cream*. Much has been published on the subject of ice cream and ice cream production and there is little we can usefully add here. We would like to briefly mention interesting new developments in the field of ice cream type products or *frozen confections*, low-cost products with a consistency and texture similar to ice cream but based on fruit pulp, juice, or flavour.

The *frozen desserts*, sold at $-20°C$ but consumed after complete thawing, are finding increasing sales in the luxuary range of products, often in family-type packs serving four to six portions.

Frozen desserts can be very similar to the chilled cabinet desserts, although there is a strong tendency towards the filled milk-type products. The basic production method also is similar, but the stability requirements are much less strict. In the frozen state, the changes in physical structure and the growth of microorganisms take place very slowly. Once thawed down the product is usually consumed within 24 hours and is normally kept in the refrigerator during that period.

However, during transport and storage frozen desserts may have to withstand a number of heat shocks or freeze–thaw cycles without the foam texture and the stabilizer system being affected.

14 Sugar confectionery

We have already mentioned that sweets are perhaps amongst the oldest aerated foods produced industrially. Such long-established articles as the "nougat de montélimar"[7] and the Italian "torrone" are clear examples of foods where aeration is essential for obtaining the typical texture. We can usefully divide confections in three groups, based on their degree of aeration. In typical light candies, marshmallows, negro-kisses and meringues, the density is reduced from 1.30 $g\,cm^{-3}$ for the nonaerated batch to below 0.45 $g\,cm^{-3}$; for very light centres it can even be reduced to 0.16 $g\,cm^{-3}$. These products show a very special texture, an extremely large volume compared to weight, a low price per unit volume and are consequently popular with the very young.

In the second group of products, the volume of air incorporated is much more restricted, a density of approximately 0.80 $g\,cm^{-3}$ is considered to be near the optimum. This is sufficiently low to produce the desired texture and to reduce sweetness and cost price, but not low enough to give the consumer the impression of buying air. Included in this group are the various nougats, the fruit chews and the chocolate coated bars.

Finally, we have in the third group those articles where the density is reduced by only 10 or 20 per cent to perhaps 1.10 or 1.20 $g\,cm^{-3}$. Even this limited aeration is, as already mentioned, sufficient to bring about important improvements in consistency, texture and appearance. Typical for this group are the fondant cream centres.

As an illustration of how aeration is used in confectionery, a more detailed discussion of the production of nougat bars is of interest. These chocolate coated bars have become enormously popular and for many years have now been the line expanding most rapidly. The centre of a nougat bar consists of a blend of sugar and corn syrup (glucose) with a moisture content of perhaps 6–10 per cent that by means of 0·3–2 per cent of a whipping agent is aerated to a density of $0·6–1·0 \text{ g cm}^{-3}$. A part of the sugar is usually present in the form of very small crystals, giving the mass its short texture. Auxiliary ingredients such as cocoa powder, fat, milk products, flavours, nuts, etc., may be present as well. The production by two-step aeration may be carried out by continuously aerating a 65 per cent sugar syrup, containing the whipping agent, to a density of $0·20 \text{ g cm}^{-3}$. The foam thus obtained is, again continuously, blended with a hot sugar–glucose mix—boiled to 126–130°C, moisture content 5–7 per cent—resulting in a heavy foam, density $0·50–0·60 \text{ g cm}^{-3}$. This heavy, hot foam is finally mixed with the auxiliary ingredients, generally including a few per cent of small sugar crystals to initiate crystallization. The finished batch, temperature approximately 80°C, is spread out in the desired thickness, allowed to cool and cut to the desired dimensions. The pieces thus formed are then chocolate coated.

15 Flavour requirements

Correct (consumer accepted) flavouring and colouring of a food spells the difference between commercial failure or success. Finding the right flavour for a new food is therefore an essential part of the development work. Extensive consumer testing will be needed.

In the case of aerated foods, the flavours chosen should meet a number of special requirements: (1) not interfere with the whipping process; (2) not adversely affect the physical stability—shelf-like; and (3) be stable against air (oxygen).

The first point particularly applies in those cases where one-step aeration is proposed and the flavours have to be added to the batch prior to aeration. Special flavours are now available for this purpose.

In dry mixes, encapsulated flavours are preferred; these now show excellent performance, both from the point of view of flavour and as far as stability is concerned.

Because of the special requirements mentioned, finding the right flavour needs, in the case of aerated foods, more than ever the full cooperation[8] of a well-established flavourhouse, preferably with experience in this particular field.

16 Conclusions

Aeration through the use of whipping agents provides a versatile tool for modifying a number of aspects in a wide range of foods. For optimal results, an accurate control over the degree of aeration is essential.

One important factor in achieving this is the use of whipping agents that show constant performance. In industrial practice, the use of Compounds, industrially produced and strictly standardized blends of adapted protein-based whipping agents with suitable hydrocolloids and other additives or auxiliary ingredients, is spreading.

These Compounds generally have a narrow field of application only, that is each has been adapted to a small group of products only. Their use allows a considerable simplification and an easy standardization of production.

References

1. Forsythe, R. H. (1960). *In* "Baking Technology and Engineering" (S. A. Matz, Ed.). AVI Publishing Co., U.S.A.
2. Baldwin, R. R. *et al.* (1967). *Poultry Science*, **46**, 1421.
3. Downs, D. E. (1948). *Confectionary Journal*, **74(880)**, 43.
4. Markley, K. S. (1951). "Soybeans and Soybean Products", Vol. 2, Academic Press, New York and London.
5. Lenderink and Co., Netherlands (1950). *Octrooischrift*, **76**, 842.
6. Knightly, W. H. (1968). *In* "Surface Active Lipids in Food". S.C.I. Monograph No. 32. Society of Chemical Industry, London.
7. Mansveldt, J. W. (1966). *Confectionary Production*, **32**, 912.
8. Selbman, H. F. and Oosterhuis, P. (1972). *Lecture* "Flavouring Aereated Confections" at Zentralfachschule der deutschen Süsswarenwirtschaft, Solingen, West Germany.

Discussion

Camina When we eat foamed foods what happens to the air; do we swallow it?

Mansvelt Yes, when consuming aerated foods a small volume of air is necessarily ingested. This volume, however, appears to be small in comparison to the volume of air or gas ingested by other means including the consumption of carbonated beverages and simple air swallowing. It is important that food foams are generally highly unstable under the conditions that occur in the stomach: high acidity, increased temperature, constant movement.

Burton Camina asked a question during the discussion of the paper presented by Mansvelt. As I commented in my own paper, large quantities of air are ingested in the normal eating process. This would also apply to foamed foods when they are eaten. The quantity of air ingested is probably very much smaller than the normal volume of air taken in during eating.

Subject Index

A

Abietic acid, 40
Adhesion of beer foams, 258
Adsorption of proteins, 243, 256
Alkaril B, 113
Amylopectin, 263
Antifoaming agents, 25, 39–47, 127–143, 237–252

B

Baking and bakery goods, 274
Beer foam, 257–271
  adhesion, 258
  cling, 258
  gushing, 250
  head retention, 261
  effect of metal ions, 265, 269
Benzene, 20
Bernoulli's law, 56
Black films, 61, 62, 64, 71, 74, 110, 117, 124, 251
Bovine serum albumin, 237
Bubble half-life, 2
  rise rate, 2, 280
  size, 217, 229
4-Butoxy-ethan-2-ol, 20
Buttercream, 273

C

β-Caesin, 237
Carbon dioxide foams, 273
Cetyl pyridinium chloride, 113
  trimethyl ammonium bromide, 33, 103, 113

Chopin Zymotachygraph, 277
Cling of beer foam, 258
Coalescence, 147–160, 228
Common black film, 33, 167, 224, 231
Contact angle, 92
Concanavalin A, 260
Copper hydrous oxide, 232
Critical micelle concentration, 75
Cyclohexane, 26

D

Decyl dimethyl phosphine oxide, 79
  methyl sulphoxide, 66, 167
Denatured proteins, 242
2,6-Dimethyl heptan-4-ol, 20
Dimethyl isobutyl carbinol (see 2,6-Dimethyl heptan-4-ol)
Disjoining pressure, 63
D.L.V.O. theory, 33, 173, 224, 231
Dodecyl polyethylene oxide, 113
  dimethylamine, 74
  trimethyl ammonium bromide, 33, 173, 224, 231
Drainage, 6, 115, 124, 199
Dynamic food foams, 273–281
  surface tension, 52
Dyspepsia, 127

E

Egg protein, 267, 273
Electrical double layer, 10, 95, 163
Electrophoresis, 164, 176, 220, 261
Electrolyte effects, 77, 92, 167, 222, 229

Index of Authors and Contributors to Discussion

Entries in heavy type refer to papers submitted for discussion, and entries in light type to discussion contributions.